THE PHILOSOPHY OF RIGHT AND LEFT

THE UNIVERSITY OF WESTERN ONTARIO
SERIES IN PHILOSOPHY OF SCIENCE

A SERIES OF BOOKS
IN PHILOSOPHY OF SCIENCE, METHODOLOGY, EPISTEMOLOGY,
LOGIC, HISTORY OF SCIENCE, AND RELATED FIELDS

VOLUME 46

THE PHILOSOPHY OF RIGHT AND LEFT

Incongruent Counterparts and the Nature of Space

Edited by

JAMES VAN CLEVE
Brown University

ROBERT E. FREDERICK
Bentley College

SPRINGER SCIENCE+BUSINESS MEDIA, B.V.

Library of Congress Cataloging-in-Publication Data

The philosophy of right and left : incongruent counterparts and the nature of space / edited by James Van Cleve, Robert Frederick.
p. cm. -- (The University of Western Ontario series in philosophy of science ; v. 46)
Includes bibliographical references and index.
ISBN 978-94-010-5661-8 ISBN 978-94-011-3736-2 (eBook)
DOI 10.1007/978-94-011-3736-2
1. Right and left (Philosophy) 2. Space and time. 3. Kant, Immanuel, 1724-1804. I. Van Cleve, James. II. Frederick, Robert. III. Series.
B105.R54P45 1990
114--dc20 90-38262

ISBN 978-94-010-5661-8

printed on acid free paper

TABLE OF CONTENTS

JAMES VAN CLEVE AND ROBERT E. FREDERICK

PREFACE

Incongruent counterparts are objects that are perfectly similar except for being mirror images of each other, such as left and right human hands. Immanuel Kant was the first great thinker to point out the philosophical significance of such objects. He called them "counterparts" because they are similar in nearly every way, "incongruent" because, despite their similarity, one could never be put in the place of the other.

Three important discussions of incongruent counterparts occur in Kant's writings. The first is an article published in 1768, 'On the First Ground of the Distinction of Regions in Space', in which Kant contended that incongruent counterparts furnish a refutation of Leibniz's relational theory of space and a proof of Newton's rival theory of absolute space. The second is a section of his Inaugural Dissertation, published two years later in 1770, in which he cited incongruent counterparts as showing that our knowledge of space must rest on intuitions. The third is a section of the *Prolegomena to Any Future Metaphysics* of 1783, in which he cited incongruent counterparts as a paradox resolvable only by his own theory of space as mind-dependent. A fourth mention in the *Metaphysical Foundations of Natural Science* of 1786 briefly repeats the *Prolegomena* point. Curiously, there is no mention of incongruent counterparts in either of the editions (1781 and 1787) of Kant's *magnum opus*, the *Critique of Pure Reason*.

Kant's arguments, whatever the final verdict on them, are highly ingenious, and evaluation of them leads one into a number of fascinating topics. These include the possibility of "higher" dimensions of space and the physicist's principle, now known to have exceptions, of parity, or right—left indifference of the laws of nature.

Interest in Kant's arguments has heightened during the last three decades. A succession of journal articles and discussions in books has appeared, some of them criticizing Kant, others defending him, and each nicely building on its predecessors. This volume brings together some of the best of this work, together with Kant's three main treatments of the topic.

Also included in this volume are three items from the nineteenth and early twentieth centuries: Möbius's note connecting Kant's problem with higher dimensions (1827), Norman Kemp Smith's discussion of the place of the three uses of incongruent counterparts in Kant's philosophy (1918), and a brief but pregnant observation from Wittgenstein's *Tractatus* (1922).

Most of the literature on incongruent counterparts focuses on Kant's argument of 1768. The issue here is whether space is an entity in its own right, as Newton believed, or merely a network of relations among material things, as Leibniz believed. Readers unfamiliar with this issue may wish to consult the *locus classicus* for the debate, which is the correspondence between Leibniz and Newton's disciple Samuel Clarke. Other background reading is listed in our bibliography.

For the most part the selections in this volume are arranged chronologically. The selections devoted mainly to the 1768 argument (space as absolute) are the first article by Gardner, the first by Van Cleve, both by Earman, both by Nerlich, and the articles by Remnant, Sklar, Walker, and Harper. (The first of the articles by Earman does not represent his current views, but it is included because it helped to shape the debate.) The articles most germane to the 1770 argument (space as intuitive) are the second by Gardner and those by Bennett, Curd, Harper, and Buroker. Finally, the articles most germane to the 1783 argument (space as mind-dependent) are those of Kemp Smith and Buroker and the second by Van Cleve.

The editors wish to thank Jonathan Bennett, John Earman, and Lewis White Beck for their advice on several points (including Bennett's suggestion for a less boring title than we had originally planned). We also wish to thank Eleanor Thum for helping prepare the manuscript, Rex Welshon for preparing the index, and Sung-Ho Chung, Teresa Ferguson, Jean Chambers, Michael Ialacci, David Martens, and John Gibbons for assisting with the bibliography.

ACKNOWLEDGMENTS

The contributions to this volume by Jill Vance Buroker and William Harper were written for the occasion, as was the second by Graham Nerlich. The other contributions are reprinted. We wish to thank the following authors and publishers for permission to reprint.

Immanuel Kant, 'On the First Ground of the Distinction of Regions in Space,' in John Handyside (trans.), *Kant's Inaugural Dissertation and Early Writings on Space* (Chicago: Open Court, 1929), pp. 19–29.

Immanuel Kant, selection from Section 15 of 'Dissertation on the Form and Principles of the Sensible and Intelligible World,' in Handyside, p. 60.

Immanuel Kant, section 13 of the *Prolegomena to Any Future Metaphysics*, trans. by L. W. Beck (Indianapolis: Bobbs-Merrill, 1950), pp. 32–4; © 1950 Macmillan Publishing Company.

August Ferdinand Möbius, 'On Higher Space,' *Der barycentrische Calcul* (Leipzig: 1827), Part 2, Chapter 1.

Norman Kemp Smith, 'The Paradox of Incongruous Counterparts,' from *A Commentary on Kant's Critique of Pure Reason* (London: Macmillan, 1918), pp. 161–66.

Ludwig Wittgenstein, 'Tractatus 6.36111,' from *Tractatus Logico-Philosophicus*, trans. by D. F. Pears and B. F. McGuinness (London: Routledge & Kegan Paul, 1961), pp. 141–2.

Peter Remnant, 'Incongruent Counterparts and Absolute Space,' *Mind*, **72** (1963), 393–99.

Martin Gardner, 'The Fourth Dimension,' Chapter 17 of *The Ambidextrous Universe* (3rd. ed.; San Francisco: W. H. Freeman, 1989); © 1989 W. H. Freeman.

Martin Gardner, 'The Ozma Problem and the Fall of Parity,' selections from Chapters 18, 20, and 22 of *The Ambidextrous Universe*; © 1989 W. H. Freeman.

Jonathan Bennett, 'The Difference Between Right and Left,' *American Philosophical Quarterly*, 7 (1970), 175–91.

John Earman, 'Kant, Incongruous Counterparts, and the Nature of Space and Space-Time,' *Ratio*, **13** (1971), 1–18.

Graham Nerlich, 'Hands, Knees, and Absolute Space,' Chapter 2 of *The Shape of Space* (Cambridge: Cambridge University Press, 1976). This is a revised version of the article of the same name that appeared in *The Journal of Philosophy*, **70** (1973), 337–51.

Lawrence Sklar, 'Incongruous Counterparts, Intrinsic Features, and the Substantiviality of Space,' *The Journal of Philosophy*, **71** (1974), 277—90.

Ralph Walker, 'Incongruent Counterparts,' from *Kant* (London: Routledge & Kegan Paul, 1978), pp. 44—51; © 1978 R. C. S. Walker.

Martin Curd, 'Showing and Telling: Can the Difference Between Right and Left Be Explained in Words?,' *Ratio*, **26** (1984), 63—69.

James Van Cleve, 'Right, Left and the Fourth Dimension,' *The Philosophical Review*, **96** (1987), 33—68.

John Earman, 'On the Other Hand. . . . : A Reconsideration of Kant, Incongruent Counterparts, and Absolute Space.' Except for minor changes this paper appears as Chapter 7 of Earman's book *World and Space-Time Enough* (Cambridge, Mass.: The MIT Press, 1989).

James Van Cleve, 'Incongruent Counterparts and Things in Themselves,' in *Proceedings: Sixth International Kant Congress*, ed. by G. Funke and Thomas M. Seebohm; © 1988 The Center for Advanced Research in Phenomenology, Inc., co-publisher (Washington, D.C.: University Press of America, Inc., 1988).

ROBERT E. FREDERICK

INTRODUCTION TO THE ARGUMENT OF 1768

Some ordinary facts about the world we live in can be readily explained by other ordinary facts. One can, for example, explain the fact that when we are facing north the sun rises on the right and not the left by appealing to ordinary facts about the rotation of the earth in its orbit about the sun. But the same is not true for other ordinary facts. It is not so easy to explain why the sky is blue and not some other color, or why water freezes at 32° F and not some other temperature.

One ordinary fact that is not readily explainable in terms of other ordinary facts is the difference between left and right hands. Although they are very similar, there is undeniably a difference between them. One can't, for instance, put a right glove on a left hand. And since there is a difference, it seems that we ought to be able to explain it. But it is not obvious what facts about the world might explain the difference, or even, perhaps, what the difference is. Exactly what is it, after all, that makes a hand left and not right?

One explanation of the apparent difference between left and right hands can be found in Kant's 1768 paper, 'On the First Ground of the Distinction of Regions of Space'. In one of his more remarkable and enduring arguments, Kant attempts to show that the only way to explain the difference between left and right hands is by supposing that there exists a thing, absolute space, such that a hand is left or right at least partly in virtue of its relation to absolute space. If his argument is correct, then he has both explained the difference between hands and shown that space cannot be, as Leibniz thought, a kind of fiction, talk of which is reducible to talk of material objects and their relations. As Kant puts it, space must have "a reality of its own, independent of the existence of all matter."[1]

For Kant's argument to be successful he must show at least three things. The first is that there is, as he says, a "real difference" between left and right hands and not a merely apparent one. The second is that no simpler alternative explanation accounts for the difference. The third is that the supposition that absolute space exists does explain the difference. Since most of the disputes about Kant's 1768 argument deal in one way or another with these three points, I will first give a preliminary reconstruction of Kant's argument and then comment briefly on each point in turn.

One way to reconstruct Kant's argument is this:

1. There is a real difference between left and right hands.
2. There is no adequate explanation of the difference that does not posit absolute space, for:
   (A) it cannot be explained by the differing relations between the parts of hands since these are the same for left and right hands, and
   (B) it cannot be explained by the relations left and right hands bear to other material objects since a hand would be either left or right if it was the only material object that existed, and
   (C) there is no other adequate explanation that does not posit absolute space.
3. The supposition that absolute space exists does adequately explain the difference.
4. Hence, granted that left and right hands exist, absolute space exists and has a reality of its own.

In this argument, as well as in his later arguments, Kant needs to begin by showing that there is a real difference between left and right hands, for if there is no difference there is nothing to be explained and the entire enterprise collapses. And it is easy enough to see the difference. We can recognize that, say, a left hand *is* left and not right even when it is not attached to a body. However, in his 1768 paper Kant's remarks about the precise nature of difference are sometimes confusing. At one point he says,

What, therefore, we desire to show is that the complete ground of the determination of the *shape* of a body [e.g. a hand] rests not merely upon the position of its parts relatively to one another, but further on a relation to universal space . . . .[2] (Emphasis added.)

This makes it sound as if hands differ in shape: left hands are left-shaped and right hands are right-shaped. But later in the paper he says,

From the common example of the two hands, it is already clear that the shape of one body can be *completely similar* to that of another, and the magnitude of their extension exactly the same, while yet there remains an *inner difference*, namely that the surface that bounds the one cannot possibly bound the other . . . .[3] (Emphasis added.)

Ordinarily one would think that if two things have completely similar shapes they have the same shape. Yet it is doubtful that Kant meant to imply this. For surely the reason the surface of a left hand cannot bound a right hand is that the surfaces of the hands differ in shape. Since it is highly unlikely that the shape of the hand differs from the shape of its surface, the shape of the hands must differ. Thus, left and right hands do not have completely similar shapes.

As Jonathan Bennett suggests in 'The Difference Between Right and Left' in this volume, perhaps what Kant meant to say is no more mysterious than this: left and right hands have completely similar shapes with the exception that one of them is left-shaped and the other is right-shaped. But is this true? Is the real difference between left and right hands a difference in shape?

In order to answer the question we first need to know something about sameness and difference in shape. If we briefly glanced at the two figures below,

(a) (b)

we might initially say that they were similar in shape but not the same. But it would repay us to look a little closer. Suppose (a) and (b) are two-dimensional figures permanently embedded in the plane of the page, and suppose we can move (b) around the page if we do not change any of the relations of distance and angle between its parts. We can rotate it about the point where its short and long shafts join, and we can move it left or right or toward the top or bottom of the page. Let us call motion of this type *rigid* motion. Note that by rotating (b) and moving it left and toward the top of the page we can superimpose it on (a) in such a way that (a) and (b) are *exactly congruent*; that is, there is no part of (a) that is not covered by a part of (b) and *vice-versa*. Objects that can be superimposed like (a) and (b) are *congruent counterparts*, and it seems we ought to say that congruent counterparts have the same shape.

Now consider the objects in the second figure.

(c) (d)

Although there is no rigid motion that will bring (c) and (d) into exact congruence, they are very similar. The length of the short and long shafts is the same for each of them, and the angle between the shafts is also the same. They are, in fact, mirror images of each other. Mirror image objects that cannot be made exactly congruent are *incongruent counterparts*. Now, do (c) and (d) have the same shape or not?

The question we are asking is whether exact congruence is both necessary and sufficient for sameness of shape. In other words, we have said that if objects like the ones above can be brought into exact congruence by some rigid motion, then they have the same shape. Assuming the discussion is restricted to objects of the same size, should we also say that if two objects have the same shape, then they can be brought into exact congruence by some rigid motion?[4]

I believe Kant would have said that exact congruence is both necessary and sufficient for sameness of shape, not only for simple line figures like the ones above but also for three-dimensional objects such as left and right hands. The case is more complex for hands, however. Left and right hands are not mirror images of each other in quite the same way that (c) and (d) are because hands are three dimensional and thus cannot be completely represented in a two-dimensional mirror image. In addition, left and right hands cannot be superimposed as we imagined (a) and (b) were superimposed. But let us expand our idea of rigid motion to include motion in three dimensions that does not change the relations of distance and angle between the parts of a thing. Now suppose laser technology has advanced to the point that we can create three-dimensional mirror holograms of objects. For instance, a three-dimensional mirror hologram of a left hand would be a three-dimensional image of the hand that is left—right reversed. It would look like an ordinary right hand. Then there seems to be no rigid motion in three dimensions that would bring a left (right) hand into exact congruence with its three-dimensional mirror hologram. They could not be made to occupy the same 'space'. I think something like this approximates Kant's idea that "the surface that bounds the one cannot possibly bound the other." So if we accept exact congruence as both necessary and sufficient for sameness of shape, and if we agree that our thought experiment with holograms captures a sense in which left and right hands are not exactly congruent, then left and right hands do not have the same shape. There is a real difference between them.

As pointed out in 'On Higher Space' by Möbius, however, there are some kinds of motions that will make a left (right) hand exactly congruent with its mirror image hologram. It will be a little easier to see how this can be done if we again consider the figures (c) and (d). Assume that in addition to two-dimensional rigid motions we can 'lift' (d) out of the plane of the page and move it around in three dimensions provided we do not change the relations of distance and angle between its parts. In order to distinguish this type of movement from motion that occurs in the plane of the page let us call it a *dimensional* motion. In general, a dimensional motion occurs when an N dimensional object is rigidly moved through an $N + 1$ dimensional space. If we lift (d), turn it over, and place it anywhere back down on the page, then we can perform some two-dimensional rigid motion that will bring it into exact congruence with (c). Alternatively, suppose we use a pair of scissors to cut out a strip of paper with both (c) and (d) on it. We then twist the strip of paper once and join the ends with glue. What we have constructed is a Möbius strip. Now, without lifting (d) or changing any relations of distance and angle between its parts, we move it once around the strip. We can call this type of motion a *Möbius* motion. Spaces in which Möbius motion is possible are usually called non-orientable spaces. If Möbius motions are not possible, then the space is orientable.[5] Oddly enough, after a Möbius motion (c) and (d) can be made exactly congruent. So under certain conditions dimensional motions in orientable spaces and Möbius motions in non-orientable spaces can make figures exactly congruent that seemed incongruent before. Should we say that these motions change the shape of the figures? Or should we say they show the figures had the same shape to begin with?

Most people would agree, I think, that dimensional and Möbius motions show that (c) and (d) have the same shape. Rigid motion does not, by itself, change the intrinsic shape of an object. Things become much more difficult, however, when we try to conceive of analogous motions for left and right hands. If we were to try to use dimensional motion to make a left hand congruent with its mirror hologram, we somehow would have to lift it and turn it over in a *fourth* spatial dimension. If we were to try to use a Möbius motion to accomplish the same thing, we would have to move the hand around a three-dimensional version of a Möbius strip. But the problem is that there is no reason to suppose that the space we inhabit either has four spatial dimensions or is a three-dimensional non-orientable space. So we cannot literally make a left hand congruent with its mirror hologram by dimensional or Möbius type motions. How, then, are dimensional and Möbius motions relevant to Kant's argument? Don't left and right hands have different shapes in the world we live in, and isn't that enough to show that there is a real difference between them?

One response is that dimensional and Möbius motions *are not* relevant. When we say that things have the same shape only if they *can* be made exactly congruent in the manner described above, what we mean is that they have the same shape only if there is some motion in the space they inhabit that will bring them into exact congruence. Since there is no motion in our actual space that will bring left and right hands into congruence with their mirror holograms, they do not have the same shape.

Another response is that dimensional and Möbius motions *are* relevant. When we say that if things *can* be made exactly congruent then they have the same shape, what we mean is that if there is a *possible* space in which some motion brings them into exact congruence, then they have the same shape. Since four-dimensional spaces and three-dimensional non-orientable spaces are possible, and since left and right hands can be made congruent by the appropriate motion in such spaces, there is no real difference in shape between left and right hands.

As I will show later, there may be a way to avoid deciding between these two positions. Thus I will leave them aside for a moment and turn to the remainder of Kant's argument. Recall that the second thing Kant needs to show is that there is no simpler adequate explanation for the difference between left and right hands. For suppose there is an alternative explanation that does not posit the existence of absolute space. Then even if Kant can also provide an explanation, Ockham's razor would require us to reject Kant's version since it multiplies entities beyond necessity. Thus, Kant must eliminate other possible explanations before he proceeds with his own. Let us turn, then, to the other explanations he considers.

In my reconstruction of Kant's argument the first alternative explanation proposed is that the difference between left and right hands can be explained by the differing relations between their parts. That is, a left hand differs from a right one in that its parts are related differently than the parts of a right hand.

Kant claimed that the relations between the parts of right and left hands are the same. If we consider only relations of distance and angle between the parts, he is correct. There are no differences of distance and angle, so we cannot use these relations to differentiate between left and right hands. In 'Kant, Incongruous Counterparts, and the Nature of Space-Time' John Earman notes, however, that hands might exemplify primitive internal relations such as standing-in-a-left-configuration or standing-in-a-right-configuration. Kant did not consider this possibility, and at first glance it may seem to be an *ad hoc* response. But it need not be since one could, for instance, change a left hand into a right one by detaching and rearranging its parts. Hence, a hand could be changed from left to right by changing the relations between its parts. So there must be some relations between the parts of hands that differentiate between left and right. If we do not find primitive left/right relations plausible, what else might they be?

In 'Right, Left, and the Fourth Dimension' James Van Cleve suggests that besides distance and angle the parts of hands are related by direction. To see that direction has something to do with the difference between left and right hands, let us try another thought experiment. Suppose evolution had taken a slightly different course — our hands do not have fingernails, and there is no difference between the skin on the palm and back of our hands. Now suppose a skilled surgeon operates on your right hand and reverses the direction the joints bend. Formerly when you held your hands together in front of your face — thumb against thumb, forefinger against forefinger, and so on — and when you closed your hands to make fists, the fingers closed in opposite directions, toward each other. After the operation when you hold your hands in the same way and close them to made fists the fingers close in the same direction, toward your right. The hand operated on, which is still on your right arm, has been converted from a normally functioning right hand to a normally functioning left hand by changing the direction the joints work.

If we assume that the joints of our hands evolved into symmetrical shapes, they are spherical for instance, then it does not seem that in these operations the surgeon alters the shape of the surface that bounds the hand. He changed the direction the joints and fingers bend, but not the hand's shape. So it seems that direction is a possibility for a relation that differentiates between hands.

An advocate of the position that handedness is some type of relation between the parts of a hand is committed to claiming that the only way to change a hand from, say, left to right is by changing the relations – either rearranging the parts of the hand or reversing the direction the joints bend. This seems to be correct as long as hands are confined to three-dimensional orientable space. But it is not in four-dimensional space or three-dimensional non-orientable space since in such spaces hands can be made exactly congruent by dimensional or Möbius motions. Hence, if the possibility of these spaces is relevant to deciding whether there is a real difference between left and right hands, the view that handedness consists in the relations between parts of hands must be mistaken.

The same objection applies to the position that handedness is a primitive relation. It does not seem plausible to suppose that by turning a hand over in a fourth dimension, or moving it around a three-dimensional non-orientable space, one would alter such primitive relations. Yet they would have to be altered if the view in question is correct. Hence, if the possibility of such spaces is relevant, it is unlikely that there are primitive relations of left or right handedness.

Even if Kant is able to show that left or right handedness is not a matter of the relations between parts of hands, there is another purported explanation of handedness that does not posit absolute space. This sort of explanation, like the one we have just considered, is compatible with the theory of space normally called 'relationism'.[6] According to advocates of relationism, space is not an entity or thing that has an independent existence or reality of its own. It is nothing other than, nothing over and above, the relations between material objects. As Lawrence Sklar says in 'Incongruous Counterparts, Intrinsic Features, and the Substantiviality of Space', "space is nothing but the collection of actual and possible spatial relations among actual and possible material objects."[7] One way to uphold relationism in the face of Kant's argument would be to argue that the difference between left and right hands can be explained by their differing relations to other actual (or possible) objects. Thus, there would be no need to posit the existence of absolute space to explain the difference.

Kant did not attempt to show that relations to other material objects cannot be used to explain the difference in handedness between *two* hands. Instead, he tried to refute relationism by arguing that under certain conditions it cannot account for the handedness of a *single* hand. The argument is this. Imagine that the only material object that exists is a human hand. It must be either a right-shaped hand or a left-shaped one. It cannot have an indeterminate shape. We already know that the relations between the parts of the hand are insufficient to account for its shape, and since there is no other material object, the hand cannot be left or right-shaped in virtue of its relation to that object. Hence, the hand must be left or right at least partly in virtue of its relation to absolute space.

An immediate problem with Kant's thought experiment is that in four-dimensional spaces and three-dimensional non-orientable spaces left and right hands do not differ in shape. If a lone hand existed in such a space it would be neither left nor right-shaped. So Kant's claim that the hand must have one shape or the other is mistaken. Thus, if the possibility of such spaces is relevant, Kant's argument presents no challenge to relationists since it depends on a false premise.

If one believes, however, that the possibility of such spaces is irrelevant to Kant's argument, then it does present a challenge to relationists. The challenge has been taken up in a variety of ways. For instance, in 'Incongruent Counterparts and Absolute Space' Peter Remnant argues that even in a space of the sort we actually inhabit a lone hand would be neither left nor right-shaped. This may seem to be an extreme position, since surely the hand *has* a shape and it must be either one shape or the other. But Remnant shows that there is no way to determine *which* shape it is. It would not help, for example, to introduce a handless human body into the space and then try to learn whether it correctly fits the left or right wrist, since we are in no better position to tell which wrist is left or right than we are to tell which hand is left or right.

In 'Hands, Knees and Absolute Space', an article that significantly changed the focus of the debate about Kant's 1768 argument, Graham Nerlich responds that Kant never intended to imply that we could discover whether the hand is left or right-shaped. Thus Remnant's argument misses the point. One way to reply to Nerlich (I do not claim it would be Remnant's reply) might be this. Kant's thought experiment requires that the lone hand exemplify one of the properties 'being right-shaped' or 'being left-shaped'. But Remnant has shown that we could not even in principle determine which property it has unless we have already determined the handedness of some other object. Hence, even though the hand may have the disjunctive property 'being either left or right-shaped' it is false (or meaningless) to claim that it exemplifies one of the disjuncts.

To evaluate this reply we would have to assess the merits of the verificationism on which it rests. Fortunately, Nerlich continues his argument with a proposal that may allow us to avoid that onerous task. The main point of Kant's lone hand example, Nerlich argues, is aimed "not at showing the hand to be a right hand or a left hand, but at showing that it is an *enantiomorph*."[8] An enantiomorph is, roughly, an asymmetrical object that *could* have an incongruent counterpart in the space in which it exists. Otherwise it is a *homomorph*. If we imagine a lone hand existing in a three-dimensional Euclidean space, it is an enantiomorph. If it is in a four-dimensional space or a three-dimensional non-orientable space, it is a homomorph.

Nerlich argues that in any space in which a lone hand could exist it has the disjunctive property of being either an enantiomorph or a homomorph. Moreover, it exemplifies one or the other of these properties. As he says, "no glimmer of sense" can be made of the idea that a lone hand exemplifies the disjunction but neither of the disjuncts. If he is correct, and if he can show that Kant's argument can be reconstructed using the enantiomorph/homomorph distinction, then we need not worry about my verificationist argument that a solitary hand is neither left nor right.

Suppose Nerlich is correct. A lone hand is either an enantiomorph or a homomorph. Then we can ask: what is it that explains the fact that a lone hand exemplifies one of these properties rather than the other? It cannot be the relations between the parts of the hand, since these are the same regardless of whether the hand is an enantiomorph or a homomorph. It cannot be the relations the hand bears to other material objects, since there are none. Hence, it can only be the relation the hand bears to absolute space. Hence, absolute space does have a reality of its own.

One advantage of Nerlich's argument is that it may avoid a potentially devastating criticism leveled against Kant's version of the argument. Earlier I mentioned that the third point Kant must establish is that positing the existence of absolute space does succeed in explaining the difference between left and right hands. There is an argument, however, that makes a persuasive case that it does not succeed. One variant of the argument is this. Assume the only thing that exists is a hand, and that neither primitive nor other types of relations serve to distinguish between hands. Then it must be some property of absolute space that makes a lone hand, say, left rather than right. Evidently the only candidate for such a property is the *handedness* of the bit of space the hand occupies. If it is a left-handed bit of space, then the hand is left; if it is a right-handed bit of space, then it is a right hand.

But what accounts for the handedness of these bits of space? One possibility is that nothing does; that is, handedness is a primitive property of bits of space, or a primitive relation between the parts of bits of space. If this is correct, however, then why couldn't we just as easily say that hands exemplify such primitive properties or relations? We would then have no need for absolute space to explain handedness. Alternatively, perhaps what explains the handedness of bits of space is something *outside* of space that bits of space are related to in some way. Even if this makes sense, which I doubt, it apparently creates an infinite regress since we can ask the same kind of questions about the relation of bits of space to this other thing.

To borrow Nerlich's phrase, this is an unlucky conclusion. Yet there may be a way to avoid it. Earman, who first devised a version of the above argument, concedes that according to Kant the handedness of a hand depends not on its relation to some bit of space, but to "space in general as a unity." In his first article in this volume Earman does not see how this helps Kant. Nerlich argues, however, that it is the properties of space considered "as a unity" that determine whether a hand is an enantiomorph or a homomorph. He writes,

Which of these ... determinate characters [being an enantiomorph or a homomorph] the hand bears depends, still, on the nature of the space it inhabits, not on other objects. The nature of this space, whether it is orientable, how many dimensions it has, is absolute and primitive.[9]

If Nerlich is correct, which Earman disputes in 'On the Other Hand . . .: A Reconsideration of Kant, Incongruent Counterparts, and Absolute Space', then he has shown that whether a hand is an enantiomorph or a homomorph depends on the properties of space in general as a unity. So absolute space does play an essential role. It explains not why a hand is left or right-shaped, but why it is an enantiomorph or a homomorph.

Another advantage of Nerlich's argument is that it may circumvent the difficult issue of the relevance of higher dimensional and non-orientable spaces. The possibility of four-dimensional orientable and three-dimensional non-orientable space seems relevant to Kant's argument since it is relevant to deciding whether hands differ in shape. But Nerlich's argument does not depend on the fact, if it is one, that hands differ in shape. It just does not matter what kinds of spaces are possible or what kind of space a hand exists in, since a hand is either a homomorph or an enantiomorph in any space in which it exists. That is all that Nerlich needs.

There is, however, an intimate link between shape and enantio/homomorphism. Let us introduce the two place predicates 'enantiomorphic relatedness' and 'homomorphic relatedness'. Two normal hands are homomorphically related in a space *S* if and only if they are congruent counterparts in *S*; they are enantiomorphically related in *S* if and only if they are incongruent counterparts in *S*. Given my earlier account of 'having the same shape', we can now say that two normal hands have the same shape in a space *S* if and only if they are homomorphically related in *S*; otherwise they have different shapes. This has an uncomfortable consequence. Suppose two normal hands are enantiomorphically related in three-dimensional Euclidean space. Then they have different shapes. Now suppose that by some philosophical magic the same hands are transported into a four-dimensional space. They are homomorphically related in this space so they have the same shape. This conflicts with the intuition that mere rigid motion can't change the shape of a thing. So something seems to have gone wrong. What is it?

I suggest that, if we wish to maintain the intuition that rigid motion can't change the shape of a thing, we must either deny that the difference between incongruent counterparts is a difference in shape or we must distinguish, as William Harper does in 'Kant on Incongruent Counterparts', between the intrinsic shape of an object and the empirical shape. We could then say that the intrinsic shape, which is defined in geometric terms, is not changed by motion, but empirical shape, which is based on the observable differences between hands, can be changed. Whether the difference between left and right is observable is also interestingly discussed by Ralph Walker in his contribution 'Incongruent Counterparts.'

Let us return to Kant and the relationists, or rather, to the relationists and Nerlich's revision of Kant's argument. Do the relationists have a response to the revised argument? Can they give an account of the enantiomorph/homomorph distinction that does not rely on absolute space? They can provided they can give a relationist account of the features of space on which enantiomorphism and homomorphism depend; that is, if they can give an explanation of the dimensionality and orientability of space.

Even if this can be done I think it is fair to say that it has not been done in any detail. Furthermore, even if it is done, relationists still face the problem of 'the fall of parity'. Parity, as well as some of its implications for the debate about absolute versus relational theories of space, is briefly discussed in the introduction to the 1770 and 1783 arguments, and is explained in more detail in 'On the Other Hand . . .' by Earman and 'The Ozma Problem and the Fall of Parity' by Martin Gardner.

To conclude this introduction I would like to present some terms (taken from 'Right, Left, and the Fourth Dimension' by Van Cleve) that may help the reader categorize the various positions discussed by the contributors. The first is *internalism*. Internalism is the view that the difference between hands can be explained by primitive relations of handedness or by some other relation internal to hands. The second term is *externalism*. Externalists argue that the difference between hands can be explained by the relations hands bear to other actual (or possible) material objects. The final term is *absolutism*. Absolutism is the view that the only way to explain handedness is by positing absolute space. In 1768 Kant, of course, was an absolutist. Internalism is represented in this volume by Earman in 'Kant, Incongruous Counterparts, and the Nature of Space-Time', and externalism by Gardner in 'The Fourth Dimension'.

There are analogs of these three positions with regard to the distinction between enantiomorphs and homomorphs. It is not necessary to adopt the same position for both distinctions. Thus Nerlich, though not a defender of Kant's absolutism about left and right, is an absolutist about enantiomorphism and homomorphism. Earman (in 'On the Other Hand . . .') and Harper (in 'Kant on Incongruent Counterparts') both oppose Nerlich's version of absolutism, defending instead internalism about enantiomorphism and homomorphism. Neither of them, however, is currently an internalist about left and right. Earman now advocates externalism (in 'On the Other Hand . . .'), and Harper defends a position that is quite similar to Kant's original absolutism.

NOTES

[1] Kant, 'On the First Ground of the Distinction of Regions in Space.' This volume, p. 28.

[2] Kant, ibid, pp. 30—31.

[3] Kant, ibid, p. 32.

[4] This is an oversimplification. Consider a two-dimensional space shaped like an hourglass. Suppose (a) is on one side of the hourglass, (b) is on the other, and the passage between the two sides is too narrow to allow us to move (b) from one side to the other by using rigid motion. Then (a) and (b) have the same shape even though they cannot be superimposed without altering the shape of space. If it seems implausible to suppose that space has an unalterable shape, imagine an impenetrable physical barrier between (a) and (b).

[5] For example, a Euclidean space of any number of dimensions is an orientable space, so Möbius motions are not possible in any Euclidean space. They are only possible in spaces with a "twisted" topological structure. For more precise definitions of orientable and non-orientable space, see the references in the bibliography.

[6] It is quite consistent to accept a relational view of space and at the same time argue that what differentiates between hands is either some relation between the parts of hands or primitive relations, and not the relations hands bear to other material objects. Earman defends just such a view in his first article.

[7] Sklar, 'Incongruous Counterparts, Intrinsic Features, and the Substantiviality of Space.' This volume, pp. 176—177.

[8] Nerlich, 'Hands, Knees, and Absolute Space.' This volume, p. 155. Emphasis added.

[9] Nerlich, ibid., p. 160.

JAMES VAN CLEVE

INTRODUCTION TO THE ARGUMENTS OF 1770 AND 1783

I

Kant's second invocation of incongruent counterparts occurs in his Inaugural Dissertation of 1770, a work written and presented on the occasion of his promotion to Professor of Philosophy at the University of Königsberg. In this work, he no longer used incongruent counterparts to show that space is an absolute being. Instead he used them to illustrate a point that he was later to defend at greater length in the Transcendental Aesthetic of the *Critique of Pure Reason*: our representation of space and spatial figures is intuitive, not conceptual. Here is how he put it in the Dissertation:

> We cannot by any sharpness of intellect describe discursively, that is by intellectual marks, the distinction in a given space between things which lie towards one quarter and things which are turned toward the opposite quarter. Thus if we take solids completely equal and similar but incongruent, such as the right and left hands (so far as they are conceived only according to extension), or spherical triangles from two opposite hemipheres, although in every respect which admits of being stated in terms intelligible to the mind through a verbal description they can be substituted for one another, there is yet a diversity which makes it impossible for the boundaries of extension to coincide. It is therefore clear that in these cases the diversity, that is the incongruence, cannot be apprehended except by pure intuition.[1]

The difference between right and left can only be grasped intuitively, through vision or some similar faculty. For readers unfamiliar with Kant, it should be pointed out that 'intuition' is the translation of the German *Anschauung* — a word that could also be translated as 'perception' or 'view'.

In 'The Difference Between Right and Left', Jonathan Bennett proposes an illuminating way of restating Kant's point. According to Bennett, Kant is claiming that the meanings of the terms 'left' and 'right' can be explained only be reference to sensorily presented examples — only by *showing*, not by *telling*. Putting the point another way, we must have recourse to an ostensive definition; a verbal definition would not suffice. Bennett calls this claim the *Kantian Hypothesis*.

The Kantian Hypothesis is closely related to what Martin Gardner calls 'the Ozma Problem' in his second paper in this volume. Gardner frames the problem this way:

> Is there any way to communicate the meaning of 'left' [to the inhabitants of some Planet X in a distant galaxy] by a language transmitted in the form of pulsating signals? By the terms of the problem we may say anything we please to our listeners, ask them to perform any experiment whatever, with one proviso: *There is to be no asymmetric object or structure that we and they can observe in common.*[2]

If the Kantian Hypothesis is correct, the answer to Gardner's question must be no. The hypothesis says that you cannot communicate the meaning of 'left' without showing, and Gardner's proviso says, in effect, that no showing is allowed.

One slight refinement of the Kantian Hypothesis is necessary. 'Right' and 'left' are two members of what we may call the *chiral* family of terms, a family that also includes 'clockwise', 'counterclockwise', the names for the various points of the compass, and the labels for the ends of magnetic poles. We *can* give a verbal definition of 'right' if we are permitted to use other chiral terms in the definition. We can say, for example, that an outstretched hand with its palm pointing up and its fingers pointing north is a right hand if and only if its thumb points east. The Kantian Hypothesis should therefore be restated as follows: we cannot give a verbal definition of any chiral term except by using *other* chiral terms; hence, if any of the chiral terms are to be understood, at least some of them must be understood ostensively.

Bennett and Gardner both defend the Kantian Hypothesis, up to a point, by showing that a wide variety of strategies for conveying our meanings of 'left' and 'right' to Planet X will not work. For example, we could tell them that if the fingers of a hand curve in the direction of rotation of a normal screw when it is being driven into a plank, the hand is a right hand if and only if its thumb points toward the plank. This obviously will not do, since for all we know, screws on Planet X are threaded in the opposite direction from that of standard screws on Earth. Any other appeal to artifacts is similarly doomed to failure. Nor do we fare any better if we appeal to biological or chemical phenomena, such as spiraling vines or molecular structures, since for all we know, these things, too, occur with the opposite handedness on Planet X. And, of course, if we were to send along a *specimen* of one of our screws or plants or chemical compounds, we would be violating the proviso; we would be resorting to a particular that we and they can observe in common.

In the end, however, neither Bennett nor Gardner defends Kant all the way. They both maintain that the Kantian Hypothesis is refuted by certain surprising findings of twentieth-century physics. Kant would have been right, they say, *but for the fall of parity*.

'The fall of parity' refers to the discovery, first announced in 1957, that parity is not conserved. I shall not attempt to say here what parity is; for that, the reader may consult 'The Ozma Problem and the Fall of Parity' by Gardner or any of several articles in our bibliography. It will be enough for our purposes to explain what it means to say that *parity is conserved* – or that it is not conserved, as the case turns out to be. The basic idea of parity conservation is this: the laws of nature are symmetric so far as right and left are concerned; they are not sensitive to any difference between right and left. This point may be explained in any of several ways. One way is to say that if a phenomenon is permitted by the laws of nature, so is its mirror image. Another way is to point out that if someone showed you a movie of an event with the film right–left reversed, you would not be able to tell, from knowledge of the laws of nature alone, that anything was wrong.

Let us pursue the movie illustration a bit further. Physicists like to use the same illustration to explain the 'time-reversibility' of all laws of nature governing basic processes in the micro-world: if you saw a movie of a micro-event run backwards, you would not be able to tell that anything was amiss. That is not to say that you could not tell the *difference* between the movie run forwards and the movie run backwards; of course you could. The point it rather that you could not tell, from knowledge of the laws of nature alone, which showing depicted events as they actually occurred. If parity is conserved, the same goes for a movie shown with the film right–left reversed: you would not be able to tell, from your knowledge of laws of nature alone, that anything was amiss. (Of course, baseball fans who saw Sandy Koufax pitching with his right arm would know that something was wrong, but they would be relying on knowledge of more than just physics.)

A third way of saying what is involved in parity conservation has been provided by Bennett:

Suppose that we have two experimental set-ups with initial states I_1 and I_2 and resultant states (arising from the initial ones in ways that can be wholly explained by basic physical laws) R_1 and R_2. The Parity principle implies that if I_1 is an enantiomorph of I_2 then R_1 is an enantiomorph of R_2.[3]

That is to say, experiments whose initial states differ only in handedness will have outcomes that differ only in handedness. The experiments as wholes will be perfect reflections of each other.

One can see, perhaps, why physicists believed for many years that parity is conserved. Who would have thought that a mere *chiral* difference (i.e., a difference in handedness) in initial states could lead to a *nonchiral* difference in results? But we now know that this is exactly what sometimes happens. Parity nonconservation was first theoretically predicted by Lee and Yang in 1956; it was experimentally verified by Wu and her associates in 1957. The reader will find a description of Wu's experiment in 'The Ozma Problem and the Fall of Parity' by Gardner. Instead of referring here to the actual experiment and its technical details, I will describe a fictitious experiment that can be understood without any knowledge of physics. Though purely make-believe, the experiment is isomorphic with real-life phenomena.

Consider the oblong boxes with one corner shaved off that Bennett uses as examples of incongruent counterparts. The boxes are just alike except for one thing: when their largest cut faces are toward you and their smallest cut faces are downmost, one of the middle-sized cut faces is to your left and the other is to your right. (See the diagram on p. 105.) Call a box of the first sort a lefty and one of the second sort a righty. Now imagine an experiment in which boxes differing only in the manner described (they are exactly the same in mass, center of gravity, etc.) are tossed in the air in identical fashion and allowed to land on a table top. Records are kept of the numbers of times boxes of each sort land with one or another side uppermost. Suppose the results are as follows: when we toss a lefty, *it usually lands large side up*, and when we toss a righty, *it usually lands small or middle side up*. Or to make the results simpler and more dramatic, suppose that lefties *always* land large side up and righties another side up.[4]

If Bennett boxes always behaved in the manner described, we would have a violation of the parity principle — parity would not be conserved. This is easily seen by reference to any of the three explanations given above. (1) If a phenomenon is permitted by the laws of nature, so is its mirror image. Not so in the case described: the mirror image of tossing a lefty and getting large side up is tossing a *righty* and getting large side up, but that never happens. (2) Knowledge of the laws of nature alone is not sufficient to let you know you are seeing a movie with the film right–left reversed. Not so in the case described: if you saw a righty landing large side up, you would know that the film was reversed. (3) If the initial states in two lawful sequences are mere enantiomorphs of each other, so are the resultant states. Not so in the case described: I_1 (= tossing a lefty) is an enantiomorph of I_2 (= tossing a righty), but R_1 (= getting a lefty with large side up) is not an enantiomorph of R_2 (= getting a righty with small or middle side up).[5]

The fall of parity bears on the Ozma Problem in an obvious way. If parity broke down in the way just described, we could communicate our meanings of 'right' and 'left' to Planet X by means of the following recipe: "Toss a Bennett box (an oblong box with one corner shaved off); if it usually lands large side up, it is a lefty; otherwise, it is a righty." This is an operational definition of the chiral terms 'lefty' and 'righty' stated wholly in nonchiral terms: 'box', 'usually', 'large', 'up', and so forth.[6]

Bennett gives the Ozma problem a slightly different twist. Instead of imagining that the inhabitants of Planet X (his 'alphans') have no notion at all about what we mean by 'left' and 'right', he imagines them to be in *error* about it: they have got our meanings of 'left' and 'right' switched around. The question then becomes, what clues could we transmit to the alphans that would enable them to discover their error? In this case, it is perhaps more natural to send a *chiral* description of the experiment. We tell them: toss a lefty and you will get large side up. They, of course, will toss a righty. But they will *not* get large side up, and at that point they will know that something is wrong.

May we conclude that the Ozma Problem is now solved? If the problem is formulated in the manner of Gardner or Bennett, the answer is yes; we have a way of conveying our meaning of 'left' and 'right' without common observation of a *particular.*[7] But is this to say that there is a way conveying our meanings of these terms without resorting to ostension? In 'Showing and Telling . . .', Martin Curd argues that the answer is no. Curd's point is that the Bennett and Gardner strategies involve a kind of ostension after all — "ostension at a distance," as he calls it. The message we send to Planet X succeeds in getting our meaning across only because it directs the attention of our listeners to an *example* of the term to be defined. Our definition is in principle like the definition "White is the color of newly fallen snow," which is in a sense ostensive even if in giving it we do not point to any snow, but simply tell our listeners where to find some.[8]

Of course, the definition just mentioned may fail of its purpose, for we cannot be sure that snow is white on Planet X, or that there is any snow there at all. What is significant about the failure of parity is that it lets us direct our hearer's attention to something we know will behave on Planet X just as it does on Earth. It lets us be sure we are observing a *type* of structure in common even when we are observing no *token* in common. Social, biological, and chemical phenomena may vary from planet to planet (owing no doubt to differences in local 'boundary conditions'), but we assume that the same basic physical laws hold throughout the universe.

In 'Kant on Incongruent Counterparts', William Harper makes a point that is related to Curd's. He argues that even if the Bennett and Gardner strategies do not involve common observation of a particular, they do require of our audience a special kind of knowledge, namely, non *de dicto* knowledge *de se*.

II

The fall of parity is arguably relevant to Kant's contentions in positive as well as negative respects. Bennett and Gardner cite it as refuting Kant's thesis of 1770, but it can also be used to bolster one of his claims of 1768. This is well brought out in the second of our selections by Earman.

Earman sums up the fall of parity this way: there are certain physical processes that have a greater probability of occurrence than their mirror images. For example, there are particles that decay significantly more often into a right-handed configuration than into a left-handed one. (See Earman's diagram on p. 246.) For a make-believe example in terms of our boxes, we could suppose that a symmetrical eight-cornered box, left on its own to 'decay' by losing a corner, more often decayed into a lefty than a righty.[9]

Earman maintains that parity-violating laws of this sort constitute an 'embarassment' for relationist theories of space. If the two decay modes are merely mirror-images of each other, and if mirror images differ only in the way allowed for by relationism, then how can nature favor one over the other?

The difficulty here may not be a difficulty for relationism as such, but it is certainly a difficulty for certain varieties of it. Consider first the variety I have called externalism.[10] Externalism says that no object is left or right by itself, but has one of these designations only in relation to other objects outside it. Two hands may be the same or different in handedness, but there is no fact about either hand alone that makes it right or left. Calling a hand 'left' only means that it differs in orientation from another hand that we have arbitrarily labelled 'right'. Thus, Kant's famous thought-experiment is rejected; a solitary hand could not be right or left. This position is represented in our volume by Gardner in 'The Fourth Dimension'; it is also well expressed in the following passage from Hermann Weyl:

> Had God, rather than making first a left hand and then a right hand, started with a right hand and then formed another right hand, he would have changed the plan of the universe *not in the first but in the second act*, by bringing forth a hand which was equally rather than oppositely oriented to the first created specimen.[11]

The challenge presented to externalism by parity-violation is this: if God cannot create a solitary left hand, *how can he create a world in which certain processes always (or even usually) have left-handed outcomes?*

The externalist may wish to answer as follows. God could not, to be sure, have made a world in which certain kinds of particles always decay into left-handed configurations, for such a world would in no way differ from a world in which they always decayed into right-handed configurations. But what he could have done, and evidently has done, is this: make a world in which the particles always decay into patterns of the *same* handedness. That is all it takes to make a parity-violating world.

Nonetheless, a puzzle remains for the externalist. God could no doubt *see to it* that certain kinds of particles always decay into configurations of the same handedness. But we need to be able to suppose that the result in question comes about through *law* rather than divine supervision. How can it be a *law* that particles always (or even usually) display decay modes of one orientation rather than another, if orientation is not intrinsic? If one particle has decayed in left-handed fashion, how does the next particle 'know' that it should do likewise? Its 'instruction' cannot be to trace a pattern of a certain intrinsic description; it can only be *to do what the first particle did.*

The problem here is not 'action at a distance', though perhaps that will trouble some. It is rather that the required laws would make ineliminable reference to *particular things*, whereas it is generally supposed to be of the essence of laws that they state relations of *kind to kind*. The problem, in short, is that it is hard to see how an externalist could regard the parity-violating differences in frequency as lawlike.

Consider now the variety of relationism portrayed by Earman. For Earman's relationist, left- and right-handed configurations differ only in what he calls 'presentation'; 'left' and 'right' are fictional absolutist descriptions of the same underlying relational reality. But that makes parity-violating phenomena hard to fathom indeed. The problem now is not that the parity-violating differences in frequency could not be lawlike; it is that they could not sensibly be supposed to occur at all. Things cannot always or usually happen one way rather than another if there is no difference between the two ways!

Earman's suggested response for the relationist is to posit intrinsic properties R^* and L^* that explain (or perhaps we should say, *make possible*) the difference in frequency of outcome. He discusses the pros and cons of this maneuver and finds nothing decisive against it. In the terms I have proposed, his suggestion is that the relationist adopt a form of internalism.

In 'Hand, Knees, and Absolute Space', Graham Nerlich argues that consideration of four-dimensional and nonorientable spaces shows that internalism (and more broadly, any view that makes left and right intrinsic properties) is untenable. The idea is that no intrinsic property can be altered by mere motion, whereas left and right *can* be altered by mere motions in the indicated spaces. Earman himself argues in similar fashion, adducing nonorientable spaces as ruling out internalism. How, then, can he recommend internalism as an escape for the relationist from the problem about parity?

Earman anticipates this question. He maintains that there is no conflict between the claim that failure of parity requires internalism and the claim that nonorientable spaces preclude it. Both can be true, *provided that worlds in which parity fails are not also worlds in which space is nonorientable*. Since parity is violated in our world, Earman concludes that our space is in fact orientable.

Drawing this conclusion may not be enough to save internalism, however. According to the original anti-internalist argument, right and left cannot be intrinsic properties if space is nonorientable; that is to say, the nonorientability of space entails the falsity of internalism. By a familiar theorem of modal logic, it follows that the *possibility* of nonorientable spaces entails the *possibility* that right and left are not intrinsic. Now let us expand the argument by adding the following plausible premise: whether a property is intrinsic or not is *no contingent matter*. That is to say, if a property is intrinsic, it is necessarily intrinsic, and if it is not intrinsic, it is not even possibly intrinsic. The original argument shows that right and left are not *necessarily* intrinsic, since they are not intrinsic in certain possible spaces. The expanded argument lets us conclude that right and left are not intrinsic *at all*, even if our actual space happens to be orientable. To put the point in a nutshell, the mere *possibility* of nonorientable spaces is all it takes to refute internalism.

If the preceding point is correct, we may now give the following argument on Kant's behalf:

(1) The fall of parity shows that externalism is false.
(2) The possibility of 'special spaces' (nonorientable or four-dimensional spaces) would show that internalism is false.
(3) The only remaining view is Kant's own absolutism.
(4) Therefore, either absolutism is correct, or special spaces are impossible.

Since Kant affirmed both limbs of this disjunction, he was right about at least one thing. He may never have entertained the idea of nonorientable spaces, but had he done so he would no doubt have maintained that space is necessarily orientable, just as he maintained that it is necessarily three-dimensional and Euclidean.

III

Kant's third use of incongruent counterparts came in the *Prolegomena to Any Future Metaphysics* of 1783. Here he enlisted them in support of Transcendental Idealism, his doctrine that space and time are not features of things in themselves, but only of appearances. He explained one consequence of Transcendental Idealism as follows in the *Critique of Pure Reason*:

> If the subject, or even only the subjective constitution of the senses in general, be removed, the whole constitution and all the relations of objects in space and time, nay space and time themselves, would vanish. As appearances, they cannot exist in themselves, but only in us.[12]

Thus, according to Transcendental Idealism, space is mind-dependent or ideal.

Two questions may be asked about the third use of incongruent counterparts. First, is the conclusion of 1783 compatible with the conclusions Kant had drawn earlier? To many readers, it has seemed that the answer is no. Second, how is the ideality of space supposed to follow from facts about incongruent counterparts? The connection is by no means obvious.

Jill Vance Buroker has dealt at length with both of these questions in her book *Space and Incongruence*, and we are pleased to be able to include here a précis she has written for this volume. In response to the first question, Buroker maintains that Kant's conclusion in 1768, that space cannot be reduced to relations among objects in space, is not repudiated, but instead presupposed in both of Kant's subsequent uses of incongruent counterparts. In 1770, the facts about space uncovered in 1768 are shown to be incompatible with Leibniz's doctrine that sensation is simply a confused mode of intellection; in 1783, they are shown to be incompatible with the doctrine that spatiality is a feature of things in themselves.

In response to the second question, Buroker calls attention to two doctrines that might have enabled Kant to argue from incongruent counterparts to idealism. The first is the *intelligibility of noumena*. According to Buroker, Kant aligns the distinction between phenomena and noumena (or appearances and things in themselves) with the distinction between sensible objects and intelligible objects. If things in themselves are intelligible objects in the requisite sense, their properties should be graspable by the intellect alone, without reliance on the faculty of sensibility. But the conclusion in 1770, reaffirmed in the *Prolegomena*, is that the difference between a right and left hand *cannot* be grasped by the intellect alone; sensible intuition is required. From this it would follow that objects like hands are not things in themselves.

The second doctrine to which Buroker refers is the *reducibility of relations*. According to this doctrine, any relational fact is derivable from a conjunction of nonrelational or qualitative facts, facts attributing qualities to objects singly. For example, the fact that apple *A* is darker in color than banana *B* might be derivable from the conjunction '*A* is red & *B* is yellow'. On one reading of Leibniz, he held that *all* relational facts are like this. Yet incongruence is *not* like this. As Kant points out, a left and a right hand, though incongruent, may be qualitatively just alike (in all relevant spatial respects). This implies that their incongruence is not derivable from their qualitative natures, for two left hands could exhibit the same qualitative features as a left and a right. Buroker sums up the argument thus: the nature of space, as discovered by Kant, conflicts with the nature of relations, as demanded by Leibniz. Since Kant agreed with the Leibnizian premise about relations, he concluded that space and its contents must be merely ideal.

Each of the doctrines identified by Buroker would make sense of Kant's argument in the *Prolegomena*, but each of them raises further interpretive questions as well. I have pursued some of these questions in 'Incongruent Counterparts and Things in Themselves'.

NOTES

[1] This volume, p. 35.
[2] This volume, p. 77.
[3] This volume, p. 127.
[4] To make the fictitious experiment genuinely isomorphic to the real-life experiment, we should keep the difference in behavior of the boxes statistical.
[5] The reader may wish to know how Bennett's schema applies in the case of Wu's experiment. If I_1 is the initial state in which the cobalt-60 nuclei are spinning in one direction around their magnetic axes, then I_2 (achievable by flipping over the surrounding magnet) would be the initial state in which the nuclei are spinning in the opposite direction. If I_1 results in the majority of electrons being emitted downward (R_1), I_2 will result in the majority of electrons being emitted upward (R_2). I_1 and I_2 are enantiomorphs of each other, but R_1 and R_2 are not.
[6] Gardner's similar definition referring to the Wu experiment is not directly a definition of 'left' and 'right', but a definition of the chiral terms 'north' and 'south' (as they apply to the poles of a magnetic field). With these understood, it is then possible to go on and define 'left' and 'right' by reference to the effects of electric currents on compass needles.
[7] In Chapter 25 of *The Ambidextrous Universe* (3rd ed.; 1989), Gardner discusses a complication that I have omitted: we could *not* communicate our meanings of 'left' and 'right' to a galaxy of time-reversed anti-matter. He goes on to argue that we could not communicate with such a galaxy *at all*, so the exception hardly matters.
[8] Roderick Firth calls this style of definition "ostensive definition by definite description." See his article 'Sense Experience' in *Historical and Philosophical Roots of Perception*, ed. by E. C. Carterette and M. P. Friedman (New York: Academic Press, 1974), pp. 4—18.
[9] We should pause to note that the phenomena cited by Earman show the need for generalizing Bennett's schema for violations of parity. That schema covers only cases in which the initial set-up is asymmetrical, hence admitting of an enantiomorph. In

Earman's cases, the initial state is symmetrical. The following schema will cover both sets of cases: parity is *conserved* only if, given any law whereby initial state *I* leads to resultant state *R*, it is also a law that a *reflection* of *I* would result in a *reflection* of *R*. Equivalently, parity is *violated* if for some states *I* and *R*, *I* leads by law to *R*, but a reflection of *I* does *not* lead to a reflection of *R*. That this schema covers both classes of cases may be seen as follows. In parity-violating cases in which *I* is symmetrical, a reflection of *I* will be of the same type as *I* itself, and the result will be the same as before — not a reflection of *R*, because it is too similar for that. In parity-violating cases in which *I* is asymmetrical, a reflection of *I* will be an enantiomorph of *I*, and the result will be a state that differs from *R* more profoundly than would any enantiomorph of it — not a reflection of *R*, because it is too different for that. The first type of case is illustrated by the symmetrical boxes that always decay into lefties; the second by the lefties that always land large side up.

[10] In 'Right, Left, and the Fourth Dimension' in this volume. For relationists, spatial designations such as 'left' and 'right' must be based exclusively on relations among material things. If the 'right-making' and 'left-making' relations are relations of a hand to other things *outside* it, we have externalism; if they are relations of part to part *within* the hand, we have internalism.

[11] Hermann Weyl, *Symmetry* (Princeton, N.J.: Princeton University Press, 1952), p. 21.

[12] *Critique of Pure Reason*, trans. by Norman Kemp Smith (New York: St. Martin's Press, 1965), A42/B59.

IMMANUEL KANT

ON THE FIRST GROUND OF THE DISTINCTION OF REGIONS IN SPACE[1]

The celebrated Leibniz enriched the sciences by many actual contributions; but he also entertained numerous greater projects, the execution of which the world has in vain awaited from him. I shall not here pronounce an opinion why this was so; whether his essays seemed to him still too incomplete — a scrupulousness peculiar to men of real worth, and one which has ever and again deprived learning of precious fragments — or whether it was with him as with the great chemists, of whom Boerhaave conjectures that they frequently announced performances as if they had them in their power, while actually they were only giving credence to their skill, believing that the execution could not miscarry if they once resolved to undertake it. At least it seems probable that a certain body of mathematical teaching, which Leibniz in anticipation entitled *Analysis Situs*,[2] never existed save in intention. Many writers, including Buffon (in dealing with the involution of Nature in germs), have lamented the loss which we thereby suffer.

I do not know exactly how far the subject which I here propose to consider is akin to that which this great man had in mind; but so far as we can judge from the words [*Analysis Situs*], I am here seeking the first philosophical ground of the possibility of that whose magnitudes Leibniz proposed to determine in a mathematical manner. For the positions of the parts of space in relation to one another presuppose the region towards which they are ordered in such relation; and this region, in ultimate analysis, consists not in the relation of one thing in space to another (which is properly the concept of position) but in the relation of the system of these positions to the absolute world-space. In anything extended the position of parts relatively to one another can be adequately determined from consideration of the thing itself; but the region towards which this ordering of the parts is directed involves reference to the space outside the thing; not, indeed, to points in this wider space — for this would be nothing else but the position of the parts of the thing in outer relation — but to universal space as a unity of which every extension must be regarded as a part.

As the explanation of these concepts is first to be found in what follows, it is not surprising if the reader as yet finds them very unintelligible; and I therefore limit myself to this one further remark, that my aim in this treatise is to investigate whether there is not to be found in the intuitive judgments of extension, such as are contained in geometry, an evident proof *that absolute space has a reality of its own, independent of the existence of all matter, and indeed as the first ground of the possibility of the compositeness of matter.*

Everybody knows how futile have been the endeavours of the philosophers, by means of the most abstract propositions of metaphysics, to settle this point once for all; and I know of no attempt, save one, to carry this out *a posteriori* as it were, that is, by means of other undeniable propositions which, though themselves lying outside the realm of metaphysics, can afford through their application in the concrete a criterion of their correctness. The one attempt, to which I have referred, was made by the celebrated Euler, the elder, in 1748, as recorded in the *History of the Royal Berlin Academy of Sciences*[3] for that year. But so far from fully achieving his purpose, he only brings to view the difficulty of assigning to the most general laws of motion a determinate meaning, should we assume no other concept of space than that obtained by abstraction from the relation of actual things. The no less notable difficulties which remain in the application of the aforesaid laws, when we endeavour to represent them in the concrete according to the concept of absolute space, are left unconsidered. The proof which I here seek should supply, not to the mechanists (as Herr Euler intended), but to the geometers themselves, a convincing ground for asserting the actuality of their absolute space, and should do so with the evidence to which they are accustomed. With this purpose in view, I make the following preparatory observations.

In physical space, on account of its three dimensions, we can conceive three planes which intersect one another at right angles. Since through the senses we know what is outside us only in so far as it stands in relation to ourselves, it is not surprising that we find in the relation of these intersecting planes to our body the first ground from which to derive the concept of regions in space.[4] The plane to which the length of our body stands perpendicular is called, in reference to us, horizontal; it gives rise to the distinction of the regions we indicate by *above* and *below.* Two other planes, also intersecting at right angles, can stand perpendicular to this horizontal plane, in such manner that the length of the human body is conceived as lying in the line of their intersection.

One of these vertical planes divides the body into two outwardly similar parts and supplies the ground for the distinction between *right* and *left*; the other, which is perpendicular to it, makes it possible for us to have the concept of *before* and *behind*. In a written page, for instance, we have first to note the difference between front and back and to distinguish the top from the bottom of the writing; only then can we proceed to determine the position of the characters from right to left or conversely. Here the parts arranged upon the surface have always the same position relatively to one another, and the parts taken as a whole present always the same outlines howsoever we may turn the sheet. But in our representation of the sheet the distinction of regions is so important, and is so closely bound up with the impression which the visible object makes, that the very same writing becomes unrecognisable when seen in such a way that everything which formerly was from left to right is reversed and is viewed from right to left.

Even our judgments about the cosmic regions are subordinated to the concept we have of regions in general, in so far as they are determined in relation to the sides of the body. All other relations that we may recognise, in heaven and on earth, independently of this fundamental conception, are only positions of objects relatively to one another. However well I know the order of the cardinal points, I can determine regions according to that order only in so far as I know towards which hand this order proceeds; and the most complete chart of the heavens, however perfectly I might carry the plan in my mind, would not teach me, from a known region, North say, on which side to look for sunrise, unless, in addition to the positions of the stars in relation to one another, this region were also determined through the position of the plan relatively to my hands. Similarly, our geographical knowledge, and even our commonest knowledge of the position of places, would be of no aid to us if we could not, by reference to the sides of our bodies, assign to regions the things so ordered and the whole system of mutually relative positions.

It is even the case that a very notable characteristic of natural organisms, which at times may even give occasion for the distinction of species, consists in the definite direction in which the arrangement of their parts is turned, a feature through which two creatures can be distinguished although they entirely agree both in size and proportion, and even in the position of their parts relatively to one another. The hairs on the crown of every man's head are turned from left to right.[5]

The hop-plant always twines round its pole from left to right; beans, however, take the opposite course. Almost all snails, only some three species excepted, have their spiral turning from left to right, that is, if we proceed from above downwards, from the apex to the mouth.[6] Since in the case of the natural existences just cited the cause of the twist or spiral lies in their very germs, this definite character remains constant in creatures of the same species without any relation to the hemisphere in which they may be found, or to the direction of the daily motion of the sun and the moon, which for us runs from left to right but for our antipodes in the opposite direction. On the other hand, when a certain revolution can be ascribed to the path of these heavenly bodies — as Mariotte[7] professes to have observed in the case of the winds, which from new to full moon tend to work round the whole compass from left to right — this motion must in the other hemisphere go from right to left, as indeed Don Ulloa considers to have been established by his observations in the South Seas.

Since the different feeling of right and left side is of such necessity to the judgment of regions, Nature has directly connected it with the mechanical arrangement of the human body, whereby one side, the right, has an indubitable advantage in dexterity and perhaps also in strength. If, therefore, we set aside individual exceptions which, like cases of squinting, cannot disturb the generality of the rule according to the natural order, all the peoples of the earth are right-handed. In mounting on horseback or striding over a ditch, the body is more easily moved from right to left than vice versa. Everywhere men write with the right hand; with it they do everything for which skill and strength are demanded. But if some investigators, e.g. Borelli and Bonnet, are to be believed, while the right hand seems to have the advantage over the left in mobility, the left has the advantage over the right in sensibility. Borelli likewise assigns to the left eye, and Bonnet to the left ear, the possession of a greater sensibility than the corresponding organ on the right side. And thus the two sides of the human body, in spite of their great outer similarity, are sufficiently distinguished by a well-marked feeling, even if we leave out of account the differing positions of the inner parts and the noticeable beat of the heart, which at every contraction strikes with its apex in oblique motion against the left side of the breast.

What, therefore, we desire to show is that the complete ground of determination of the shape of a body rests not merely upon the position of its parts relatively to one another, but further on a relation to the universal space which geometers postulate — a relation, however, which is such that it cannot itself be immediately perceived. What we do perceive are those differences between bodies which depend exclusively upon the ground which this relation affords. If two figures drawn upon a plane are equal and similar, they can be superimposed. But with physical extension and also with lines and surfaces that do not lie in one plane, the case is often quite different. They can be perfectly equal and similar, yet so different in themselves that the boundaries of the one cannot be at the same time the boundaries of the other. A screw which winds round its axis from left to right will not go into a threaded cylinder whose worm goes from right to left, although the thickness of the stem and the number of turns in an equal length correspond. Two spherical triangles can be perfectly equal and similar, and yet not allow of superposition. But the commonest and clearest example is to be found in the limbs of the human body, which are symmetrically disposed about its vertical plane. The right hand is similar and equal to the left, and if we look at one of them alone by itself, at the proportions and positions of its parts relatively to one another and at the magnitude of the whole, a complete description of it must also hold for the other in every respect.

When a body is perfectly equal and similar to another, and yet cannot be included within the same boundaries, I entitle it the incongruent counterpart of that other. To show its possibility, take a body which is not composed of two halves symmetrically disposed to a single intersecting surface, say a human hand. From all points of its surface draw perpendiculars to a plane set over against it, and produce them just as far behind the plane as these points lie in front of it; the extremities of the lines so produced, if connected, then compose the surface and shape of a physical body which is the incongruent counterpart of the first; i.e., if the given hand is the right, its counterpart is the left. The image of an object in a mirror rests upon the same principle; for it always appears just as far behind the mirror as the object lies in front of its surface, and so the mirrored image of a right hand is always a left. If the object itself consists of two incongruent counterparts, as does the human body when divided by a vertical section from front to back, its image is congruent with it, as can easily be seen by allowing it in thought to make a half turn; for the counterpart of the counterpart of an object is necessarily congruent with the object.

The above considerations may suffice for understanding the possibility of spaces which are completely equal and similar and yet incongruent. We now proceed to the philosophical application of these concepts. From the common example of the two hands, it is already clear that the shape of one body can be completely similar to that of another, and the magnitude of their extension exactly the same, while yet there remains an inner difference, namely that the surface which bounds the one cannot possibly bound the other. Since this surface bounds the physical space of the one but cannot serve as boundary to the other, however one may turn and twist it, this difference must be such as rests upon an inner ground. This inner ground cannot, however, depend on any difference in the mode of connection of the parts of the body relatively to one another; for, as can be seen from the examples adduced, in this respect everything may be completely identical in the two cases. Nevertheless, if we conceive the first created thing to be a human hand, it is necessarily either a right or a left, and to produce the one a different act of the creating cause is required from that whereby its counterpart can come into being.

Should we, then, adopt the conception held by many modern philosophers, especially in Germany, that space consists only in the outer relations of the parts of matter existing alongside one another, in the case before us all actual space would be that which this hand occupies. But since, whether it be right or left, there is no difference in the relations of its parts to one another, the hand would in respect of this characteristic be absolutely indeterminate, i.e., it would fit either side of the human body, which is impossible.

Thus it is evident that instead of the determinations of space following from the positions of the parts of matter relatively to one another, these latter follow from the former. It is also clear that in the constitution of bodies differences are to be found which are real differences, and which are grounded solely in their relation to absolute, primary space. For, only through this relation is the relation of bodily things possible. Since absolute space is not an object of an outer sensation, but a fundamental concept which first makes all such sensations possible, it further follows that whatsoever in the outline of a body exclusively concerns its reference to pure space, can be apprehended only through comparison with other bodies.

A reflective reader will accordingly regard as no mere fiction that concept of space which the geometer has thought out and which clear-thinking philosophers have incorporated into the system of natural philosophy. There is, indeed, no lack of difficulties surrounding this concept, if we attempt to comprehend its reality — a reality which is sufficiently intuitable to inner sense — through ideas of reason. This difficulty always arises when we attempt to philosophise on the first data of our knowledge. But it reaches its maximum when, as in this case, the consequences of an assumed concept [that of spatial relations as subsequent to and dependent on the relations of bodies to one another] contradict the most obvious experience.

NOTES

[1] [This treatise was published in 1768. For a statement and discussion of Kant's changing views with regard to the argument here developed see Vaihinger, *Commentar zu Kant's Kritik der reinen Vernunft*, Bd. II, pp. 518 ff.; Kemp Smith, *Commentary to Kant's Critique of Pure Reason*, pp. 161 ff.; this volume, pp. 43—48.]

[2] [The programme of this new science Leibniz outlines in a fragment published by Gerhardt, *Leibnizens mathematische Schriften*, Bd. V, pp. 178—83.]

[3] [Euler there published his *Refléxions sur l'espace et le temps*.]

[4] [Kant returned to this subject in 1786 in his treatise, *Was heisst: sich im Denken orientiren?*]

[5] [Cf. Walter Kidd, *The Direction of Hair in Animals and Man*, 1903, pp. 76—77: "Over the posterior fontenelle the familiar whorl or 'crown' is always present and may be to the right or left of the middle line, seldom quite in the middle line, and it is double in a certain number of persons, one whorl lying on each side of the middle line."]

[6] [The facts, as now known, are more complicated than Kant here suggests. Not only are there many examples among climbing plants both of right-handed (i.e., following the sun, proceeding from left to right) and of left-handed twining stems, but the direction of turning is not always constant throughout a natural order of plants, and sometimes is reversed even in successive internodes of the same stem. In Bittersweet, a plant in our hedgerows, individuals are occasionally found to twine in opposite directions. As regards molluscs, about seventeen genera have spiral shells which are normally sinistral; in about fourteen genera the majority of the species are dextral, but in each genus there are several species with sinistral shells. Some four genera contain species of which the individuals seem to be indifferently either dextral or sinistral. Also there are about two hundred species in which sinistral individuals are exceptionally found among the normal dextral forms. For this information I am indebted to my colleague, Professor J. D. Ashworth.]

[7] [See Kant's treatise on the theory of the winds, *Werke*, Berlin ed., Bd. I, p. 502.]

IMMANUEL KANT

SELECTION FROM SECTION 15 OF *DISSERTATION*

C. *The concept of space is thus a pure intuition*, since it is a singular concept. It is not put together from sensations, but is the fundamental form of all outer sensation. This pure intuition can be readily observed in the axioms of geometry, and in every mental construction of postulates or of problems. For that space has not more than three dimensions, that there is but one straight line between two points, that from a given point in a plane surface with a given straight line as radius a circle can be described, etc., are not inferred from any universal notion of space, but can only be discerned[1] in space in the concrete. We cannot by any sharpness of intellect describe discursively, that is, by intellectual marks, the distinction in a given space between things which lie towards one quarter, and things which are turned towards the opposite quarter. Thus if we take solids completely equal and similar but incongruent, such as the right and left hands (so far as they are conceived only according to extension), or spherical triangles from two opposite hemispheres, although in every respect which admits of being stated in terms intelligible to the mind through a verbal description they can be substituted for one another, there is yet a diversity which makes it impossible for the boundaries of extension to coincide. It is therefore clear that in these cases the diversity, that is, the incongruence, cannot be apprehended except by pure intuition. Hence geometry employs principles which not only are unquestioned and discursive, but which are such as fall under the mind's direct observation. *Evidence* in demonstrations (meaning thereby the *Clearness* of assured knowledge, so far as this clearness can be likened to that of sense) is found in geometry not merely in the highest degree, but is found there alone of all the pure sciences. Geometrical evidence is thus the model for, and the means of attaining, all evidence in the other sciences. For since geometry contemplates the relations of space, the concept of which contains in itself the very form of all sensual intuition, there can be nothing clear and perspicuous in things perceived by outer sense except through the mediation of the intuition which that science is occupied in contemplating. Further, geometry does not demonstrate its universal propositions by apprehending the object through a universal concept, as is done in matters of reason, but by submitting it to the eyes in a singular intuition, as is done in matters of sense.[2]

NOTES

[1] [*cerni*]

[2] I here pass over the proposition that space must necessarily be conceived as a continuous quantum, since that is easily demonstrated. Owing to this continuity it follows that the simple in space is not a part but a limit. But a limit in general is that in a continuous quantum which contains the ground of its limits [*limes;* elsewhere, throughout, the term used is *terminus*]. A space which is not the limit of a second space is complete, i.e., a solid. The limit of a solid is a surface, of a surface a line, of a line a point. Thus there are three sorts of limits in space, just as there are three dimensions. Of these limits two, surface and line, are themselves spaces. The concept of limit has no application to quanta other than space and time.

PROLEGOMENA, SECTION 13

§ 13. Those who cannot yet rid themselves of the notion that space and time are actual qualities inherent in things in themselves may exercise their acumen on the following paradox. When they have in vain attempted its solution and are free from prejudices at least for a few moments, they will suspect that the degradation of space and time to mere forms of our sensuous intuition may perhaps be well founded.

If two things are quite equal in all respects as much as can be ascertained by all means possible, quantitatively and qualitatively, it must follow that the one can in all cases and under all circumstances replace the other, and this substitution would not occasion the least perceptible difference. This in fact is true of plane figures in geometry; but some spherical figures exhibit, notwithstanding a complete internal agreement, such a difference in their external relation that the one figure cannot possibly be put in the place of the other. For instance, two spherical triangles on opposite hemispheres, which have an arc of the equator as their common base, may be quite equal, both as regards sides and angles, so that nothing is to be found in either, if it be described for itself alone and completely, that would not equally be applicable to both; and yet the one cannot be put in the place of the other (that is, upon the opposite hemisphere). Here, then, is an internal difference between the two triangles, which difference our understanding cannot describe as internal and which only manifests itself by external relations in space. But I shall adduce examples, taken from common life, that are more obvious still.

What can be more similar in every respect and in every part more alike to my hand and to my ear than their images in a mirror? And yet I cannot put such a hand as is seen in the glass in the place of its original; for if this is a right hand, that in the glass is a left one, and the image or reflection of the right ear is a left one, which never can take the place of the other. There are in this case no internal differences which our understanding could determine by thinking alone. Yet the differences are internal as the senses teach, for, notwithstanding their complete equality and similarity, the left hand cannot be enclosed in the same bounds as the right one (they are not congruent); the glove of one hand cannot be used for the other. What is the solution? These objects are not representations of things as they are in themselves and as some mere[1] understanding would know them, but sensuous intuitions, that is, appearances whose possibility rests upon the relation of certain things unknown in themselves to something else, namely, to our sensibility. Space is the form of the external intuition of this sensibility, and the internal determination of every space is possible only by the determination of its external relation to the whole of space, of which it is a part (in other words, by its relation to the outer sense). That is to say, the part is possible only through the whole, which is never the case with things in themselves, as objects of the mere understanding, but which may well be the case with mere appearances. Hence the difference between similar and equal things which are not congruent (for instance, two symmetric helices) cannot be made intelligible by any concept, but only by the relation to the right and left hands which immediately refers to intuition.

NOTE

1 [In German, *pure*. The clause is meant ironically.—L.W.B.]

AUGUST FERDINAND MÖBIUS

ON HIGHER SPACE[1]

§139, page 181. If, given two figures, to each point of one corresponds a point of the other so that the distance between any two points of one is equal to the distance between the corresponding points of the other, then the figures are said to be *equal and similar.*

. . .

§140, pages 182—183. *Problem. — To construct a system of n points which is equal and similar to a given system of n points.*

Solution. Let A, B, C, D, . . . , be the points of the given system, and A', B', C', D', . . . , the corresponding points of the system to be constructed. We have to distinguish three cases according as the points of the first set lie on a line, or in a plane, or in space.

. . .

Finally, if the given system lies in space, then A' is entirely arbitrary, B' is an arbitrary point of the spherical surface which has A' for center and AB for radius, C' is an arbitratry point of the circle in which the two spherical surfaces drawn from A' with AC as radius and from B' with BC as radius intersect, and D' is one of the two points in which the three spherical surfaces drawn from A' with AD, from B' with BD, and from C' with CD as radii intersect. In the same way as D' will also each of the remaining points, for example, E', be found, only that of the two common intersections of the spherical surfaces drawn from A', B', C', with AE, BE, CE as radii, that one is taken which lies on the same side or on the opposite side of the plane $A'B'C'$ as D', according as the one or the other is the case with the corresponding points in the given system.

For the determination of A' therefore no distance is required, for the determination of B' one, for the determination of C' two, and for the determination of each of the remaining $n - 3$ points three. Therefore in all

$$1 + 2 + 3(n - 3) = 3n - 6$$

distances are required.

Remark. — Thus only for the point D', and for none of the remaining points, are we free to choose between the two intersections on the three spherical surfaces falling on opposite sides of the plane $A'B'C'$. These two intersections are distinguished from each other in this way, that looking from one the order of the points A', B', C' is from right to left, but from the other from left to right, or, as also we can express it, that the former point lies on the left, the latter on the right of the plane $A'B'C'$. Now according as we choose for D' the one or the other of these two points, so also will the order formed be the same or different from that in which the point D appears from the points A, B, C. In both cases are the systems A, B, C, D, . . . , and A', B', C', D', . . . indeed equal and similar, but only in the first case can they be brought into coincidence.

It seems remarkable that solid figures can have equality and similarity without having coincidence, while always, on the contrary, with figures in a plane or systems of points on a line equality and similarity are bound with coincidence. The reason may be looked for in this, that beyond the solid space of three dimensions there is no other, none of four dimensions. If there were no solid space, but all space relations were contained in a single plane, then would it be even as little possible to bring into coincidence two equal and similar triangles in which corresponding vertices lie in opposite orders. Only in this way can we accomplish this, namely by letting one triangle make a half revolution around one of its sides or some other line in its plane, until it comes into the plane again. Then with it and the other triangle will the order of the corresponding vertices be the same, and it can be made to coincide with the other by a movement in the plane without any further assistance from solid space.

The same is true of two systems of points A, B, . . . and A', B', . . . on one and the same straight line. If the directions of AB and $A'B'$ are opposite, then in no way can a coincidence of corresponding points be brought about by a movement of one system along the line, but only through a half revolution of one system in a plane going through the line.

For the coincidence of two equal and similar systems, A, B, C, D, . . . and A', B', C', D', . . . in space of three dimensions, in which the points D, E, . . . and D', E', . . . lie on opposite sides of the planes ABC and $A'B'C'$, it will be necessary, we must conclude from analogy, that we should be able to let one system make a half revolution in a space of four dimensions. But since such a space cannot be thought, so is also coincidence in this case impossible.

NOTE

[1] (From *Der barycentrische Calcul*, Leipzig, 1827, part 2, Chapter 1.)

NORMAN KEMP SMITH

THE PARADOX OF INCONGRUOUS COUNTERPARTS

The purpose, as already noted, of the above sections II. to IV., as added in the second edition, is to afford 'confirmation' of the ideality of space and time. That being so, it is noticeable that Kant has omitted all reference to an argument embodied, for this same purpose, in §13 of the *Prolegomena*. The matter is of sufficient importance to call for detailed consideration.[1]

As the argument of the *Prolegomena* is somewhat complicated, it is advisable to approach it in a light of its history in Kant's earlier writings. It was to his teacher Martin Knutzen that Kant owed his first introduction to Newton's cosmology; and from Knutzen he inherited the problem of reconciling Newton's mechanical view of nature and absolute view of space with the orthodox Leibnizian tenets. In his first published work[2] Kant seeks to prove that the very existence of space is due to gravitational force, and that its three-dimensional character is a consequence of the specific manner in which gravity acts. Substances, he teaches, are unextended. Space results from the connection and order established between them by the balancing of their attractive and repulsive forces. And as the law of gravity is merely contingent, other modes of interaction, and therefore other forms of space, with more than three dimensions, must be recognised as possible.

"A science of all these possible kinds of space would undoubtedly be the highest enterprise which a finite understanding could undertake in the field of geometry."[3]

In the long interval between 1747 and 1768 Kant continued to hold to some such compromise, retaining Leibniz's view that space is derivative and relative, and rejecting Newton's view that it is prior to, and pre-conditions, all the bodies that exist in it. But in that latter year he published a pamphlet[4] in which, following in the steps of the mathematician, Euler,[5] he drew attention to certain facts which would seem quite conclusively to favour the Newtonian as against the Leibnizian interpretation of space. The three dimensions of space are primarily distinguishable by us only through the relation in which they stand to our body. By relation to the plane that is at right angles to our body we distinguish 'above' and 'below'; and similarly through the other two planes we determine what is 'right' and 'left,' 'in front' and 'behind.' Through these distinctions we are enabled to define differences which cannot be expressed in any other manner. All species of hops — so Kant maintains — wind themselves around their supports from left to right, whereas all species of beans take the opposite direction. All snail shells, with some three exceptions, turn, in descending from their apex downwards, from left to right. This determinate direction of movement, natural to each species, like the difference in spatial configuration between a right and a left hand, or between a right hand and its reflection in a mirror, involves in all cases a reference of the given object to the wider space within which it falls, and ultimately to space as a whole. Only so can its determinate character be distinguished from its opposite counterpart. For as Kant points out, though the right and the left hand are *counterparts*, that is to say, objects which have a common definition so long as the arrangement of the parts of each is determined in respect to its central line of reference, they are none the less inwardly *incongruent*, since the one can never be made to occupy the space of the other. As he adds in the *Prolegomena*, the glove of one hand cannot be used for the other hand. This inner incongruence compels us to distinguish them as different, and this difference is only determinable by location of each in a single absolute space that constrains everything within it to conform to the conditions which it prescribes. In three-dimensional space everything must have a right and a left side, and must therefore exhibit such inner differences as those just noted. Spatial determinations are not, as Leibniz teaches, subsequent to, and dependent upon, the relations of bodies to one another; it is the former that determine the latter.

"The reason why that which in the shape of a body exclusively concerns its relation to pure space can be apprehended by us only through its relation to other bodies, is that absolute space is not an object of any outer sensation, but a fundamental conception which makes all such differences possible."[6]

Kant enforces his point by arguing that if the first portion of creation were a human hand, it would have to be either a right or a left hand. Also, a different act of creation would be demanded according as it was the one or the other. But if the hand alone existed, and there were no pre-existing space, there would be no inward difference in the relations of its parts, and nothing outside it to differentiate it. It would therefore be entirely indeterminate in nature, *i.e.* would suit either side of the body, which is impossible.

This adoption of the Newtonian view of space in 1768 was an important step forward in the development of Kant's teaching, but could not, in view of the many metaphysical difficulties to which it leads, be permanently retained; and in the immediately following year — a year which, as he tells us,[7] "gave great light" — he achieved the final synthesis which enabled him to combine all that he felt to be essential in the opposing views. Though space is an absolute and preconditioning source of differences which are not conceptually resolvable, it is a merely subjective form of our sensibility.

Now it is significant that when Kant expounds this view in the *Dissertation* of 1770, the argument from incongruous counterparts is no longer employed to establish the absolute and pre-conditioning character of space, but only to prove that it is a pure non-conceptual intuition.

"Which things in a given space lie towards one side, and which lie towards the other, cannot by any intellectual penetration be discursively described or reduced to intellectual marks. For in solids that are completely similar and equal, but incongruent, such as the right and left hand (conceived solely in terms of their extension), or spherical triangles from two opposite hemispheres, there is a diversity which renders impossible the coincidence of their spatial boundaries. This holds true, even though they can be substituted for one another in all those respects which can be expressed in marks that are capable of being made intelligible to the mind through speech. It is therefore evident that the diversity, that is, the incongruity, can only be apprehended by some species of pure intuition."[8]

There is no mention of this argument in the first edition of the *Critique*, and when it reappears in the *Prolegomena* it is interpreted in the light of an additional premiss, and is made to yield a very different conclusion from that drawn in the *Dissertation*, and a *directly opposite conclusion* from that drawn in 1768. Instead of being employed to establish either the intuitive character of space or its absolute existence, it is cited as evidence in proof of its subjectivity. As in 1768, it is spoken of as strange and paradoxical, and many of the previous illustrations are used. The paradox consists in the fact that bodies and spherical figures, conceptually considered, can be absolutely identical, and yet for intuition remain diverse. This paradox, Kant now maintains[9] in opposition to his 1768 argument, proves that such bodies and the space within which they fall are not independent existences. For were they things in themselves, they would be adequately cognisable through the pure understanding, and could not therefore conflict with its demands. Being conceptually identical, they would necessarily be congruent in every respect. But if space is merely the form of sensibility, the fact that in space the part is only possible through the whole will apply to everything in it, and so will generate a fundamental difference between conception and intuition.[10] Things in themselves are, as such, unconditioned, and cannot, therefore, be dependent upon anything beyond themselves. The objects of intuition, in order to be possible, must be merely ideal.

Now the new premiss which differentiates this argument from that of 1768, and which brings Kant to so opposite a conclusion, is one which is entirely out of harmony with the teaching of the *Critique*. In this section of the *Prolegomena* Kant has unconsciously reverted to the dogmatic standpoint of the *Dissertation*, and is interpreting understanding in the illegitimate manner which he so explicitly denounces in the section on *Amphiboly*.

"The mistake ... lies in employing the understanding contrary to its vocation transcendentally [*i.e.* transcendently] and in making objects, *i.e.* possible intuitions, conform to concepts, not concepts to possible intuitions, on which alone their objective validity rests."[11]

The question why no mention of this argument is made in the second edition of the *Critique* is therefore answered. Kant had meantime, in the interval between 1783 and 1787,[12] become aware of the inconsistency of the position. So far from being a paradox, this assumed conflict rests upon a false view of the function of the understanding.[13] The relevant facts may serve to confirm the view of space as an intuition in which the whole precedes the parts;[14] but they can afford no evidence either of its absoluteness or of its ideality. In 1768 they seem to Kant to prove its absoluteness, only because the other alternative has not yet occurred to him. In 1783 they seem to him to prove its ideality, only because he has not yet completely succeeded in emancipating his thinking from the dogmatic rationalism of the *Dissertation*.

As already noted,[15] Kant's reason for here asserting that space is intuitive in nature, namely, that in it the parts are conditioned by the whole, is also his reason for elsewhere describing it as an Idea of Reason. The further implication of the argument of the *Prolegomena*, that in the noumenal sphere the whole is made possible only by its unconditioned parts, raises questions the discussion of which must be deferred. The problem recurs in the *Dialectic* in connection with Kant's definition of the Idea of the unconditioned. In the Ideas of Reason Kant comes to recognise the existence of concepts which do not conform to the reflective type analysed by the traditional logic, and to perceive that these Ideas can yield a deeper insight than any possible to the discursive understanding. The above rationalistic assumption must not, therefore, pass unchallenged. It may be that in the noumenal sphere all partial realities are conditioned by an unconditioned whole.

Concluding Paragraph.[16] — The wording of this paragraph is in keeping with the increased emphasis which in the *Introduction* to the second edition is given to the problem, how *a priori* synthetic judgments are possible. Kant characteristically fails to distinguish between the problems of pure and applied mathematics, with resulting inconsecutiveness in his argumentation.

NOTES

1 Upon this subject cf. Vaihinger's exhaustive discussion in ii. p. 518 ff.

2 *Gedanken von der wahren Schätzung der lebendigen Kräfte* (1747).

3 *Op. cit.* § 10. Cf. above, p. 117 ff.

4 *Von dem ersten Grunde des Unterschiedes der Gegenden im Raume.*

5 Euler, *Réflexions sur l'espace et le temps* (1748). Vaihinger (ii. p. 530) points out that Kant may also have been here influenced by certain passages in the controversy between Leibniz and Clarke.

6 *Loc. cit.*, at the end.

7 In the Dorpater manuscript, quoted by Erdmann in his edition of the *Prolegomena*, p. xcvii *n.*

8 §15 C.

9 So also in the *Metaphysical First Principles of Natural Science* (1786), *Erstes Hauptstück, Erklärung* 2, *Anmerkung* 3.

10 Cf. above, p. 105.

11 A 289 = B 345.

12 More exactly between the writing of the *Metaphysical First Principles* (in which as above noted the argument of the *Prolegomena* is endorsed) and 1787.

13 Cf. A 260 ff. = B 316 ff. on the *Amphiboly of Reflective Concepts.*

14 The *Dissertation* cites the argument only with this purpose in view. And yet it is only from the *Dissertation* standpoint that the wider argument of the *Prolegomena* can be legitimately propounded.

15 Above, pp. 96—8, 102 *n.* 4; below, pp. 390—1.

16 B 73.

LUDWIG WITTGENSTEIN

TRACTATUS 6.36111

6.36111 Kant's problem about the right hand and the left hand, which cannot be made to coincide, exists even in two dimensions. Indeed, it exists in one-dimensional space

- - - o ——— × - - × ——— o - - - -

a *b*

in which the two congruent figures, *a* and *b*, cannot be made to coincide unless they are moved out of this space. The right hand and the left hand are in fact completely congruent. It is quite irrelevant that they cannot be made to coincide.

A right-hand glove could be put on the left hand, if it could be turned round in four-dimensional space.

PETER REMNANT

INCONGRUENT COUNTERPARTS AND ABSOLUTE SPACE[1]

Immanuel Kant is usually associated with the theory that space is not a thing in itself, nor a property of things in themselves, but a "form of sensibility" by means of which we organize our sensations. This theory, however, was only the last of several strenuous attempts by Kant to arrive at a satisfactory account of the nature of space.

In 1768, thirteen years before the appearance of the *Critique of Pure Reason*, Kant published a paper entitled "On the first ground of the distinction of regions in space",[2] in which he attempted to prove that "absolute space has a reality of its own, independent of the existence of all matter" (this volume, p. 28). The argument employed for this purpose seems to have been entirely his own invention, and is characteristically odd and ingenious.

I

Many natural objects come in pairs, the members of which are "perfectly equal and similar" to one another and yet cannot be "included within the same boundaries"; such pairs Kant calls "incongruent counterparts" (this volume, p. 31). As examples he mentions snails whose shells twist opposite ways, screws of the same overall dimensions but threaded in opposite directions, and, most familiar of all, pairs of human hands precisely alike except for one being right and the other left. Given any pair of incongruent counterparts one of the pair is, so to speak, the mirror image of the other.

Now when two objects are incongruent counterparts, Kant continues, "if we look at one of them alone by itself, at the proportions and positions of its parts relatively to one another and at the magnitude of the whole, a complete description of it must also hold for the other in every respect" (this volume, p. 31). In other words, a description which restricts itself to an account of the way in which the various parts of such an object are related to each other, without making reference to the relations between the object and anything else, will do equally well as a description of the incongruent counterpart of that object. The incongruence of the two objects will be revealed only when one is compared with the other or when both are considered in relation to some appropriate third thing.

This point is crucial to the whole subsequent argument and deserves some elaboration. Suppose I am presented with a comprehensive description of the ways in which the various parts of a human hand are inter-related, an account which restricts itself entirely to the intrinsic properties of the hand and neither refers to nor presupposes any relationships between the hand and anything else. Then the description will include such information as that the line across the palm from the tip of the thumb to a given point on the middle of the wrist is three-quarters of the length of the line from that point to the tip of the middle finger, and that there is a thirty degree angle between the two lines. But it will not include such information as that the line from fingertip to wrist is eight inches long — since this would presuppose comparison between the hand and a standard of length — nor will it include such information as that the line from wrist to thumb is to the right of the line from wrist to middle finger, or that the line from wrist to thumb can be reached by rotating the line from wrist to middle finger through thirty degrees in a clockwise direction — since in either case such information would presuppose a comparison between the hand and some standard of direction. However, unless I am given *some* such information about the directions from one line or point to others it is obvious that I will be unable to say whether what is being described is a left hand or a right hand. (Nor, although Kant apparently fails to notice, will there be any indication of the size of the object. Only when we have been given information about the relations between two things, or about the relations of each to some third thing, can we tell whether they are in fact counterparts — that is, whether in addition to being of similar structure they are of the same size.)

It is on the basis of considerations such as these that Kant concludes that for any pair of incongruent counterparts a description of the intrinsic properties of either member of the pair will serve equally well as a description of the other. I do not entirely share Kant's confidence in the reasoning underlying this conclusion; however I intend to assume that he is correct on this point and then attempt to show that his subsequent argument breaks down nonetheless. I shall also overlook the fact that the kind of descriptions I have been talking about can be arrived at only by someone, physical body and all, confronting the object in question and examining it. Whether, under these circumstances, it is legitimate to speak of a description which takes into account nothing but the intrinsic properties of a single object, seems extremely doubtful.

"We now proceed to the philosophical application of these concepts" (this volume, p. 32). A left hand and a right hand, however alike they may be in many respects, are obviously different in one important respect. No amount of turning and twisting them will enable us to fit one into the space occupied by the other — we cannot, for example, put a left hand into a right-hand glove. Now if Kant is right this difference between right hands and left hands does not consist in any difference in the intrinsic arrangement of the parts of the two hands since in this respect they are identical. In what does it consist then? In view of the fact that the difference in question only reveals itself when we consider each counterpart in relation to other objects there is a temptation to conclude that the difference between a left hand and a right hand *consists* in the different relationships in which each stands to other physical things.

However Kant's next move is designed to rule out any such convenient solution and, along with it, the Leibnizian doctrine that "space consists only in the outer relations of the parts of matter existing alongside one another . . ." (this volume, p. 32). Let us suppose that there exists nothing in the universe but a single human hand. Then no description of the "outer relations of the parts of matter existing alongside one another" will indicate whether this is a left hand or a right hand since such a description will be entirely restricted to the intrinsic properties of the hand. Nevertheless the hand must be either right or left since, if we were to suppose it indeterminate and were then to suppose a handless human body brought into existence, we should have to conclude that the hand would fit equally appropriately onto either wrist — and this is obviously absurd.

The hand must therefore be either a right hand or a left hand even though no other physical object exists. Which it is cannot be determined by any of its own intrinsic properties nor, since no other object exists, by any relationship between itself and other physical things. Kant concludes that whether the hand is left or right must depend upon the relation between it and absolute space, and in general that "the complete ground of determination of the shape of a body rests not merely upon the position of its parts relatively to one another, but further on a relation to the universal space which geometers postulate — a relation, however, which is such that it cannot itself be immediately perceived". And he adds that "what we do perceive are those differences between bodies which depend exclusively upon the ground which this relation affords" (this volume, p. 31).

Kant abandoned his belief in a Newtonian absolute space almost immediately after the pubiication of the paper which I have been summarizing, but he continued to maintain that the shapes of physical objects are determined by their relations to the whole system of space and that these relations cannot be perceived but must be constructed out of perceptible relations among physical things. He also continued to employ the notion of incongruent counterparts in subsequent arguments about the nature of space. In the *Inaugural Dissertation* of 1770, by which time he had decided that "space is not something objective and real . . . but subjective and ideal" (Handyside, p. 61), he calls up the peculiarities of incongruent counterparts in the process of showing that "geometry does not demonstrate its universal propositions by apprehending the object through a universal concept . . . but by submitting it to the eyes in a singular intuition. . . ." (*Ibid.*) And although incongruent counterparts do not appear in the *Critique of Pure Reason*, they crop up again in the *Prolegomena* (Sec. 13; this volume, pp. 37—38) where they are used to prove that spatially extended things must be appearances, rather than things in themselves, and that space is the form of outer intuition.

II

To return to the question of absolute space, put as briefly as possible Kant's argument is as follows:

1. Suppose that the universe contains nothing but a single human hand — then, since left hands and right hands are exact, but incongruent, counterparts of each other, no information about the interrelations of its parts will indicate whether *this* hand is left or right.

2. Next suppose that a handless human body comes into existence — the hand will then fit properly, with the thumb up when the palm is placed against the chest, on either the right wrist or the left wrist.

3. Suppose that it fits onto the right wrist — then, if we take it for granted that the body's coming into existence produced no change in the hand, it is *and was* a right hand.

4. But since its right-handedness, before the body came into existence,

cannot have consisted in any intrinsic property of the hand itself nor in any relation between the hand and any other physical object, it must have consisted in an empirically undetectable relation between the hand and universal space.

5. Therefore space cannot itself be constituted of relations between physical objects but must exist absolutely, independent of the existence of matter.

The argument is highly ingenious but it is difficult to regard its conclusion with enthusiasm. Not only is the concept of absolute space itself suspect — an "empty figment of reason" which "pertains to the world of fable", as Kant himself later called it (Handyside, p. 62) — but in addition it is not at all clear what properties space is supposed to possess in order that right hands and left hands should stand in different relations to it; presumably some sort of pervasive asymmetry. However I shall leave these questions aside and concentrate on the argument itself.

Consider, to begin with, another set of incongruent counterparts: two similar but opposite scalene triangles. For convenience let us call one right and the other left, depending on the position of the shortest side when the triangle is standing on its longest side. If we suppose the triangles to be movable then it is obvious that no amount of sliding them about on the surface upon which they are placed will bring one into coincidence with the other. However as soon as we permit ourselves to rotate one of the triangles through the third dimension they can readily be brought into coincidence.

Now if we were to suppose such a triangle, made out of plywood perhaps, to be the sole occupant of the universe we would be unable to say whether it was left or right. Or, more precisely, we would have to say that which it was would remain indeterminate until a physical surface had been brought into existence and we had decided which way up to place the triangle on this surface.

There is some analogy with the case of the single human hand, but there also seem to be important differences. We cannot choose whether the hand shall be left or right simply by deciding to put it on this wrist or that; the relation between the hand and the body seems to be pre-determined in a way that the relation between the triangle and the surface obviously is not. Nor does there seem to be any sense in Wittgenstein's suggestion that, just as two-dimensional objects can be rotated into their incongruent counterparts through the third dimension, so can three-dimensional objects be rotated through the fourth dimension. (*Tractatus*, 6.36111) Some three-dimensional objects *can* be "rotated" into their incongruent counterparts — a left-hand glove can be made into a right-hand glove simply by turning it inside out — but this has nothing to do with the fourth dimension and is in virtue of the fact that a glove, by reason of its physical structure, is a somewhat degenerate kind of three-dimensional object. However, I have no idea at all what it means to speak of rotating a fully qualified three-dimensional object through the fourth dimension.

Again let us suppose that nothing exists but a single human hand. If we now suppose that, instead of a handless body, a sphere comes into existence then an account of the relations between the various parts of the hand and the sphere will not enable us to say whether the hand is right or left. If a second object is to reveal the handedness of the first object it must itself be asymmetrical in some respect. So, instead of a sphere, let us suppose that what comes into existence is a second human hand. Now obviously a description of the relations between the two hands will enable us to say whether they are congruent or incongruent, and if we already know whether the second hand is right or left we will then be able to say which the first hand is.

But there's the rub — unless we are provided with some information about at least one of the hands, in addition to a description of the spatial relations between them, then although we will be able to say whether or not they are incongruent we will still be unable to say whether a given one of them is left or right. That this is so can readily be shown. Suppose that the two hands are incongruent; then, to distinguish the first hand from the second, let us suppose that the first hand is black and the second hand white; and finally suppose the two hands placed together in some definite relationship, say palm to palm with finger-tips touching. Then if the black hand is left we will have one arrangement and if it is right we will have a different arrangement. However since the one arrangement will be the incongruent counterpart of the other and since we are assuming that the two hands now constitute the total contents of the universe, there will be absolutely no difference between the description of the one arrangement and of the other.

Now, with reference to Kant's own example, let us suppose that the handless human body which comes into existence has only one arm. If its arm is a right one and the hand fits onto it then it will follow that the hand is a right hand. But to such a body there will correspond its incongruent counterpart, a one-armed body with its arm on the opposite side. And if Kant is correct in maintaining that, for any pair of incongruent counterparts, a description of the intrinsic properties of either will apply equally well to the other, then precisely the same considerations apply to one-armed bodies as to hands: given nothing but a single one-armed body it would be impossible to say whether it was right-armed or left-armed and given two incongruent bodies it would be impossible to say which was which. By the same token, given nothing but a hand and a one-armed body onto which it fits, it would be impossible to say whether we were dealing with a right hand and a right-armed body or with the incongruent counterparts of these.

There is no essential difference between this example, with its one-armed bodies, and Kant's example in which the body is a normal two-armed one. However, one of the things that makes Kant's account confusing is that, externally at any rate, a two-armed body and its mirror image are not incongruent counterparts — roughly speaking they are congruent. We can remedy this difficulty by supposing that a "standard" body has a green right arm and a red left arm. It will then have an incongruent counterpart with colours reversed. Now suppose that we change the sequence in Kant's example and imagine that the first thing to come into existence, in an otherwise empty universe, is a handless human body with one arm red and the other arm green. Is it a standard body with its right arm green, or is it the incongruent counterpart of a standard body with its left arm green? If Kant is correct there will be no way of deciding from a description of the way in which the various parts of the body are related to each other. Next suppose that a hand comes into existence and that it fits onto the green arm. Then if we already know that it is a right hand we know that the body must be standard, or if we already know that the body is standard we know that the hand must be right. But when all there is in the universe is a hand and a handless body, then even though it is quite determinate which arm the hand belongs on, it remains completely indeterminate whether this is a right arm or a left arm and consequently indeterminate whether the hand is right or left.

We can now see where Kant's own argument goes wrong: it involves the inconsistency of maintaining that it is impossible to say of a hand, considered entirely in isolation from everything else, whether it is right or left, while assuming that it would be possible to say of a handless body, considered by itself, which was its right side and which its left.

Kant seems to have been led into this blunder by his rather curious idea that human beings have something like an innate capacity to distinguish their left sides from their right sides – "the two sides of the human body, in spite of their great outer similarity, are sufficiently distinguished by a well-marked feeling, even if we leave out of account the differing positions of the inner parts and the noticeable beat of the heart. . . ." (this volume, p. 30) – and that we distinguish different regions in space by projecting this distinction into the outside world. But even if this were so it would be easy to imagine the existence of an incongruent counterpart of the normal human body in which the well-marked feelings of leftness and rightness occurred in the reverse of their normal relationship. However, contrary to Kant's opinion, we must first learn to apply the distinction of left and right to physical objects, our own bodies included, in terms of various publicly perceptible features which they possess, and only subsequently do we learn to associate differences in bodily feeling with this distinction. If we were not surrounded with obviously asymmetrical things and processes the distinction between left and right would have no application.

I conclude that Kant is mistaken in thinking that he has demonstrated the existence of absolute space. It is an essential premiss of his argument that where two objects are incongruent counterparts, one of the other, there is no difference whatever in the way in which the parts of each are internally inter-related; with the help of this premiss he concludes that the difference between incongruent counterparts consists fundamentally in the fact that they stand in different relationships to space itself. Although I am not entirely convinced of the truth of the premiss I am unable to show that it is false. If it *is* false then obviously Kant's argument collapses immediately. However if it is true then, in a universe which contained nothing but a single hand, it would not just be empirically undecidable whether that hand were left or right; it would be strictly indeterminate. In other words, the situation would not after all be so different from that of the plywood triangle. Just as I can choose whether the triangle shall become a left one or a right one by deciding which way to lay it down on a surface, so, if I were creating a universe, I could choose whether the hand with which I had begun was to become left or right by deciding next to create a standard body or its incongruent counterpart – a standard universe or its incongruent counterpart. And, as I hope has already become clear, in this context the word "standard" is nothing more than an expository convenience; it implies no ontological seniority in terms of which the hand could still be said to be really one thing as opposed to the other.

NOTES

[1] Originally read at the 1961 Congress of the Canadian Philosophical Association.

[2] *Cf. Kant's Inaugural Dissertation and Early Writings on Space*, trans. John Handyside, Open Court, 1929. Unless otherwise indicated all references are to this work.

MARTIN GARDNER

THE FOURTH DIMENSION

Immanuel Kant, the great German philosopher of the eighteenth century, was the first eminent thinker to find a deep philosophical significance in mirror imagery. That an asymmetric object could exist in either of two mirror-image forms seemed to Kant both puzzling and mysterious. Before discussing some of the implications Kant drew from left-right asymmetry, let us first see if we can recapture something of the mood in which he approached this topic.

Imagine that you have before you, on a table, solid models of the enantiomorphic polyhedrons shown in Figure 1. The two models are *exactly alike* in all geometrical properties. Every edge of one figure has a corresponding edge of the same length on the other figure. Every angle of one figure is matched by a duplicate angle on the other. No amount of measurement or inspection of either figure will disclose a single geometrical feature not possessed by the other. They are, in a sense, identical, congruent figures. Yet clearly they are *not* identical!

This is how Kant expressed it, in section 13 of his famous *Prolegomena to All Future Metaphysics*: "What can more resemble my hand or my ear, and be in all points more like, than its image in the looking-glass? And yet I cannot put such a hand as I see in the glass in the place of its original. . . ."

That two objects can be exactly alike in all properties, yet unmistakably different, is certainly one reason why the looking-glass world has such an eerie quality for children and for primitive people when

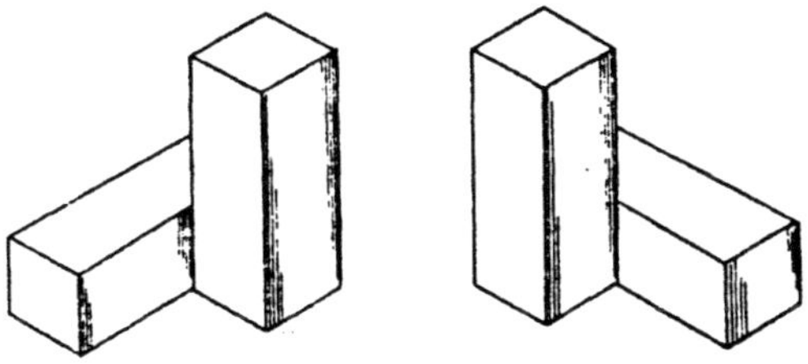

Fig. 1. Enantiomorphic polyhedrons.

they encounter it for the first time. Of course the major source of spookiness is simply the appearance behind the glass of a world that looks as real as the world in front, yet is completely illusory. If you want to puzzle and fascinate a small child, stand him in front of a large wall mirror at night, in a dark room, and hand him a flashlight. When he shines the flashlight into the mirror the beam goes straight into the room behind the glass and illuminates any object toward which he aims it. This strong illusion of a duplicate room is spooky enough, but it grows even spookier when one becomes aware of the fact that everything in the duplicate room "goes the other way." It is the *same* room, yet it isn't.

Exactly what Kant made of all this is a tangled, technical, controversial story. During the past few decades Kant has been so mercilessly pilloried by Bertrand Russell and other leading philosophers of science that readers on the sidelines are apt to think of Kant as a woolly-brained metaphysician who had little comprehension of mathematics and science. The fact is that Kant was well-trained in the science and mathematics of his day. He began his career as a lecturer on physics, and most of his early writings were on scientific topics. Like Alfred North Whitehead, he turned from mathematics and physics to the construction of a metaphysical system only in his later years. Whatever one may think of his final conclusions, there is no denying the importance of his ground-breaking contributions to the philosophy of modern science.

Kant's first published paper, 'Thoughts on the True Estimation of Living Forces' (1747), contains a remarkable anticipation of *n*-dimensional geometry. Why, he asks, is our space three-dimensional? He concludes that somehow this is bound up with the fact that forces such as gravity move through space, from a point of origin, like expanding spheres. Their strength varies inversely with the square of the distance. Had God chosen to create a world in which forces varied inversely with the *cube* of the distance, a space of four dimensions would have been required. (Similarly, though Kant did not mention it, forces in 2-space, moving out from a point source in expanding circles, would vary only inversely with the distance.) Kant here adopted a view of space that had been put forth a century earlier by Gottfried Wilhelm von Leibnitz, the great German philosopher and mathematician. Space has no reality apart from material things; it is nothing more than an abstract, mathematical description of relations that hold between objects. Although the notion of a fourth dimension had occurred to mathematicians, it had been quickly dropped as fanciful speculation of no possible value. No one had hit on the fact that an asymmetric solid object could (in theory) be reversed by rotating it through a higher space; it was not until 1827, eighty years after Kant's paper, that this was first pointed out by August Ferdinand Möbius, the German astronomer for whom the Möbius strip is named. It is surprising, therefore, to find Kant writing as early as 1747: "A science of all these possible kinds of space [spaces of more than three dimensions] would undoubtedly be the highest enterprise which a finite understanding could undertake in the field of geometry." He adds, "If it is possible that there are extensions with other dimensions, it is also very probable that God has somewhere brought them into being; for His works have all the magnitude and manifoldness of which they are capable." Such higher spaces would, however, "not belong to our world, but must form separate worlds."

In 1768, in a paper 'On the First Ground of the Distinction of Regions in Space,' Kant abandoned the Leibnitzian view of space for the Newtonian view. Space is a fixed, absolute thing — the "ether" of the nineteenth century — with a reality of its own, independent of material objects. To establish the existence of such a space, Kant turned his attention toward what he called "incongruent counterparts" — asymmetric solid figures of identical size and shape but opposite handedness, such as snail shells, twining plants, hair whorls, right and left hands. The existence of such twin objects, he argued, implies a Newtonian space. To prove it, he made use of a striking thought experiment, which can be stated as follows.

Imagine that the cosmos is completely empty save for one single human hand. Is it a left, or a right hand? Since there are no intrinsic, measurable differences between enantiomorphic objects, we have no basis for calling the hand left or right. Of course, if you imagine yourself looking at the hand, naturally you will see it as either left or right, but that is equivalent to putting yourself (with your sense of handedness) into 3-space. You must imagine the hand in space to be completely removed from all relationships with other geometrical structures. Clearly, it would be as meaningless to say that the hand is left or right as it would be to say it is large or small, or oriented with its fingers pointing up or down.

Suppose now that a human body materializes in space near the hand.

The body is complete except for both hands; they have been severed at the wrist and are missing. It is evident that the hand will not fit both wrists. It will fit only one — say the left wrist. Therefore it is a left hand. Do you see the paradox confronting us? If it proves to be a left hand, by virtue of fitting the left wrist, it must have been a left hand *before* the body appeared. There must be some basis, some ground, for calling it "left" even when it is the sole object in the universe. Kant could see no way of providing such a ground except by assuming that space itself possessed some sort of absolute, objective structure — a kind of three-dimensional lattice that could furnish a means of defining the handedness of a solitary, asymmetric object.

A modern reader familiar with n-dimensional geometry should have little trouble seeing through the verbal confusion of Kant's thought experiment. In fact, Kant's error was effectively exposed by an episode in Johnny Hart's syndicated comic strip called *B.C.*, in newspapers of July 26, 1963. One of Hart's cavemen has just invented the drum. He strikes a log with a stick held in one hand and says, "That's a left flam." Then he hits the log with a stick in his other hand and says, "That's a right flam."

"How do you know which is which?" asks a spectator.

The drummer points to the back of one hand and replies, "I have a mole on my left hand."

Let us see how this relates to Kant's error. Imagine that Flatland contains nothing but a single, flat hand. It is true that it is asymmetrical, but it is meaningless to speak of it as left or right if there is no other asymmetric structure on the plane. This is evident from the fact that we in 3-space can view the hand from either side of the plane and see it in either of its two mirror-image forms. The situation changes if we introduce a handless Flatlander and *define* "left" as, say, the side on which his heart is located. This by no means entails that the hand was "left" or "right" before introducing the Flatlander, because *we can introduce him in either of two enatiomorphic ways*. Place him in the plane one way, the hand becomes a left hand. Turn him over, place him the other way, and the hand becomes a right hand — "right" because it will fit the wrist on the side opposite the heart.

Does this mean that the hand alters its handedness, or that the Flatlander's heart magically hops from one side of his body to the other? Not at all. Neither the hand nor the Flatlander changes in any respect. It is simply that their relations to each other in 2-space are changed. It is all a matter of words. "Left" and "right" are words which mean, as Humpty Dumpty said, whatever we want them to mean. The solitary hand can be labeled with either term. So can the sides of a solitary Flatlander. It is only when the two asymmetric objects are present in the same space, and a choice of labels has been made with respect to one, that labels applied to the other cease to be arbitrary.

It is the same in 3-space. Not until we introduce the handless body, with the understanding that "left" is the side the heart is on, do we have a basis for deciding what to call the hand. If the body is "turned over" by rotating it through 4-space, the hand's label automatically changes. Suppose we first label the solitary hand, calling it, say, a "right" hand. When the body appears, its "right" wrist will be, by simple definition, the wrist on which the hand fits. The important point is that the initial choice of terms is wholly arbitrary. Hart's caveman who chose to call one hand "left" because it had a mole on it was making a completely rational first step in defining handedness. The humor of the strip lies in the way the caveman phrased his reply. Instead of saying that he knew the difference between left and right flams because he had a mole on his left hand, he should have said: "Because I have decided to call 'left' the hand that has a mole on it." There is nothing paradoxical about such a situation, therefore no need to introduce Newton's absolute space.[1]

Actually, even a fixed, Newtonian ether is no help in providing a label for the solitary hand unless the structure of space itself is somehow asymmetrical. If the hand floats inside a spherical, cylindrical, or conical cosmos, or in an infinite space crisscrossed with the lines of a cubical lattice, we are no better off than before. If the cosmos has the shape of one enormous human hand, the situation changes. We could call the cosmic hand "right" (or "plus" or "Yin"); then, if the solitary human hand is of opposite handedness, we are forced to call it "left" (or "minus" or "Yang"). We could also define the hand's handedness on the basis of an asymmetric "grain" in space, a submicroscopic lattice of geodesics (straightest possible paths) like the asymmetric lattice of quartz or cinnabar. In later chapters we will see that such speculations are now of the highest interest in connection with recent discoveries about the asymmetric behavior of certain elementary particles.

Kant himself soon realized that his thought experiment proved nothing. In later, more mature reflections he combined the views of Newton and Leibnitz into a novel synthesis of his own, intimately bound up with his transcendental idealism. Newton was right, he argued, in regarding space as independent of material bodies, but Leibnitz was also right in denying reality to space. Space is independent of bodies precisely because it is not real; it is ideal, subjective, a mode by which we view a transcendent reality utterly beyond our comprehension.

Space and time are like the two lenses in a pair of glasses. Without the glasses we could see nothing. The actual world, the world external to our minds, is not directly perceivable; we see only what is transmitted to us by our space-time spectacles. The *real* object, what Kant called the *Ding-an-sich* or Thing-in-itself, is transcendent, beyond our space-time, completely unknowable. ("The solution of the riddle of life in space and time lies outside space and time," writes Ludwig Wittgenstein in his *Tractatus Logico-Philosophicus*, 6.4312.) We experience only our sensory perceptions: what we see, hear, feel, smell, taste. These perceptions are, in a sense, illusions. They are shaped and colored by our subjective sense of space and time, as the color of an object is influenced by colored glasses or the shape of a shadow is influenced by the surface on which it falls.

> Space is a swarming in the eyes; and time,
> A singing in the ears.[2]

"What then is the solution?" Kant asks in his *Prolegomena*. "These [mirror-image] objects are not presentations of things as they are in themselves, and as the pure understanding would cognize them, but they are sensuous intuitions, *i.e.,* phenomena, the possibility of which rests on the relations of certain unknown *things in themselves* to something else, namely, our sensations."

In trying to get at the meaning of statements made by philosophers who lived many generations ago, it is sometimes worth the risk to try to rephrase the statements in current terminology and in the light of current knowledge. Of course, it is highly speculative. Nevertheless, I think that if Kant were alive today he would make his point somewhat as follows.

Eighteenth-century mathematicians, as we have seen, had not yet discovered that Euclidean geometry could be extended to any number of dimensions. A straight line, one foot in length, is a one-dimensional figure. In two dimensions the corresponding figure is a square, one foot on a side. In three dimensions it is a cube, one foot on a side. This can be generalized by adding as many new dimensions as one wishes. A hypercube is a cube, one foot on a side, which extends in four directions, each direction at right angles to the other three. The mathematician can work out the geometrical properties of such a cube. There is no reason why a four-dimensional world could not exist, containing material hypercubes, or for that matter a world of five dimensions or six or seven. The hierarchy is endless. At each level the geometry is Euclidean – as valid and consistent as the familiar plane and solid Euclidean geometry taught in high school.

Mathematical techniques can uncover the properties of figures in these higher Euclidean spaces, but our minds are firmly trapped in a Euclidean 3-space, which is united with the single onrushing arrow of time. We find it impossible to conceive of a thing existing without extension in three spatial dimensions and duration in the one dimension of time. Perhaps with the right sort of training, or in some future age when the mind of man has evolved into a more powerful tool, one might learn to think in four spatial dimensions. At present we cannot do so. We see the world through our space-time spectacles: one lens is one-dimensional time, the other is three-dimensional space. We cannot visualize in our brain the structure of a hypercube or any other 4-space structure. We can only visualize 3-space structures that endure – which move along the single track of time.

Suppose, however, that there is a transcendent world, a world of 4-space, inaccessible to our senses and beyond our powers to imagine. How would a hyperperson, in such a hyperworld, view two solid asymmetric objects such as the polyhedrons in Figure 1 that are mirror images of each other? The mathematician can give a clear and unambiguous answer: The polyhedrons would appear identical, each superposable on the other.

To understand this, imagine yourself looking down on a world of 2-space and seeing the two asymmetric shapes shown in Figure 2. Flatlanders living on the plane would be just as puzzled by those two

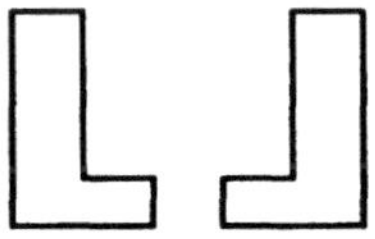

Fig. 2. Enantiomorphic polygons.

figures as Kant was puzzled by his ears and their mirror reflections. How can two figures be so alike, the Flatlanders ask themselves, and yet be nonsuperposable? We who live in 3-space can understand. They *are* alike. It is only because the poor Flatlanders are trapped in 2-space, seeing things only through their 2-space Euclidean lenses, that they cannot see that the two shapes *are* superposable. We can prove that they are simply by picking one up, turning it over, and fitting it, point for point, on the other. If we return the reversed figure to the plane, next to the other one, the two figures will be seen by the Flatlanders as identical in every respect, including their handedness. Since the Flatlanders cannot conceive of 3-space, they will think a miracle has occurred. A rigid, asymmetric object has been changed to its mirror image! Yet we have done nothing to the object. We have not stretched, damaged, or altered it in any way. We have only altered its orientation in 2-space — its position relative to other objects in that space.

The two asymmetric polyhedrons in Figure 1 are similarly identical and superposable. It is only because we cannot see them through the transcendent spectacles of 4-space that we think they are not alike. If we could rotate one of them through hyperspace — turn it over, so to speak, through a fourth dimension — we would have a pair of congruent polyhedrons of the same handedness.

Kant did not, of course, express such views. Nevertheless, I think that if one makes a serious, well-informed attempt to put himself into the center of Kant's final vision of existence, he will find it not frivolous to suppose that Kant might have argued in this way had the mathematical knowledge of the twentieth century been available to him.

Leibnitz also had, I am persuaded, an intuitive grasp of the then-as-yet-undiscovered higher Euclidean spaces. He once considered the question of what would happen if the entire universe were suddenly reversed so that everything in it became its mirror image. He concluded that nothing would happen. It would be meaningless to say such a reversal had occurred, because there would be no way one could detect such a change. To ask why God created the world this way and not the other is to ask, Leibnitz said, "a quite inadmissible question."

When we view this question in the light of the various levels of Euclidean space, we see at once that Leibnitz is right. To "reverse" an entire Flatland on a sheet of paper, all we need do is turn the paper over and view the figures from the other side. We do not even have to turn the paper. Imagine a Flatland on a vertical sheet of glass standing in the center of a room. It is, say, a left-handed world when you view it from one side of the glass. Walk around the glass, you see it as a right-handed world.

EXERCISE 11. *When Mrs. Smith started to push open the glass door at the entrance to the bank, she was puzzled to see the word TUO printed on the door in large black letters. What does the word mean?*

Flatland itself does not change in any way when you view it from another side. The only change is in the spatial relation, in 3-space, of Flatland and you. In precisely the same way, an inhabitant of 4-space could view one of our kitchen corkscrews from one side and see a right-handed helix, then change his position and see the same corkscrew from the other side as a left-handed helix. If he could pick up one of our corkscrews, turn it over, and replace it in our continuum, it would seem to us a miracle. We would see the corkscrew vanish then reappear in reflected form.

Enantiomorphic objects are identical not only in all metric properties; they are also topologically identical. Even though a right-handed knot in a closed loop cannot be deformed into a left-handed one, the two are topologically equivalent. Very young children seem to grasp this more readily than adults. Jean Piaget and Bärbel Inhelder, in their book *The Child's Conception of Space* (Humanities Press, 1956) report on strong experimental evidence that children actually recognize topological properties before they learn to recognize Euclidian properties of shape, including the distinction between left and right forms. When asked to copy a triangle, for example, very young children often draw a circle. The angles and sides of the triangle are less noticeable to them than the property of being a closed curve. They will see no difference between colors that go in a certain order clockwise around a circle and a circle on which the same colors go counterclockwise in the same order. Their untrained minds seem to sense that the two circles are identical: not that they realize that one can be turned over to become like the other, but rather that they see no difference to begin with. This may explain why even strongly right-handed children so often print letters backward, or sometimes entire words.

Perhaps our minds are potentially more flexible than Kant suspected. Our inability to visualize 4-space structures such as the hypercube may be due solely to the fact that all our memories are derived from experiences in a 3-space world. With suitable training toys, could a child learn to think in 4-space pictures? The question has been discussed seriously by a number of mathematicians and of course it is a familiar science-fiction gimmick, notably in Lewis Padgett's much anthologized tale 'Mimsy Were the Borogoves'.

Are there mirror-image forms among the hypersolids of 4-space — that is, shapes identical in all respects except their handedness? Yes, this duality exists on every level. In one dimension, figures are mirrored by a point; in two dimensions, by a line; in three dimensions, by a plane; in four dimensions, by a solid. And so on for the higher spaces. In every space of n dimensions the "mirror" is a "surface" of $n - 1$ dimensions. In every space of n dimensions an asymmetric figure can be made to coincide with its reflection by rotating it through a space of $n + 1$ dimensions. Perhaps our imaginary twentieth-century Kant would put it this way: only the "pure understanding" of God Himself, who stands outside space and time, would see all pairs of enantiomorphic structures, in all spaces, as identical and superposable.

H. G. Wells was the first to base a science-fiction story on the reversal of an asymmetric solid structure by turning it around in 4-space. In 'The Plattner Story', one of Wells's best, a young chemistry instructor named Gottfried Plattner explodes a mysterious green powder that blows him straight into 4-space. What he sees during the nine days that he lives in the dark "Other World," with its huge green sun and unearthly inhabitants, you will have to discover for yourself by reading Wells's story. It can be found in a collection, *28 Science Fiction Stories*, by H. G. Wells (Dover, 1952). After nine days in 4-space, Plattner slips on a boulder, the bottle of green powder explodes in his pocket, and he is blown back into 3-space. But his body has been turned over. His heart is now on the right. He writes a mirror script with his left hand.[3]

The drifting, mute figures in Wells's 4-space are the souls of those who once lived on earth. This notion that departed souls inhabit a higher space was a common one in the spiritualist circles of Wells's day; from time to time mediums actually were asked to change an asymmetric object to its mirror image as proof they were in genuine contact with 4-space inhabitants. Henry Slade, a clever American medium who was world-famous in the late nineteenth century, claimed that his controls had the power of moving objects in and out of 4-space during his séances. One of his favorite tricks was to produce knots in unknotted closed loops of rope, a feat that (barring trickery) could be explained only by assuming that part of the rope had been passed through a higher space. A German astronomer and physicist named Johann Carl Friedrich Zöllner, a remarkably stupid fellow who was incredibly ignorant of conjuring methods, fell completely for Slade's elementary brand of magic. Zöllner wrote an unintentionally hilarious book called *Transcendental Physics* in which he defended Slade's exploits against the charges of fraud.[4]

To obtain definitive, irrefutable proof of Slade's contact with spirits in 4-space, Zöllner once proposed that the medium reverse some dextrotartaric acid so that it would rotate a plane of polarized light to the left instead of the right. He also brought Slade a number of shell with conical helices that twisted right or left, to see if Slade could convert them to their mirror images. Such feats would surely have been as simple as tying a knot by passing part of a rope through 4-space, but from a conjuring standpoint they presented difficulties. Slade would have had to obtain some levotartaric acid, which could be synthesized only in a laboratory and was hard to come by, and it would have been even more difficult for him to find shells that were exact duplicates but of opposite handedness to the shells given him. As might be expected, neither of these crucial experiments succeeded. Of course, this made not the slightest dent in the hard shell of Zöllner's faith.

Is it possible that someday science will find evidence that a higher space is more than just a mathematical abstraction or the wild speculation of spiritualists and occultists? It is possible, though at present there are no more than tantalizing hints. The four-dimensional continuum of relativity is one in which 3-space is combined with time and handled mathematically as a non-Euclidean geometry of four dimensions. This is not at all the same thing as a 4-space consisting of four spatial coordinates. On the other hand, many cosmological models have been devised in which 3-space actually curves through 4-space in a way that could, in principle, be tested. Einstein, for instance, once proposed a cosmic model in which an astronaut could set out in any direction and if he traveled far enough, in the straightest possible line, he would return to his starting point. In this model our world of 3-space is treated as the hypersurface of an enormous hypersphere. Going around it would be comparable to a Flatlander's trip around the surface of a sphere.

In other cosmic models the hypersurface twists through 4-space in a manner analogous to such 2-space surfaces as the Klein bottle and the projective plane. These are closed, one-sided surfaces, without edges, which twist on themselves in a way similar to the way a Möbius strip twists. For example, if you suppose every point on a sphere is joined to every point exactly opposite it on the other side (you cannot imagine this; it has to be worked out mathematically), you have a model of what topologists call projective 3-space. An astronaut making a round trip through projective 3-space would return in reflected form, like H. G. Wells's Plattner.

To understand how the astronaut would be reversed, the following simple experiment is instructive. Cut two paper strips exactly alike, put one on top of the other, then (treating them as a single strip) make a half-twist and join the ends in the manner shown in Figure 3. The model you have formed is not the familiar Möbius strip, but the space *between* the two strips is.[5] The paper may be thought of as a covering for a Möbius surface of zero thickness. Now cut two small swastikas from a piece of dark-colored paper. Put both cutouts inside the double Möbius band, keeping them in place with paper clips as shown. The two swastikas must be placed side by side with the same handedness. Free one from its clip and slide it once around the Möbius surface, sliding it between the "two" strips until it is back where it was originally.

Examine the two swastikas. You will see at once that the cutout that made the round trip has changed its handedness. The two swastikas are no longer superposable. Of course, if you slide the cutout around once more it will recover its former handedness. This same sort of reversal would occur to an astronaut in 3-space if he made a round trip through a cosmos that twisted through 4-space in a manner analogous to the twist in a Möbius surface.

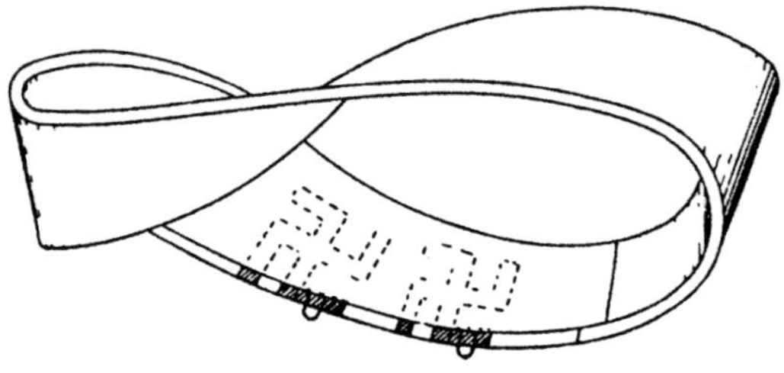

Fig. 3. An experiment with a double Möbius band.

EXERCISE 12. *Figure 4 is a picture of a Klein bottle — a one-sided surface without edges. If an asymmetric Flatlander lived on such a surface (remember, it must be thought of as having zero thickness), would it be possible for him to make a trip around his cosmos in such a way that he would return in a form that was reversed with respect to his surroundings?*

Fig. 4. Model of a Klein bottle.

NOTES

[1] See Peter Remnant's paper on 'Incongruent Counterparts and Absolute Space', *Mind*, vol. 73, July 1963, p. 393–99 and this volume, pp. 51–59, in which Kant's thought experiment is analyzed, with conclusions essentially the same as those given here. For English versions of Kant's two early papers on space, see *Kant's Inaugural Dissertation and Early Writings on Space*, translated by John Handyside (Open Court, 1929). The thought experiment is discussed by Norman Kemp Smith in a section headed 'The Paradox of Incongruous Counterparts', in *A Commentary on Kant's Critique of Pure Reason* (Macmillan and Co., Ltd., 1918, pp. 161–66) this volume, pp. 43–47; and in Hans Vaihinger's earlier German commentary on the same work, vol. 2, pp. 518ff.

[2] These two lines are from Canto 2 of 'Pale Fire', a beautiful poem by Vladimir Nabokov that is the heart of his bizzare novel of the same name. The poem is supposedly written by Nabokov's invented poet, John Francis Shade. As a joke, in the first edition of this book I credited the lines only to Shade and listed only Shade's name in the index. Nabokov returned the joke in his novel *Ada* (note the palindrome), where the action takes place on Anti-Terra, a kind of mirror image of our earth. On page 542 Nabokov repeats the same two lines, then adds that they were written by "a modern poet, as quoted by an invented philosopher ('Martin Gardiner') in *The Ambidextrous Universe* . . . ."

[3] For two amusing later stories about a man reversed in 4-space (both more up-to-date

in their science than Wells's pioneer yarn), see 'Technical Error' by Arthur C. Clarke (in Clarke's *Reach for Tomorrow*, Ballantine, 1956), and 'The Heart on the Other Side' by George Gamow (in *The Expert Dreamers*, Frederik Pohl, ed., Doubleday, 1962).

[4] Zöllner's book, first published in Germany in 1879, was later translated into English and issued in many editions. Sir Arthur Conan Doyle devotes a chapter to the defense of Slade in his *History of Spiritualism* (George H. Doran, 1926). A good discussion of Slade's methods of cheating will be found in Section 2 of the *Proceedings of the American Society for Psychical Research, Inc.*, vol. 15, 1921, in an article by Walter F. Prince on 'A Survey of American Slate-Writing Mediumship'. For more on this shameless mountebank consult John Mulholland, *Beware Familiar Spirits* (Scribners, 1938) and Harry Houdini, *A Magician Among the Spirits* (Harper, 1924).

[5] Actually, there are not two strips but only one! For a discussion of some of the puzzling properties of this double Möbius band see Chapter 7 of my *Scientific American Book of Mathematical Puzzles and Diversions* (Simon and Schuster, 1959), reissued as *Hexaflexagons and Other Mathematical Diversions* (Chicago: University of Chicago Press, 1988).

MARTIN GARDNER

THE OZMA PROBLEM AND THE FALL OF PARITY

A. THE OZMA PROBLEM

On controversial scientific questions for which there is a scarcity of empirical data, scientific opinion sometimes shifts back and forth like the changing fashions of women's clothes. The skirt is low in one decade, high in the next, then back down it goes again. When I was in college it was fashionable among astronomers to think that planets were extremely rare in the universe, on the theory that the earth was the result of an improbable collision or near approach of two suns. Quite possibly (it was believed) life in the cosmos is confined to our solar system, perhaps even to the earth. Today, informed opinion has swung the other way. Astronomers now suspect that planets are extremely *common* in the universe. Perhaps there are billions of them in our galaxy alone, millions of which may support intelligent life. If so, it seems likely that inhabitants of some of these planets, with a knowledge of science equal to or in advance of our own, may be trying to communicate with other planets.

On this assumption Project Ozma was started in 1960. A powerful radio telescope at Green Bank, West Virginia, was pointed toward various suns in the galaxy in a systematic search for radio messages from another world. Frank D. Drake, the radio astronomer who directed the project, is a long-time admirer of L. Frank Baum and his Oz books. He named the project for Ozma, the ruler of Baum's mythical utopia. It is an appropriate name. The location of Oz is unknown. Its inhabitants are "humanoid" but not necessarily "meat people" like us (witness the Tin Woodman and the Scarecrow.) Moreover, Oz is surrounded on all sides by the impassable Deadly Desert, which destroys anyone who so much as touches one grain of its sand. One of Baum's characters, the Nome King, has a servant called the Long Eared Hearer. The ears of this "nome" are several feet across. By placing one of them on the ground he can hear sounds thousands of miles away. Frank Drake's radio telescope is his Long Eared Hearer. It listened patiently for some type of coded signal, perhaps a repetition of a simple sequence of numbers, which could come only from an intelligent source that understood the universal laws of mathematics. The prospect of hearing such a signal is indeed an Ozzy one! It is hard to estimate the shattering effect such a signal would have on our self-centered, earthbound ways of thinking.

What should we do if we hear such a signal? Physicist Chen Ning Yang (we will hear more about him later) has made one suggestion: "Don't answer!" Such a response seems unlikely. Already, mathematicians and logicians are busy at work on step-by-step procedures by which two planets could slowly build up a common language for talking to each other. In 1962 Hans Freudenthal, a Dutch mathematician, published part 1 of an ambitious work called *Lincos: Design of a Language for Cosmic Intercourse.* There is no doubt whatever that coded pulses could be used for fluent communication. Once contact was made, it would be a simple matter to transmit detailed pictures. In crudest form it would only be necessary to divide a rectangle into thousands of tiny square units, like a sheet of graph paper, then transmit a binary code of ones and zeros indicating which unit squares — scanning the rectangle from top to bottom, left to right — should be blacked in. Better pictures, perhaps even moving TV pictures, could later be transmitted by the use of scanning beams. The long time intervals involved (it takes more than four years for a radio signal to reach the star nearest earth) introduce complications, but no one doubts that it would be only a matter of time until the two planets would be communicating with each other as easily, or almost as easily, as two nations on earth that speak different languages.

Did the reader notice the use of the phrase "left to right" in describing how that picture rectangle is to be scanned? Unless the inhabitants of the distant planet — we will call it Planet X for short — scan their rectangle from left to right, they will produce a picture which is a mirror image of the one we intend to transmit. How can we let them know what we mean by the phrase "left to right"?

Assume we have already established fluent communication with Planet X by means of a language such as Lincos and by the use of pictures. We have asked them to scan their rectangles from "top to bottom" and from "left to right." There is no possibility of their misinterpreting what we mean by "top to bottom." "Top" is the direction away from the center of a planet, "bottom" is toward the planet's center. "Front and back" is no problem either. But having established the meanings of up, down, front, back, how do we make clear our understanding of that third pair of directions, left and right? How can we be sure, when we transmit a picture of, say, what we call a right-handed helix, they receive a picture of a helix with the same handedness? If they have taken "left to right" in the same sense that we use the phrase, the pictures will match, but if they are scanning the other way, our picture of a right helix will be reproduced on Planet X as a left helix. In brief, how can we communicate to Planet X our meaning of left and right?

It is a puzzling question. Although an old problem, it has not yet been given a name.[1] I propose to call it the Ozma problem. To state it precisely: Is there any way to communicate the meaning of "left" by a language transmitted in the form of pulsating signals? By the terms of the problem we may say anything we please to our listeners, ask them to perform any experiment whatever, with one proviso: *There is to be no asymmetric object or structure that we and they can observe in common.*

Without this proviso there is no problem. For example, if we sent to Planet X a rocket missile carrying a picture of a man labeled "top," "bottom," "left," "right," the picture would immediately convey our meaning of "left." Or we might transmit a radio beam that had been given a helical twist by circular polarization. If the inhabitants of Planet X built antennas that could determine whether the polarization was clockwise or counterclockwise, a common understanding of "left" could easily be established. Or we might ask them to point a telescope toward a certain asymmetric configuration of stars and to use this stellar pattern for defining left and right. All of these methods, however, violate the proviso that there must be no common observation of a particular asymmetric object or structure.

Is it possible to transmit instructions for drawing a geometric design or graph of some sort that would explain to them what we mean by *left*? After considering it for a while, you can easily convince yourself that the answer is no. Every asymmetric pattern has both right and left forms. Until we and Planet X have a common understanding of left and right, there is no way to make clear which of the two patterns we have in mind. We could, for instance, ask them to draw a picture of a Nazi swastika, then define right as the direction toward which the top arm of the swastika points. Unfortunately, we have no way of telling them what we mean by a Nazi swastika. The swastika can spiral either way. Until we have agreed on left and right, we cannot give unambiguous instructions for drawing the swastika correctly.

Perhaps the field of chemistry would furnish a method of defining left and right. Could we explain to Planet X how to identify a crystal such as quartz or cinnabar that twisted polarized light a certain way? Yes, but even if they found such a crystal on their planet, the specimen would be of no help, [. . .] an optically active crystal can be of either handedness. Without a prior understanding of left and right, we would have no way of knowing the handedness of any particular crystal specimen they might find or grow in their laboratories.

The same ambiguity applies to all optically active stereoisomers. Every chemical compound capable of twisting polarized light — that is, every compound with atoms arranged asymmetrically in the molecule — also has both left and right forms. We could easily come to an understanding with Planet X about what we meant by an asymmetric form of tartaric acid, but if they succeeded in finding or synthesizing it we would not know whether they had obtained it in the right or the left form.

How about the asymmetry of carbon compounds in living tissues? We learned in an earlier chapter that all amino acids in living organisms on the earth are left-handed, and all helices of protein and nucleic acid are right-handed. If the inhabitants of Planet X are made of carbon compounds, perhaps they too contain protein and nucleic acid helices, and of course if they have proteins they also have amino acids. Could we not define left and right in terms of the structure of such asymmetric carbon compounds? No, we could not. It is entirely accidental that our carbon compounds have their particular handedness. So far as we know there is no reason why every carbon compound in every living thing on earth could not, if evolution had taken a different turn at the beginning, have gone the other way. Without a prior understanding of right and left, we could not know whether *their* amino acids were right- to left-handed.

Assume that their planet, like earth, is rotating on an axis. Is there any way this rotation could be used as a basis for defining left? The direction of rotation of the earth can be demonstrated by means of a heavy weight suspended by a long fine wire and swinging slowly back and forth. The device is known as a Foucault pendulum, after Jean Bernard Léon Foucault, the French physicist who first demonstrated it, in Paris in 1851. The swinging weight's intertia keeps the direction of its swing constant in relation to the stars while the planet rotates beneath it. In the Northern Hemisphere a Foucault pendulum rotates clockwise; in the Southern Hemisphere it rotates the other way. But how could we explain to Planet X what we mean by North and South Hemispheres? We could not say: stand on your equator, facing the direction your planet rotates, and the Northern Hemisphere will be on your left. That would presuppose an understanding of "left." Unless we could make clear to Planet X which hemisphere was which, the Foucault pendulum would be no help. The same is true of the various asymmetric phenomena that are the result of a planet's Coriolis forces. We could not say: fire a missile from the equator toward your North Pole and you will see it deviate in the direction we call "right." Such a statement would be ambiguous unless we had previously agreed on which pole was "north." This we could not do without an agreement on what we meant by left and right.

Perhaps Planet X has a magnetic field with north and south poles that correspond closely to the poles of the planet's axis of rotation. Would that be of any help? No. In the first place, we do not know yet the cause of a planet's magnetic field. Presumably it is related in some way to a planet's rotation, but we cannot say with assurance that what we call a north magnetic pole is always associated with the end of the axis of rotation that is on the left when you face the direction of rotation. It may be on the right. The sun always rotates in the same direction, but [. . .] every now and then the magnetic poles of the sun do a peculiar flip-flop; the north pole becomes the south pole and vice versa. The moon, which rotates slowly (one rotation for each revolution around the earth), apparently has *no* magnetic poles. We have no grounds, at present, for guessing how the magnetic poles of Planet X would be placed with repect to the direction of the planet's rotation. Even if we *did* know how they were placed, it still would not help us define left and right, as we will see in the next chapter.

One possibility remains: the asymmetric phenomena associated with electrical and magnetic forces. To take the most familiar example, the magnetic lines of force surrounding a current go around the current in a counterclockwise direction if you face the direction of current flow. In the nineteenth century, when it was thought that current flowed through a wire from positive to negative poles of a battery, this asymmetry was expressed by what physicists called the right-hand rule. If you grasped a wire with your right hand, its thumb pointing along the wire from positive to negative poles, your fingers would curl around the wire in the direction of the magnetic lines of force. Today we know that the current actually flows the opposite way. The motion of free electrons, which produces the wave pulse that is the electric current, goes from the negative pole of a battery to the positive. In this book we adopt the practice of physicists who prefer the convention of a "left-hand rule."

Exactly what does a physicist mean when he says that if you curl your left fingers around a wire, thumb pointing in the direction of current flow, the fingers will point in the direction of the current's magnetic field? He means that if you put a magnetic needle near the wire, the north pole of the needle will always point in a direction counterclockwise around the wire as you face the direction of current flow. Figure 1 shows how the magnetic needle behaves when placed at various positions around a wire carrying a current moving in the direction of the arrow.

Here we have a simple, striking instance of asymmetry. We could

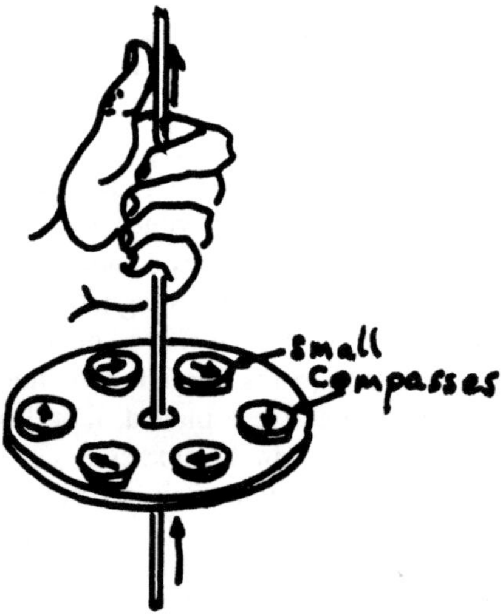

Fig. 1. The left-hand rule for determining the direction of a magnetic field surrounding an electrical current.

explain to the inhabitants of Planet X exactly how to make a battery by mixing certain chemicals and inserting metals in the liquid to provide positive and negative poles. Once we and planet X agreed on the direction of current flow along a wire (there is no difficulty in agreeing on this) could we not then say: put a magnetic needle above the wire, face the direcion the current moves, and the north end of the needle will point in the direction that we on earth call left?

Here, surely, is a simple experiment that provides a clear, unambiguous, operational definition of left and right. No?

No. The experiment would do the trick only if we have some unambiguous way of telling Planet X which end of the needle is the end we call north. Alas, there is no way of communicating this necessary information without first having a common understanding of left and right. To understand why this is so, we must first understand the fundamentals of the modern theory of magnetism. This will be the task of the following chapter. [Editor's note: In the next chapter, here omitted, Gardner explains that there are no intrinsic features distinguishing one end of a compass needle from the other. Our labeling of the ends is purely conventional. Moreover, until the fall of parity, our conventions could be conveyed only in ways that presuppose our concept of left and right — for example, by saying that the north pole of a compass needle is the pole that is attracted by the pole of the earth's axis that is thumbward when the fingers of the right hand curl in the direction of the earth's rotation.]

B. PARITY

If you had asked a physicist in 1950 for a solution to the Ozma problem, you would have been told: there *is* no solution. There is no way, he would have said, to communicate the meaning of left and right to the intelligent beings on Planet X without turning their attention toward a particular asymmetric structure — a configuration of stars, a beam of circularly polarized light, or the like — which both we and they could observe in common. There is no experiment, involving any of the known laws of nature, that can provide an operational definition of left and right.

When something in nature always remains the same, physicists like to express the invariance by a conservation law. For example: the law of the conservation of mass-energy states that the total amount of mass-energy in the universe never changes. Mass is one form of energy (in accordance with Einstein's famous formula, $E = mc^2$), and there is never an increase or loss in mass-energy. The conservation law that implies the universe's fundamental, never-changing mirror symmetry – its lack of bias for left or right in its basic laws – is the law of the conservation of parity.

The term *parity* was first used by mathematicians to distinguish between odd and even numbers. If two integers are both even or both odd, they are said to have the same parity. If one is even and the other odd, they are said to have opposite parity. The term came to be applied in many different ways to any situation in which things fall neatly into two mutually exclusive classes that can be identified with odd and even integers. For a simple illustration, place three pennies in a row on the table, each head-side up. Now turn the coins over, one at a time, taking them in any order you please, but make an *even* number of turns. You will find that no matter how many turns you make – 2, 74, 3,496, any even number – you are sure to end with one of the following four patterns:

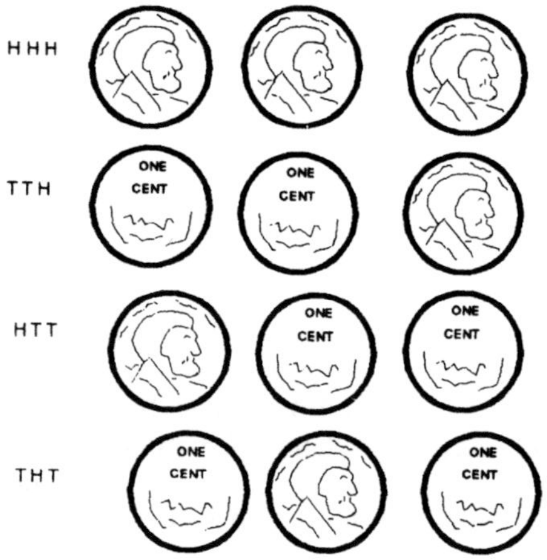

Place the three pennies, all heads up, in a row again. This time make an *odd* number of turns, taking the coins in any order you please. You are sure to end with one of the next patterns shown.

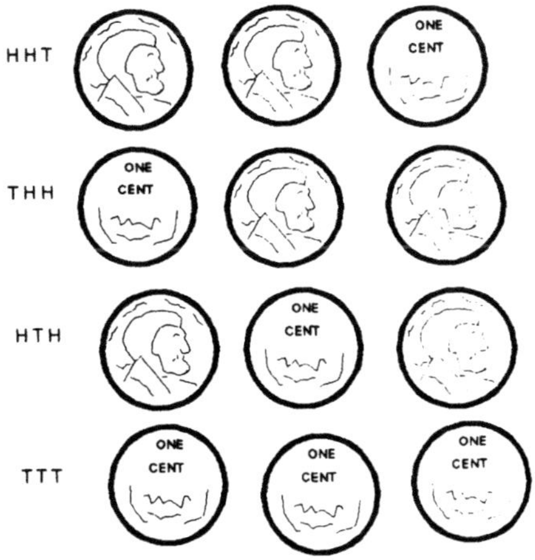

The first set of patterns can be said to have even parity, the second set an odd parity. Experiment will show that the parity of a pattern is conserved by any even number of turns. If you start with an even pattern and make say, 10 turns, the final pattern is sure to be even. If you start with an odd pattern and make 10 turns you are sure to end with an odd pattern. On the other hand, any pattern changes its parity if you make an odd number of turns.

Many tricks with cards, coins, and other objects exploit these principles. For example, ask someone to take a handful of coins out of his pocket and toss them on a table. While your back is turned, he turns over coins at random, one at a time, calling out "Turn" each time he reverses a coin. He stops when he pleases, covers one coin with his hand. You turn around and tell him whether the hidden coin is heads or tails.

The method is a simple application of what mathematicians call a "parity check." Before you turn your back, count the number of heads and remember whether it is an even or odd number. If he makes an even number of turns you know that the parity of the heads remains the same. An odd number of turns changes the parity. Knowing the parity, a simple count of the heads showing, after you turn around, will tell you whether the hidden coin is heads or tails. To vary the trick, you can have him cover two coins and tell him whether they match or not.

EXERCISE 14. *Place six drinking glasses in a row, the first three brim up, the next three brim down. Seize any pair of glasses, one in each hand, and simultaneously reverse both glasses. (That is, if a glass is brim down it is turned brim up, and vice versa). Do the same with another pair of glasses. Continue reversing pairs as long as you please. Is it possible to end with all six glasses upright? With all six upside down? Can you prove your answers mathematically?*

The concept of parity is applied to rotating figures in 3-space in the following manner. Consider the rotating cylinder drawn with solid black lines in Figure 2. Its structure can be described by reference to a coordinate system of three mutually perpendicular axes, traditionally labeled x, y, z as shown. The position of any point on the cylinder is given by an ordered set of three numbers. The first number is the point's distance, measured along the x-axis, from a plane passing through the center of the coordinate system and perpendicular to the x-axis. The second number is the distance of the point, measured in similar fashion along the y-axis. The third number is the distance on the z-axis.

The cylinder drawn with dotted lines is the figure that results when

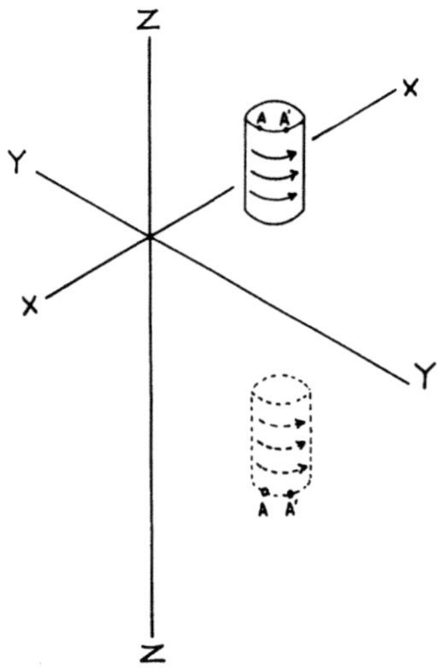

Fig. 2. A rotating cylinder has even parity.

all the z coordinate numbers, in the triples that designate the cylinder's points, have been changed in sign from plus to minus. Note that as the upper cylinder rotates in the direction of the arrows, point A on its upper edge moves from A to A'. The positions of A and A' on the dotted cylinder show that it is rotating in the same direction. True, the cylinder has been turned upside down by this transformation, but since the ends of the cylinder are indistinguishable, the upper and lower cylinders (including their spins) are superposable. In short, the entire system remains unchanged by the change of the sign for all z numbers.

Consider now the rotating cone drawn with solid lines in Figure 3. Below it is the cone that results when the z coordinate numbers are changed from plus to minus. Are the two figures superposable? No, they are mirror images of each other. If you turn the top cone upside down so that it coincides, point for point, with the bottom cone, then the spins will be in opposite directions. And if you turn the cones so that their spins coincide, the cones will point in opposite directions. The rotating cone is an asymmetric system possessing handedness.

It is not hard to see that any symmetric system in 3-space remains unchanged by a change in the sign of any one coordinate. Such systems are said to have an even parity. Asymmetric systems are transformed to

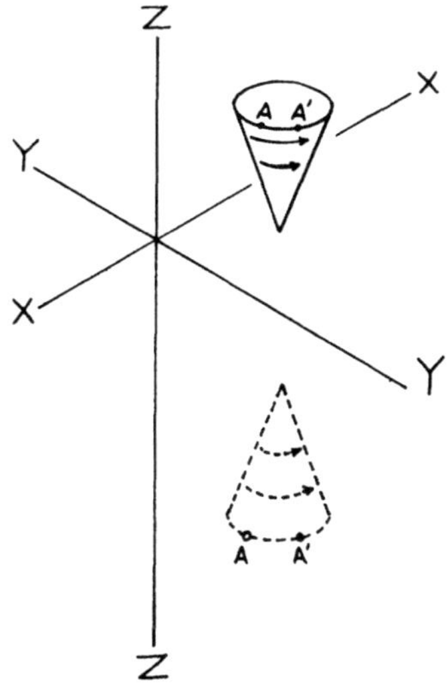

Fig. 3. A rotating come has odd parity.

mirror images by a change in the sign of one coordinate. Such systems are said to have an odd parity. The three coordinates, each of which can be plus or minus, behave in a manner somewhat like the three pennies, each of which can be heads or tails. If the system is asymmetrical, any *odd* number of sign changes has the same effect as changing one sign: it mirror-reflects the system. If you change the signs of all three axes, the system is reflected, because 3 is an odd number. Each single change produces a mirror reflection, but if a mirror reflection is reflected, you are back where you started. Every even number of sign changes leaves the system unaltered with respect to left and right. (This is why two trick mirrors [. . .] give unreversed images; they reverse *two* axes of the coordinate system.) Every odd number of sign changes transforms it to its mirror image. Of course if the system is symmetrical (has even parity) then any number of sign changes, odd or even, leaves the system unchanged.

Physicists found it useful, in the 1920s, to apply these mathematical concepts to the wave functions that describe the elementary particles. Each function contains *x, y,* and *z* space coordinate numbers. If a change in the sign of one (or all three) coordinate numbers leaves the function unaltered, the function is said to have even parity. This is indicated by assigning to the functions a quantum number of +1. A function that changes its sign by a change in the sign of one (or all three) coordinate numbers is said to have odd parity. This is indicated by a quantum number of −1.

Theoretical considerations (such as the left-right symmetry of space itself) as well as experiments with atomic and subatomic particles indicated that, in any isolated system, parity was always conserved. Suppose, for example, that a particle with even (+1) parity breaks down to two particles. The two new particles can both have even parity or both have odd parity. In either case the sum of the parities is even because an even number plus an even number is even, and an odd number plus an odd number is even. To say the same thing differently, the *product* of the two parity numbers is +1. (+1 times +1 is +1, and −1 times −1 is also +1.) The final state of the system has a total parity of +1. Parity is conserved. If an even particle should break down into two particles, one even and the other odd, the total parity of the final state would be odd. (An even number plus an odd number is an odd number, or +1 times −1 is −1.) Parity would not be conserved.

It is important to realize that we are no longer dealing with simple geometrical figures in 3-space but with complex, abstract formulas in quantum mechanics. It is impossible to go into more technical details about the exact meaning of parity conservation in quantum theory or the many ways in which it turned out to be a useful concept. Fortunately, the implications are not hard to understand. In 1927 Eugene P. Wigner was able to show that parity conservation rests squarely on the fact that all the forces involved in particle interactions are free of any left-right bias. In other words, any violation of parity would be equivalent to a violation of mirror symmetry in the basic laws that describe the structure and interaction of particles. Physicists had long known that mirror symmetry prevails in the macroworld of whirling planets and colliding billiard balls. The conservation of parity suggests that this mirror symmetry extends down into atomic and subatomic levels. Nature, apparently, is completely ambidextrous.

This does not mean that asymmetry cannot turn up in the universe in all sorts of ways. It only means that anything nature does in a left-handed way she can do just as easily and efficiently in a right-handed way. For example, our sun moves through the galaxy in such a direction that the earth's motion with respect to the galaxy is along a helical path. Here is a clear instance of astronomical asymmetry. But this asymmetry is merely an accident in the evolution of the galaxy. Other planets, orbiting other suns, no doubt trace helical paths of opposite handedness. Our bodies have hearts on the left. Again, no fundamental asymmetry in natural law is involved. The location of the human heart is an accident in the evolution of life on this planet. In theory a person could be constructed with a heart on the right; in fact [. . .] such persons actually exist. Here we have an instance of an asymmetric structure that exists in both left and right forms, but one form is extremely rare. The parity conservation law does not say that mirror images of asymmetric structures or moving systems must exist in equal quantities. It merely asserts that there is nothing in nature's laws to prohibit the possible existence of both types of handedness.

Physicists sometimes explain the mirror symmetry of the universe in this way. Imagine a motion picture taken of any natural process. The film is mirror-reflected and projected on a screen so that you see a reversed movie of what actually occurred. Is it possible to examine this reflected motion picture and tell if it has been reversed? No, said the physicist in the 1940s, it is not. Of course, we could recognize at once that it was reversed if we saw in the film any man-made asymmetric structures, such as printed letters or numbers, the face of a clock, and so on. But we are concerned only with the fundamental processes of nature, uncontaminated by the artificial asymmetry introduced by living things. Perhaps we are watching drops of oil falling into water, or a chemical reaction taking place. There is no way, physicists said in the 1940s, that we can tell if such a film has been reversed.

If we took a motion picture showing the growth of left-handed crystals from a left-handed compound, it is true that a reversal of the film would show right-handed crystals being formed. But unless we had advance information, we would have no way of knowing that we were not watching an unreflected motion picture of the growth of right-handed crystals from a right-handed compound. Suppose we paint the north end of a magnetic needle red, then take a color motion picture of the needle-and-wire experiment that shocked Mach. The reversed picture would, it is true, show the red end pointing the wrong way. But if we saw such a picture without having previously been told how it was made, we could assume that someone had painted the *south* end of the needle red and all would be well. If magnets do not have their poles labeled *N* and *S*, or distinguished in some other way, a reflected picture of an experiment involving them does not provide any clue by which one can be sure the film has been reversed.

All this is, of course, just another way of stating the Ozma problem. If an experiment could be performed that violated the law of parity, that showed a basic preference of nature for either right or left, we would immediately have a solution to the Ozma problem. We would simply explain to the scientists of Planet X how to set up such an experiment. From its asymmetric twist we and they could easily arrive at a common understanding of left and right.

C. THE FALL OF PARITY

In the previous chapter [Editor's Note: omitted from this volume] we spoke of the four fundamental forces (or "interactions" as physicists prefer to say) that govern the universe. In decreasing order of strength they are: nuclear force, electromagnetism, weak force, and gravity. As we learned, electromagnetism and the weak force are now seen as manifestations of a single electroweak force. There are good reasons to believe that the superweak force will soon be united to the strong, and perhaps all four will be unified by supergravity.

The strong force is the force that holds together the protons and neutrons in the nucleus of an atom. It is often called the "binding force" of the nucleus. Electromagnetism is the force that binds electrons to the nucleus, atoms into molecules, molecules into liquids and solids. Gravity, as we all know, is the force with which one mass attracts another mass; it is the force chiefly responsible for binding together the substances that make up the earth. Gravitational force is so weak that unless a mass is enormously large it is extremely difficult to measure. On the level of the elementary particles its influence is negligible.

The remaining force, the force involved in weak interactions, is the force about which the least is known. That such a force must exist is indicated by the fact that in certain decay interactions involving particles (such as beta-decay, in which electrons or positrons are shot out from radioactive nuclei), the speed of the reaction is much slower than it would be if either nuclear or electromagnetic forces were responsible. By "slow" is meant a reaction of, say, one ten-billionth of a second. To a nuclear physicist this is an exceedingly lazy effect — about a ten-trillionth the speed of reactions in which nuclear force is involved. To explain this lethargy it has been necessary to assume a force weaker than electromagnetism but stronger than the extremely weak force of gravity.

The "theta-tau puzzle," over which physicists scratched their heads in 1956, arose in connection with a weak interaction involving a "strange particle" called the *K*-meson. (Strange particles were called "strange" because they did not seem to fit in anywhere with any of the other particles.) There appeared to be two distinct types of *K*-mesons. One, called the theta meson, decayed into two pi mesons. The other, called the tau meson, decayed into three pi mesons. Nevertheless, the two types of *K*-mesons seemed to be indistinguishable from each other. They had precisely the same mass, same charge, same lifetime. Physicists would have liked to say that there was only one *K*-meson; sometimes it decayed into two, sometimes into three pi mesons. Why didn't they? Because it would have meant that parity was not conserved. The theta meson had even parity. A pi meson has odd parity. Two pi mesons have a total parity that is even, so parity is conserved in the decay of the theta meson. But *three* pi mesons have a total parity that is odd.

Physicists faced a perplexing dilemma with the following horns:

1. They could assume that the two *K*-mesons, even though indistinguishable in properties, were really two different particles: the theta meson with even parity, the tau meson with odd parity.
2. They could assume that in one of the decay reactions parity was not conserved.

To most physicists in 1956 the second horn was almost unthinkable. As we [have seen], it would have meant admitting that the left-right symmetry of nature was being violated; that nature was showing a bias for one type of handedness. The conservation of parity had been well established in all "strong" interactions (that is, in the nuclear and electromagnetic interactions). It had been a fruitful concept in quantum mechanics for thirty years.

In April 1956, during a conference on nuclear physics at the University of Rochester in New York, there was a spirited discussion of the theta-tau puzzle. Richard P. Feynman raised the question: is the law of parity sometimes violated? In corresponding with Feynman, I received some of the details behind this historic question. They are worth putting on record.

The question had been suggested to Feynman the night before by Martin Block, an experimental physicist with whom Feynman was sharing a hotel room. The answer to the theta-tau puzzle, said Block, might be very simple. Perhaps the lovely law of parity does not always hold. Feynman responded by pointing out that if this were true, there would be a way to distinguish left from right. It would be surprising, Feynman said, but he could think of no way such a notion conflicted with known experimental results. He promised Block he would raise the question at next day's meeting to see if anyone could find anything wrong with the idea. This he did, prefacing his remarks with, "I am asking this question for Martin Block." He regarded the notion as such an interesting one that, if it turned out to be true, he wanted Block to get credit for it.

Chen Ning Yang and his friend Tsung Dao Lee, two young and brilliant Chinese-born physicists, were present at the meeting. One of them gave a lengthy reply to Feynman's question.

"What did he say?" Block asked Feynman later.

"I don't know," replied Feynman. "I couldn't understand it."

"People teased me later," writes Feynman, "and said my prefacing remark about Martin Block was made because I was afraid to be associated with such a wild idea. I thought the idea unlikely but possible, and a very exciting possibility. Some months later an experimenter, Norman Ramsey, asked me if I believed it worthwhile for him to do an experiment to test whether parity is violated in beta decay. I said definitely yes, for although I felt sure that parity would *not* be violated, there was a possibility it would be, and it was important to find out. 'Would you bet a hundred dollars against a dollar that parity is not violated?' he asked. 'No. But fifty dollars I will.' 'That's good enough for me. I'll take your bet and do the experiment.' Unfortunately, Ramsey didn't find time to do it then, but my fifty dollar check may have compensated him slightly for a lost opportunity."

During the summer of 1956 Lee and Yang thought some more about the matter. Early in May, when they were sitting in the White Rose Cafe near the corner of Broadway and 125th Street, in the vicinity of Columbia University, it suddenly struck them that it might be profitable to make a careful study of all known experiments involving weak interactions. For several weeks they did this. To their astonishment they found that although the evidence for conservation of parity was strong in all strong interactions, there was no evidence at all for it in the weak. Moreover, they thought of several definitive tests, involving weak interactions, which would settle the question one way or the other. The outcome of this work was their now-classic paper "Question of Parity Conservation in Weak Interactions."

"To decide unequivocally whether parity is conserved in weak interactions," they declared, "one must perform an experiment to determine whether weak interactions differentiate the right from the left. Some such possible experiments will be discussed."

Publication of this paper in *The Physical Review* (October 1, 1956) aroused only mild interest among nuclear physicists. It seemed so unlikely that parity would be violated that most physicists took the attitude: Let someone else make the tests. Freeman J. Dyson, a physicist now at the Institute for Advanced Study in Princeton, writing on "Innovation in Physics" (*Scientific American*, September 1958) had these honest words to say about what he called the "blindness" of most of his colleagues: "A copy of it [the Lee and Yang paper] was sent to me and I read it. I read it twice. I said, 'This is very interesting,' or words to that effect. But I had not the imagination to say, 'By golly, if this is true is opens up a whole new branch of physics.' And I think other physicists, with very few exceptions, at that time were as unimaginative as I."

Several physicists were prodded into action by the suggestions of Lee and Yang. The first to take up the gauntlet was Madam Chien-Shiung Wu, a professor of physics at Columbia University and widely regarded as one of the world's leading physicists. She was already famous for her work on weak interactions and for the care and elegance with which her experiments were always designed. Like her friends Yang and Lee, she, too, had been born in China and had come to the United States to continue her career.

The experiment planned by Madam Wu involved the beta-decay of cobalt-60, a highly radioactive isotope of cobalt which continually emits electrons. In the Bohr model of the atom, a nucleus of cobalt-60 may be thought of as a tiny sphere that spins like a top on an axis labeled north and south at the ends to indicate the magnetic poles. The beta-particles (electrons) emitted in the weak interaction of beta-decay are shot out from both the north and the south ends of nuclei. Normally, the nuclei point in all directions, so the electrons are shot out in all directions. But when cobalt-60 is cooled to near absolute zero (−273 degrees on the centigrade scale), to reduce all the joggling of its molecules caused by heat, it is possible to apply a powerful electromagnetic field that will induce more than half of the nuclei to line up with their north ends pointing in the same direction. The nuclei go right on shooting out electrons. Instead of being scattered in all directions, however, the electrons are now concentrated in two directions: the direction toward which the north ends of the magnetic axes are pointing, and the direction toward which the south ends are pointing. If the law of parity is not violated, there will be just as many electrons going one way as the other.

To cool the cobalt to near absolute zero, Madam Wu needed the facilities of the National Bureau of Standards in Washington, D.C. It was there that she and her colleagues began their historic experiment. If the number of electrons divided evenly into two sets, those that shot north and those that shot south, parity would be preserved. The theta-tau puzzle would remain puzzling. If the beta-decay process showed a handedness, a larger number of elecrons emitted in one direction than the other, parity would be dead. A revolutionary new era in quantum theory would be under way.

At Zurich, one of the world's greatest theoretical physicists, Wolfgang

Pauli, eagerly awaited results of the test. In a now-famous letter to one of his former pupils, Victor Frederick Weisskopf (then at the Massachusetts Institute of Technology), Pauli wrote, "I do *not* believe that the Lord is a weak left-hander, and I am ready to bet a very high sum that the experiments will give symmetric results."

Whether Pauli (who died in 1958) actually made (like Feynman) such a bet is not known. If he did, he also lost. The electrons in Madam Wu's experiment were *not* emitted equally in both directions. Most of them were flung out from the south end; that is, the end toward which a majority of the cobalt-60 nuclei pointed their south poles.

At the risk of being repetitious, and possibly boring readers who see at once the full implication of this result, let us pause to make sure we understand exactly why Madam Wu's experiment is so revolutionary. It is true that the *picture* (Figure 4) of the cobalt-60 nucleus, spinning in a certain direction around an axis labeled *N* and *S*, is an asymmetric structure not superposable on its mirror image. But this is just a picture. As we have learned, the labeling of *N* and *S* is purely conventional. There is nothing to prevent one from switching *N* and *S* on all the magnetic fields in the universe. The north ends of cobalt-60 nuclei would become south, the south ends north, and a similar exchange of poles would occur in the electromagnetic field used for lining up the nuclei. Everything prior to Madam Wu's experiment suggested that such a switch of poles would not make a measurable change in the experimental situation. If there were some intrinsic, observable difference between poles — one red and one green, or one strong and one

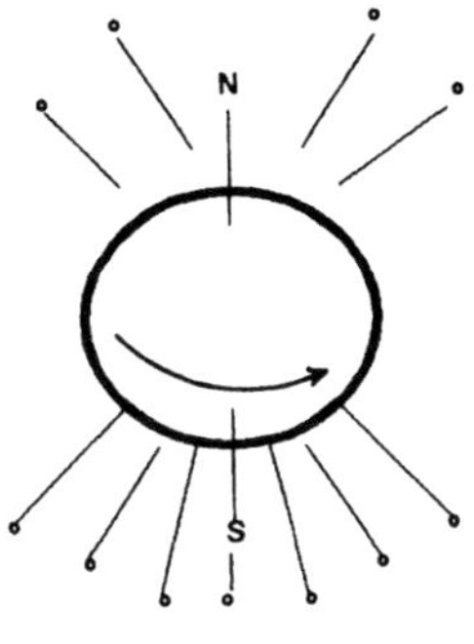

Fig. 4. An electron is more likely to be flung out from the south end of a cobalt-60 nucleus than from its north end.

weak — then the labeling of N and S would be more than a convention. The cobalt-60 nuclei would possess true spatial asymmetry. But physicists knew of no way to distinguish between the poles except by testing their reaction to other magnetic axes. In fact, as we have learned, the poles do not really exist. They are just names for the opposite sides of a spin.

Madam Wu's experiment provided for the first time in the history of science a method of labeling the ends of a magnetic axis in a way that is not at all conventional. *The south end is the end of a cobalt-60 nucleus that is most likely to fling out an electron.*

The nucleus can no longer be thought of as analogous to a spining sphere or cylinder. It must now be thought of as analogous to a spinning cone. Of course, this is no more than a metaphor. No one has the slightest notion at the moment of why or how one end of the axis is different, in any intrinsic way, from the other. But there is a difference! "We are no longer trying to handle screws in the dark with heavy gloves," was the way Sheldon Penman of the University of Chicago put it (*Scientific American*, July 1961); "we are being handed the screws neatly aligned on a tray, with a little searchlight on each that indicates the direction of its head."

It should be obvious now that here at long last is a solution to the Ozma problem — an experimental method of extracting from nature an unambiguous definition of left and right. We say to the scientists of Planet X: "Cool the atoms of cobalt-60 to near absolute zero. Line up their nuclear axes with a powerful magnetic field. Count the number of electrons flung out by the two ends of the axes. The end that flings out the most electrons is the end that we call 'south.' It is now possible to label the ends of the magnetic axis of the field used for lining up the nuclei, and this in turn can be used for labeling the ends of a magnetic needle. Put such a needle above a wire in which the current moves aways from you. The north pole of this needle will point in the direction we call 'left.' "

We have communicated precisely and unambiguously to Planet X our meaning of the word *left*. Neither we nor they will be observing in common any single, particular asymmetric structure. We will be observing in common a universal law of nature. In the weak interactions, nature herself, by her own intrinsic handedness, has provided an operational definition of left and right.

NOTE

[1] I do not know who was the first to give this problem explicitly as one of communication. It is, of course, implied in Kant's discussion of left and right, and many later philosophers allude to it. This is how William James puts it in his chapter on "The Perception of Space" in *Principles of Psychology*, 1890:

> "If we take a cube and label one side *top*, another *bottom*, a third *front*, and a fourth *back*, there remains no form of words by which we can describe to another person which of the remaining sides is *right* and which *left*. We can only point and say *here* is right and *there* is left, just as we should say *this* is red and *that* blue."

James's way of presenting the problem is probably based on his reading of a similar presentation by Charles Howard Hinton in the first series of his *Scientific Romances* (George Allen & Unwin, 1888). Hinton (we will meet him again later) believed that he had taught himself to think in 4-space images by building models with cubes that had been colored in various ways. In discussing these cubes (page 220) he gives a clear statement of what I am calling the Ozma problem.

JONATHAN BENNETT

THE DIFFERENCE BETWEEN RIGHT AND LEFT

I. THE "PARADOX" ABOUT RIGHT AND LEFT

Kant seems to have been the first to notice that there is something peculiar about the difference between right and left, but he failed to say exactly what the peculiarity is. His clearest account of the matter is in his inaugural lecture:[1]

> We cannot describe [in general terms] the distinction in a given space between things which lie towards one quarter, and things which are turned towards the opposite quarter. Thus if we take solids which are completely equal and similar but incongruent, such as the right and left hands . . . although in every respect which admits of being stated in terms intelligible to the mind through a verbal description they can be substituted for one another, there is yet a diversity which makes it impossible for their boundaries to coincide. (15 C; this volume, p. 35.)

One can see roughly what Kant's point is. Take two coins which differ only in their spatial positions: any description of one in general terms also fits the other; but then it is also true that "their boundaries coincide" or, as Kant says elsewhere, that "each can be replaced by the other in all cases and all respects, without the exchange causing the slightest recognizable difference." For example, if I tell you that I earned *this* coin and stole *that*, then shuffle them and show them to you again, you cannot re-identify the one I earned unless you have tracked one of them through the shuffle.

A left and a right hand are more different than this. If I showed you two detached hands which differed only as right and left, told you that I was given *this* one and stole *that*, then shuffled and reproduced them, you could re-identify the stolen one without having tracked either through the shuffle. The two hands would be qualitatively different as well as numerically distinct; it would not be true that "each can be replaced by the other . . . without the exchange causing the slightest recognizable difference"; for example, a glove which fitted one would not fit the other. And yet, Kant thinks, this difference between the two hands cannot be "stated in terms intelligible to the mind through a verbal description": he says that it is a qualitative difference which cannot be captured in language.

That is false. We can state in language what the difference is between the two hands, for we can describe one as "a right hand" and the other as "a left hand." If you did not see them before the shuffle, you can still identify the stolen one if someone tells you "The right hand is the one he stole" — this being meant and understood not as saying what arm the hand used to grow on but rather as describing the hand itself, as saying what *kind* of hand it is.

That refutes what Kant says, taken dead literally. Behind what he actually says, though, there is a less vulnerable claim about the meanings of "left" and "right" and their equivalents in other languages. It is the claim that one could explain the meanings of these words only by a kind of showing — one could not do it by telling. That is the claim I am going to explore.

Still, there was a point in skirmishing with Kant on the basis of a ploddingly literal reading of his words. He uses several unsatisfactory formulations like the one I have attacked, and these help him to think he can report something surprising — in one place he calls it a "paradox" — about the right/left distinction. "Two things can differ qualitatively, although the difference cannot be expressed in words" — that would indeed be surprising if it were not false! Again, Kant says that two hands whose boundaries do not coincide may nevertheless be "completely equal and similar," which would be astonishing if it were true. Other writers, too, have offered one-sentence formulations of what they suppose to be obviously a "problem" about right and left. Reichenbach, for instance, refers to "the problem of the existence of equal and similarly shaped figures that cannot be superimposed" (p. 109; see also Caird, p. 166). Taking "similarly shaped" to mean "having the same shape," that would be a problem indeed; but if you think that your hands have the same shape, just try putting a glove first on one and then, without turning it inside out, on the other.

When Kant and others say that a left and a right hand have "similar shapes" or the like, perhaps they mean — as any mathematician would mean — that the hands do not differ in shape *except* to the extent that one is a right hand and the other a left. Then what they say is true. But now where is the "problem" or "paradox"? My two hands differ only as right and left; but they do differ in that way, so *of course* a single glove won't fit both. Why should I find this surprising or paradoxical?

There is indeed a peculiarity about the right/left distinction. But it does not lie on the surface: philosophical work will be needed to dig it out and lay it bare, and so it could not possibly afford a simple, immediate surprise of the sort Kant thought he had in store for us.

II. KANT'S USES OF THE "PARADOX"

The real peculiarity of the left/right distinction, as well as being more elusive than Kant realized, has a different kind of philosophical interest from any that he found in it. He tried to argue from it to some of his larger philosophical views, but without much success. His major attempt of this kind is definitively treated in a paper by Remnant; his minor ones are hardly worth discussing. Since Kemp Smith has fully described the roles which the left/right matter plays throughout Kant's writings (this volume, pp. 43—48), I need only to sketch them. This section and the next are not presupposed by the rest of the paper.

The relevant background facts are these. (1) Kant was a transcendental idealist, i.e., he held a certain view about the analysis of spatial concepts — *any* spatial concepts. (2) He took sides in the dispute about absolute versus relative space, i.e., the dispute about whether the concept of spatial location is more or less basic than that of spatial relations between things. (1) concerns the analysis of the basic spatial concepts, whatever they may be; whereas (2) concerns which spatial concepts are basic, whatever their further analysis might be. Yet we are told by Weyl:

> Kant finds the clue to the riddle of left and right in transcendental idealism; (P. 84.)

and by Russell:

> Right and left hands, spherical triangles, etc. . . . show, as Kant intended them to show, the essential relativity of space; (§150.)

and by Smart:

> Kant supported the absolute theory of space. In particular he thought that the relational theory could not do justice to the difference between a left hand and a right hand. (P. 6.)

These conflicting accounts of Kant's intentions reflect the instability of those intentions themselves. Kant's first discussion of left and right was in a little paper in 1768. He returned to the topic briefly in his inaugural lecture of 1770. In the first *Critique* in 1781 he took over much of that lecture almost verbatim, but made no mention of left and right. In the *Prolegomena* of 1783, intended as a popular summary of the *Critique*, he resurrected left/right and gave it a short section to itself. But then in the second edition of the *Critique* (1787), in which several new arguments and emphases are borrowed from the *Prolegomena*, the left/right matter once more disappears from sight. Kant seems to have been genuinely unsure whether he could draw philosophical conclusions from his point about the right/left distinction.

He also wavered in his views about *what* conclusions he could draw. Although he did not firmly enough distinguish (1) the issue over transcendental idealism from (2) the issue about absolute versus relative space, it is not too misleading to say: in 1768 he used the left/right matter to support the absolute theory of space; in 1783 he took it to support transcendental idealism; while in 1770 he adduced it in support of a doctrine which is not quite either of these though it arguably entails both.

In short, Kant could not decide which if any of his doctrines about space can draw strength from special facts about the right/left distinction. I am sure none of them can.

III. PROLEGOMENA § 13

Behind Kant's words in the inaugural lecture I have detected the claim that an explanation of the meanings of "right" and "left" requires *showing*, i.e., demands an appeal to sensorily presented examples. I shall call this claim the Kantian Hypothesis. It may not be what Kant "really meant" when he wrote about right and left, but it is the best we can get from him. In defense of this contention I shall examine *Prolegomena* §13, which is Kant's longest and most detailed treatment of the matter, and also, I believe, his last. When examined carefully, this passage can be seen to amount to a series of pointers toward the Kantian Hypothesis. This is not a bad thing to amount to; and really my only criticism is that in *Prolegomena* §13 Kant purports to be expressing, not merely pointing toward, the peculiarity of the right/left distinction. (In the final sentence I make two corrections which the translator accepts. The numbers are for subsequent reference.)

[One would have thought that] if two things are [1] completely the same in all points that can be known at all about each separately (in all determinations belonging to quantity and quality), it must follow that each can be replaced by the other in all cases and all respects, without the exchange causing the slightest recognizable diference. This is in fact the case with plane figures in geometry; but various spherical figures show, notwithstanding this [2] complete inner agreement, an outer relation such that one cannot be replaced by the other. For example two spherical triangles on opposite hemispheres which have an arc of the equator as their common base can be completely equal, in respect of sides as well as angles, so that [3] nothing is found in either, when it is described alone and completely, which does not also appear in the description of the other (on the opposite hemisphere). Here then is an *inner* difference between the two triangles which [4] no understanding can show to be inner and which only reveals itself through the outer relation in space. But I will quote more usual cases which can be taken from ordinary life.

What can be more like my hand or my ear, and more equal in all points, than its image in the mirror? And yet I cannot put such a hand as is seen in the mirror in the place of its original: for if the original was a right hand, the hand in the mirror is a left hand, and the image of a right ear is a left ear, which could never serve as a substitute for the other. Here are [5] no inner differences that any understanding could think; and yet the differences are inner so far as the senses tell us, for the left hand cannot be enclosed in the same boundaries as the right (they cannot be congruent) notwithstanding all their mutual equality and similarity; the glove of one hand cannot be used on the other. ... We cannot make the difference between similar and equal but yet incongruent things (e.g. spirals winding opposite ways) [6] intelligible by any concept whatsoever, but only by their relation to the right and left hand, which immediately involves intuition.

I have omitted Kant's "solution." The question I want to answer is: What is his problem?

The problem, as Kant sees it, is that a certain plausible proposition is false. (My addition of "One would have thought that" at the start of the passage, though it wrongly makes Kant explicit about this, must be legitimate. Without it, Kant asserts something which he immediately proceeds to deny.) The proposition in question has the form

$$(x)\,(y)(Fxy \rightarrow Gxy).$$

Kant says that, although this is plausible, there are in fact values of x and y such that $(Fxy \mathbin{\&} \sim Gxy)$; and to solve his problem will be to explain this surprising fact. *Our* problem is to discover what F and G are.

There is no difficulty about G. Gxy is the statement that x can be replaced by y "without the exchange causing the slightest recognizable difference." Thus Gxy is true if x and y are newly minted coins from the same die, and false if they are a normal pair of hands, i.e., a pair differing only as right and left.

The search for *F* is embodied in the question: What does Kant think he can say about a normal pair of hands from which one might naturally, though wrongly, infer that they could not be told apart? We can safely pin everything on the one example of a pair of hands, for it is universally agreed that in this area Kant's examples stand or fall together.

He expresses *Fxy* in six different ways. Here are two of them:

(1) *x* and *y* "are completely the same in all points that can be known at all about each separately (in all determinations belonging to quantity and quality)."

(3) When *x* is "described alone and completely," its description is the same as *y*'s.

To describe something "alone and completely" is presumably to say everything about it except how it relates — spatially and otherwise — to other things. But then is (3) true of a normal pair of hands? In describing one of the pair "completely" we can use a phrase which does not fit the other, namely "a right hand" — taking this to express a fact not about which arm it grows on but about its shape, e.g., about which sort of glove will fit it. To exclude this, Kant must say that if we use "right" in describing a hand we are not describing it "alone": the phrase "a right hand," he must say, is covertly relational, and not merely in the attenuated way in which any description, e.g., "a small hand," is covertly relational. This is not obviously true, and the only arguments I can find to support it stem from the Kantian Hypothesis.

If (1) is not also to amount to a pointer toward the Kantian Hypothesis, the phrase "determinations [= properties] belonging to quantity and quality" must be turned to account. But it cannot be. The difference between a left and a right hand *is* "qualitative" in any plain sense of the word; and Kant's technical sense of "quality" in the *Critique* is too unclear to help us here.

Here are Kant's other four ways of expressing *Fxy*:

(2) There is a "complete inner agreement" between *x* and *y*.

(4) The "inner difference" between *x* and *y* is one which "no understanding can show to be inner."

(5) Between *x* and *y* there are "no inner differences that any understanding could think."

(6) "We cannot make the difference between [*x* and *y*] intelligible by any concept whatsoever."

We must presume (2) to be a careless contraction of (4) or (5). Otherwise, Kant is saying that between *x* and *y* there is (2) a "complete inner agreement" and also (4) an "inner difference." So (2) can be ignored.

(5) and (6) go together. For Kant, "the understanding" is the faculty of "concepts": to be thought by the understanding is to be brought under, thought through, or made intelligible by, concepts. So (5) and (6) both say that a right hand need not fall under any concepts which do not equally apply to a left hand, which is tantamount to denying that there is any concept of rightness-as-distinct-from-leftness. Since "right as distinct from left" is a meaningful description, why should Kant deny that there is a concept corresponding to it? His only hint at an answer is in his remark, at the end of the passage, that we can explain the right/left difference only in a way "which immediately involves intuition [= sense-experience]." But this is — and so (5) and (6) are just unargued pointers toward — the Kantian Hypothesis.

Whereas (5) and (6) say that the understanding cannot show or express *what the left/right difference is*, (4) says that it cannot show *that the difference is an inner one*, implying that one could show this only with the aid of "intuition" or sense-experience. To assess this, we must know what an "inner" difference is. It seems to be just a difference in respect to something other than spatial location or orientation — a difference in respect to some property that a thing can carry around with it. This yields the wanted result that there is an inner difference between a pair of normal hands, and not between two new coins from the same die. It also fits my example in Sect. I above: if two things are to be separately re-identifiable after a shuffle, without being tracked through it, what is needed is precisely some "inner" difference between them, i.e., some difference of the kind that can be carried through a shuffle.

So (4) seems to say that someone who has grasped what the difference is between a right and a left hand must make a further appeal to experience if he is to grasp that one hand cannot be made congruent with the other just by moving it around. This is in fact correct; for there are mathematically possible spaces in which a right hand could, by sheer travel, become a left hand; and if our space is not of such a kind, that is an empirical fact about it and in that sense a fact which can be known only by appeal to experience. But it is not credible that that is the point Kant was trying to make in (4). I am sure that what he says about showing (4) *that the difference is inner* is meant to follow from what he says about showing (5, 6) *what the difference is*. When he says at (5) in the quoted passage:

Here are no inner differences that any understanding could think; and yet the differences are inner so far as the senses tell us,

isn't it clear that he is simply failing to distinguish "what the inner difference is" from "that the difference is inner"? If he is, and if that explains (4), then the latter goes the same way as (5) and (6) – toward the Kantian Hypothesis.

So *Prolegomena* §13 does, to its great credit, yield the Kantian Hypothesis. But that is all it yields; and it does not make clear just what the force of the Hypothesis is, or why it is true. There remains work to be done.

Before we go on with it, there are two footnotes to the claim that Kant was the first philosopher to notice that right/left is peculiar.

In a letter to Clarke, Leibniz says that God could have no reason for choosing (a) the way things are in fact arranged in space rather than (b) an arrangement "preserving the same situation of bodies among themselves" and differing from (a) only in "changing East into West"; whence he infers that (a) and (b) are not really different (p. 26). He probably thinks of (b) as the world's being rotated through 180°, changing north into south as well as east in west. Still, all he actually says is "changing East into West"; so he could be envisaging a systematic left/right switch, or mirror-image transformation, in which case he has anticipated something like Kant's point. I find the latter reading implausible. It credits Leibniz with introducing an original philosophical insight in an incredibly offhand way, and arguing from it – without first explaining or defending it – even though he could further his main argument much less vulnerably with the rigid-rotation version of (b). Also, when he reverts to this matter in his next letter to Clarke he clearly construes it in the rigid-rotation rather than the mirror-transformation manner (p. 37). Kant's thoughts about right and left, however, grew out of his disagreements with Leibniz, and the east/west remark may well be what put him on the track.

The 11th century Arab philosopher Ghazali has a better claim to have anticipated Kant's insight:

The highest sphere moves from east to west and the spheres beneath it in the opposite direction, but everything that happens in this way would happen equally if the reverse took place, i.e., if the highest sphere moved from west to east and the lower spheres in the opposite direction. For all the same differences in configuration would arise just as well. Granted that these movements are circular and in opposite directions, both directions are equivalent; why then is the one distinguished from the other, which is similar to it? (Quoted in Averroës, Vol. I, p. 30.)

(I am indebted to George F. Hourani for calling this passage to my attention.)

IV. ENANTIOMORPHISM

It is a nuisance that, when we want to use "a left hand" to mean something about the hand's shape, what sort of glove will fit it, etc., the phrase can also mean "a hand that grows on a left arm." In either meaning it applies to just the same objects, but that is a mere contingency. For this reason, and for others that will emerge shortly, hands are not the best example of the relationship we are interested in. I prefer these two boxes:

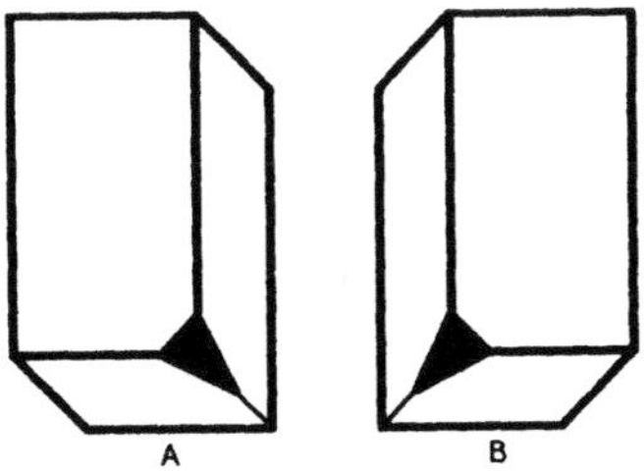

In Kantian language, these differ as "things which lie towards one quarter and things which are turned toward the opposite quarter." Such pairs are sometimes called "incongruous counterparts," which means that (a) their boundaries do not coincide, and that (b) one of them looks just as the other would in a mirror. If the sliced-off corners were restored, (a) would be false and the boxes would not be "incongruous"; if just one had is corner restored, or if one were bigger, (b) would be false and the boxes would not be "counterparts."

The mathematical term for two things which are thus related is *enantiomorphs* ("having contrary shapes"). I shall sometimes use this word instead of the longer "incongrous counterparts," but not to mark any distinction.

It is time to confess that my paper's real topic is not right/left as such, but rather enantiomorphism, or the difference between incongrous counterparts. The right/left distinction can bear the whole weight of the difference between any pair of enantiomorphs: that is, any such pair can be so described that a "right"/"left" switch turns a description of either into a description of the other. It this section I shall show how such descriptions work, to show that in discussing incongruous counterparts it is *convenient but not essential* to use "right" and "left" or some other pair of terms which similarly refer to the two sides of the human body.

If the two boxes *A* and *B* are to be described by the use of "right" and "left," without anything's being assumed, it apparently cannot be done more simply than this:

A: When (1) the line from its small cut to its small uncut face runs the same way as the line from your feet to your head, and (2) the line from its large cut to its large uncut face runs the same way as the line from your back to your front, then (3) the line from its middling cut to its middling uncut face runs the same way as the line from your right side to your left side.

B: Switch "left" and "right" in the above description of *A*.

The following would be simpler, but they make assumptions:

A: When (1) its small cut face it downmost and (2) its large cut face is toward you, then (3) its middling cut face is to your right.

B: Replace "right" by "left" in the above description of *A*.

Those simpler versions are accurate if you are on your feet and facing the box, or on your head with your back to it. They are wrong if you are on your feet with your back to the box, or on your head facing it.

What the longer descriptions make explicit is that we use "right" and "left" to express the difference between an object and its incongruous counterpart by fixing directions along two of the object's dimensions and then employing "right" and "left" to make the required distinction in the third dimension. (Here and throughout I ignore the mathematically sound but entirely unhelpful remark – e.g., in Wittgenstein, 6.36111 – that in a fourth spatial dimension *A* could be flipped over so as to become congruous with *B*.) To discriminate *A* from *B* by reference to the human body in this way, we need to be able to pick out three axes of the human body and to be able to distinguish the two directions along each axis. It is harder to distinguish directions along the left/right axis than along either the head/feet or the back/front axis; but this fact, which connects with our being broadly and superficially left/right symmetrical, is irrelevant to the use of human bodies to discriminate *A* from *B*. My first description of *B* above could just as accurately have ordered a "head"/"feet" or a "front"/"back" switch in the long description of *A*.

We can also use "right (side)" and "left (side)" to distinguish the two sorts of hand, and not through the contingency about which sort of hand grows on which side (I now use a self-explanatory shorthand):

Left hand: When thumb → little-finger runs with back → front, and wrist → fingertips runs with feet → head, then palm → knuckles runs with right-side → left-side.

Right hand: Switch "right" and "left" in the above description of the left hand.

But the two sorts of hand can be distinguished without reference to human flanks, just so long as we have some pair of enantiomorphs – e.g., the two boxes – to use as a standard:

Left hand: When thumb → little-finger runs with large-cut → large-uncut face of *A*, and wrist → fingertips runs with small-cut → small-uncut face of *A*, then palm → knuckles runs with middling-cut → middling-uncut face of *A*.

Right hand: Replace "*A*" by "*B*" in the above description of the left hand.

It is commonly believed that the distinction between a pair of enantiomorphs, when properly spelled out, must refer to the "point of view" of an "observer"; but this is false if it goes beyond the general point that any empirical distinction must, qua empirical, have a possible observer lurking in the conceptual background. The idea seems to be that we should describe *A* like this: "When the line from its small cut to its small uncut face runs the same way as the line from the observer's feet to his head . . . etc.". But if a human body is used in describing *A*, why should it be an observer's body? A corpse would serve as well.

In any case, human bodies are not needed at all. It is sometimes said that we can distinguish enantiomorphs only because our bodies are asymmetrical in at least two dimensions, but this is false too. If our bodies were symmetrical about a point we could still make the distinction we now make in terms of "right" and "left," the one exemplified by *A* and *B*; only we should have to express it in terms of something other than the sides of our bodies. Perhaps it is worth a paragraph to explain how this might be done.

Traveling from Ridge toward Lougheed, I must turn left at a certain corner to reach the University. If humans were spherical I might be told which way to roll at that corner by reference to the box *A*:

> If (1) small-cut → small-uncut face of *A* runs with ground → sky, and (2) large-cut → large-uncut face of *A* runs with turning-corner → Lougheed, then (3) middling-cut → middling-uncut face of *A* runs with the next part of your journey.

That may seem to compare ill with the instructions I can in fact be given:

> If at that corner you (1) stand (2) facing Lougheed, (3) you must turn left before proceeding;

but this, though briefer, is not logically simpler. It spells out into:

> If you so orientate yourself that (1) feet → head runs with ground → sky, and (2) back → front runs with turning-corner → Lougheed, then (3) right-side → left-side runs with the next part of your journey.

Also, it is routine work to construct definitions of "*A*-turn" and "*B*-turn" which would let us describe a route unambiguously and quite briefly by specifying where the spherical traveler should make an *A*-turn and where a *B*-turn. I have heard it insisted that if our bodies were spherical we could not remember the difference between *A* and *B*, or between *A*-like boxes and *B*-like boxes, or between *A*-turns and

B-turns; but I know of no principles in the epistemology of spherical rational animals which could justify this claim.

Failure to grasp the conventions underlying our use of "left" and "right" has generated the mildly famous "mirror problem": why does a mirror reverse left/right but not up/down? Martin Gardner (pp. 29–31) presents the only clear account I know of the solution to this: the answer to "Why does a mirror . . . etc.?" is *It doesn't*! Your image in a normal mirror is a visual representation of an incongruous counterpart of your body, and we conventionally describe this sort of relationship as a "left/right reversal." But this convention does not pick out one dimension as privileged over the other two: it is merely a natural and convenient way of expressing the fact of enantiomorphism in a case where each member of the enantiomorphic pair has – like a normal human body – a superficial over-all bilateral symmetry. (Of course an object which was precisely and totally bilaterally symmetrical could not have an enantiomorph.) If we are to describe what an ordinary mirror does, in a way which really does select one axis of the body in preference to the other two, then we must say this: if you face the mirror, it reverses you back/front; if you stand side-on to it, it reverses you left/right; if you stand on it, it reverses you up/down. These facts, once they are properly described, do not offer a problem. They are explained by routine optics. For some deeper aspects of this matter, see the paper by Pears.

V. "WHAT IS THE DIFFERENCE?"

I am going to test the Kantian Hypothesis that the difference between right and left – by which I really mean "the difference between anything and its enantiomorph" – can be explained only by showing and not by telling. Now, there is one way of taking this in which it is obviously false, the following being a counter-example:

> If you have a man on one side of you and a woman on the other, then you have *either* a man on your left and a woman on your right *or* a man on your right and a woman on your left, depending upon which side each is on.

Or we can tell someone what the difference is between the boxes *A* and *B* by giving him a mathematical description of each (the two descriptions will differ only in that one will have a minus-sign before each value for x), and telling him that of these two descriptions one fits A and the other fits B.

In ways like these we can explain the *difference*: we can say what distinction is marked by "right" and "left," or what kind of difference there is between a pair of incongruous counterparts, without saying anything about how to tell which is which. Analogously, someone might learn what "the difference between" blue and green is by being told that sunny skies characteristically have one of these colors and well-watered grass the other. Confronted with two shirts, say, he would then be in a position to say "I know what the difference between these is — one is blue and the other green"; but he would not be able to say which is blue and which green.

When Kant says — in episode (4) of the long passage — that between two incongruous counterparts there is "an inner difference which no understanding can show to be inner," he may mean that one could not explain in general terms "what the difference is" even in this attenuated sense. If so, he is surely wrong. (Thus Weyl, p. 80. But Weyl errs in thinking that this is Kant's only point.)

The Kantian Hypothesis that I want to discuss says that we must use sensorily presented instances — must resort to showing — if we are to explain the *direction* of the left/right distinction, i.e., to explain which is which. I shall for brevity's sake go on using the phrase "the difference between," but always intending it in this which-is-which manner. In my use, someone does not know the difference between right and left unless he knows which is his right side and which his left; and we have not told someone what the difference is between A and B unless we have equipped him to pick out A as distinct from B.

VI. TACTICS

A good way of examining how something could be explained is to consider how someone could discover that he has it wrong. So I shall invent someone — call him an Alphan — whose grasp of English is perfect except that he gives to "right" the meaning of "left" and vice versa. We have to see how he could learn of his mistake.

For a contrast case I shall take someone — call him a Betan — whose grasp of English is perfect except that he has switched the meanings of some other pair of spatial expressions. The Betan's mistake concerns the word "between": he gives to the form "x is between y and

z" the meaning we give to "y is between x and z". (He thinks that the thing asserted to be between the other two is the thing whose name occurs between the names of the other two: any English sentence containing the form "x is between y and z" is a kind of *picture* of what the Betan thinks it means.) The contrasts I shall draw between the Alphan and the Betan will not depend at all upon special features of betweenness — e.g., that it is a triadic relation, or that it concerns order rather than shape or size. Essentially the same contrasts could be drawn if the Betan had switched the meanings of "large" and "small," "inside" and "outside," "round" and "square," or any one of dozens of others pairs of spatial expressions. Nor does it matter that the Betan has not switched a pair of *words*. Pretend that English also contains "botween," defined by "'y is botween x and z' = 'x is between y and z'," and then think of the Betan as having switched the meanings of "between" and "botween."

Let us ask how the Alphan and the Betan can discover their respective semantic errors. In seeing how the two cases differ we shall see that the Kantian Hypothesis is nearly true.

If the Alphan encounters a statement using "right" or "left" which he knows to be false given the meanings he attaches to those words, but which might for all he knows be true if their meanings were switched, he may guess that the speaker or writer is mistaken or lying. As such cases pile up, however, the Alphan ought to conclude that *he* had made an error — a semantic one. Similarly, the Betan will realize his mistake about "between" if he encounters enough statements which he knows to be false on his understanding of them but which might for all he knows be true on the other relevant interpretation, i.e., the one which is in fact correct.

I shall take these to be the only ways in which either man can discover his error. Any corrective force that verbal definitions have can be expressed in the pattern of correction I have described, and it will make for clarity if everything is brought under the one pattern.

So our question about each man is: what true statements will he, interpreting them in his mistaken way, think to be false? The inquiry is not a psychological one. The intellectual responses of the Alphan and the Betan are dramatic embodiments of logical relations, so we credit them both with maximum alertness, intelligence, retentiveness, and so on.

VII. ADMISSIBLE EVIDENCE

Here are some boring ways of correcting the Alphan. Say to him "I am now touching your right shoulder," while touching his right shoulder. Say to him "Your right shoulder is the one with the birthmark," when he knows which of his shoulders has a birthmark and it is indeed his right shoulder. Say to him "As I stood facing Boulogne, I had Dover on my left and Folkestone on my right," and give him a map of Europe or a look at Europe.

All these correct him by applying "right" and "left" to particular bits of the world of which he has relevant independent knowledge — from his own observation of those particulars, or from inspecting maps or pictures or statues of them. It is obvious — and the Kantian Hypothesis does not deny — that the Alphan can be corrected in ways like these, as indeed can the Betan. What the Hypothesis says, in effect, is that if we rigorously exclude all such references to particulars which are also known through observation, the Betan can still be corrected while the Alphan cannot. If we are to test the Hypothesis, therefore, we must deprive both men of any statements referring to particulars which they know about from observation.

We must also ban all English statements about particulars which the Alphan or Betan knows about from hearsay in languages other than English. Any attempt to capitalize on the Alphan's correct grasp of some pair of non-English synonyms of "right" and "left" would merely force us to redirect our inquiry — making us ask about his grasp of those other words rather than of "left" and "right" — without altering the inquiry's fundamental nature.

So the English statements encountered by the Alphan or Betan are to say *nothing relevant* about any particular things or places or situations regarding which he has *any relevant* knowledge from any source other than what he reads in English. The word "relevant" here means "relevant to his semantic mistake," and it isn't always clear whether something is relevant in this way. Rather than constantly watching for hidden relevances, let us exclude more: the English statements encountered by the Alphan or the Betan are to say *nothing at all* about any particulars regarding which he knows *anything at all* from any source other than what he reads in English. This will be much easier to handle, and it cannot affect the validity of our results: anything allowed in by the weaker exclusion but kept out by the stronger must, *ex hypothesi*, be irrelevant to the matter in hand.

Think of each man as receiving an account, written in English, of some part of reality about which he knows nothing from any other source (and, in the meantime, forget that this involves his receiving ink-samples about which his correspondent might make comments in English). It is crucial that they are to know *nothing* about the described part of reality other than (a) general truths about it which hold true of all reality, and (b) truths about it in particular, or about particulars in it, which they learn simply from what they read in English. They can be in a position to say of something they observe, "This is *a thing of the kind* the Englishman was referring to when he wrote . . . ," but never to say "This is *the thing* the Englishman was referring to when he wrote . . ." They must not even be in a position to relate particulars described to them in English with particulars presented in any other way, apart from merely comparing them. So they must not be in a position to say "This rain was caused by the atomic explosion the Englishman wrote about" or "The mountain the Englishman wrote about is 7,568 miles NNE of my village." It follows that among the things they must not know about the part of reality described to them in English is where it is in relation to themselves.

The line of exclusion I am drawing is not arbitrary or willful. There is a good reason for depriving both Alphan and Betan of any independent knowledge, however remote and relational, of any particular they read about in English. Everything thus excluded is either irrelevant to our inquiry or else logically on a continuum with the trivial case where we touch the Alphan's shoulder while saying "I am now touching your right shoulder."

Even with all this excluded, the Alphan and Betan call still encounter millions of uses "left" and "right," or of "between." And they may still be able to judge some of what they read to be false; for one can reject a statement about a particular of which one has no independent knowledge, on the grounds that it conflicts with a generalization which one knows to be true. I heard the BBC say that 9,000 civilians would be evacuated from Aden within a year, at the rate of 500 per month: without investigating Aden I was entitled to reject *that* — the thing is logically impossible. In Shelley's *The Cenci*, a torturer says of his intended victim:

As soon as we
Had bound him on the wheel, he smiled on us,
As one who baffles a deep adversary;
And holding his breath, died.

I wasn't there; but I know that this report is false — Marzio cannot have committed suicide by holding his breath, because that is a physiological impossibility.

Of those two examples, one concerns a logical generalization, and the other a contingent, broadly causal generalization. I shall use this dichotomy in what follows.

VIII. LOGICAL CLUES: THE BETAN

There are countless "logical clues" to the Betan's error — that is, countless true statements which, interpreted according to his semantic error, will come out logically false. Here are two examples, with the Betan's pictures indicated in brackets:

(a) "I sat between a silent old bore and a talkative young bore [I-bore-bore]. Since there were only two bores present, I resented having one on each side of me."

(b) "Since Baltimore is between Washington and New York [B-W-NY], and we were flying in a straight line, we passed over New York first, then Baltimore, then Washington."

These bring the Betan's correct understanding of "each" and "side," and of "straight" and "first" and "then," into logical conflict with his incorrect reading of the form "*x* is between *y* and *z*." With no independent knowledge of the dinner or of the flight, he nevertheless knows that there is something amiss with each statement or with his understanding of it.

Those statements are logical clues for the Betan only because he does understand all the other words correctly. Perhaps, then, we can shield him from logical clues to his error about "between" by supposing that he errs also about other words such as "straight" and "each," and that these other errors *match* his mistake about "between." Can we do this? Can we credit him with a set of semantic errors which dovetail together so that no true statement will give him a logical clue to his having misunderstood "between" or any of the other words in the set?

("Between" can conflict with itself, because the Betan would equate "*x* is between *y* and *z*" with "*z* is between *y* and *x*" but not with "*x* is between *z* and *y*." But since that is a special feature of "between," and would not obtain for most of the examples I might have taken as contrasts to right/left, I cannot avail myself of it. The Betan is in enough trouble anyway.)

The first point to notice is that dozens of words have direct meaning-connections with "between." To remain sheltered from logical clues to his error about "between," the Betan must err about the meanings not just of the words I have mentioned but also of "symmetrical," "lopsided," "middle," "pinch," "trapped," "separated," and many more.

Also, it is hard to see *what* semantic error we must suppose him to make in each case. In (b), for instance, will he give to the sentence "We passed over New York first, then Baltimore, then Washington" the meaning we give to "We passed over Baltimore first, then Washington, then New York"? It is not clear what underlying semantic error, concerning what word(s) or phrase(s), could generate that reading of the sentence.

Finally, if he is to have no logical clue to any of his semantic errors, then each error with which we initially credit him will presumably have to be matched by yet others, these in their turn by others again, and so on outward. I can't illustrate this in detail because, as just noted, I can't say what semantic error is required in any single case; but I am sure that if we could specify a semantic error which would produce a "match" in a given case, it would be one which could remain unclued only if matched by further errors. For example, if we try to draw (a)'s sting by supposing the Betan to make a matching mistake about the word "each," then we must protect the latter mistake from statements which connect "each" with such words as "both" and "two" and "neither" and so on. The Betan's semantic errors, in short, must ramify until they infect his understanding of most words in the language — and far beyond the point where we could still say that he does, with certain exceptions, understand English.

The proposed revision in our account of the Betan is, therefore, impossible.

IX. LOGICAL CLUES: THE ALPHAN

What logical clues could the Alphan have to his error about the meanings of "right" and "left"? That is, what true statements might he read which, on his interpretation of them, would be logically false? Perhaps these would do:

(a) "As I stood on the deck facing forward, a gun to my right fired a short burst. It was the starboard oerlikon."

(b) "As a pitcher he is a southpaw — he can't pitch at all with his right hand."

Confronted with either of these, the Alphan would smell a rat — provided he understood "starboard" and "southpaw" correctly.

Can we protect him from any such logical clues by crediting him with matching semantic errors?

It is encouraging that so few words are involved. Indeed, the only certain examples I can find — apart from ones drawn from very limited dialects — are "port" and "starboard," "southpaw," the words for the four points of the compass, and a few cricketing terms. Also "clockwise" and "anticlockwise," *if* it is contingent that most clock-hands move clockwise. I have doubtless missed some, but not many.

Still, the language could have been otherwise. Screws might be called "standard" and "nonstandard" according to how they have be rotated to be driven in, a righthanded golf club might be called "a hogan" and a lefthanded one "a charles," and so on. Let us pretend, as we easily can, that hundreds of English words are thus meaning-connected with the left/right distinction: *now* can we shield the Alphan from logical clues by the "matching errors" move?

Easily! In each case we know exactly what the matching semantic error must be, namely a simple switch — of the meanings of "port" and "starboard," "hogan" and "charles," and so on. Furthermore, these errors need not ramify and infect words which are not directly meaning-connected with "right" and "left." The initial set of switches completes the whole job, leaving the Alphan with no source of logical clues to his error about "right" and "left" or to any of his compensating semantic errors.

Some we can, for example, comfortably suppose that he begins with his mistake about "right" and "left" and is smoothly seduced by it into his other mistakes without ever having, so far as meaning-relationships are concerned, the faintest hint that he has gone astray. The analogous supposition about the Betan collapses in chaos.

That, then, is my first contrast between "left"/"right" and "between"

– indeed, I believe, between "left"/"right" and any pair of spatial terms which is not equivalent to the left/right distinction. Our terminology for the left/right distinction, unlike any other part of our spatial terminology, has an extremely simple internal logical structure and is thoroughly insulated from the rest of the language. It is for those two reasons that good dictionaries, which do not define "between" as "the normal relation of the mouth to the nose and chin," or "round" as "the normal shape of the pupil of a human eye," do perforce define "right" in terms of "that hand which is normally the stronger of the two."

X. CONTINGENT CLUES: THE BETAN

I now drop logical clues to ask what "contingent clues" either of our men could have to his semantic error. That is, what true statements can he read which, interpreted as he will interpret them, conflict with contingent generalizations which he knows to be true?

Here are some contingent clues for the Betan, again with his pictures indicated in brackets:

(a) "James stood between a snow-clad mountain and me [James-mountain-me]: I could see him perfectly."

(b) "Finding myself between a sheer cliff and the oncoming tide [me-cliff-tide], I was naturally afraid that I should be drowned."

(c) "My brother flung himself between the gun and my body [brother-gun-me], so that the bullet hit him instead of me."

Let us see whether the Betan can evade the force of all such contingent clues, in the following way. Each time he reads a statement which, on his understanding of it, conflicts with a generalization which he has hitherto accepted, he concludes that the generalization does not hold true in the part of the world described in the statement (call it "England"). This would enable him to think that the statement is true on his interpretation of it, and is therefore not evidence that he has made a semantic mistake. It does not matter that he would be silly to try to neutralize each contingent clue by supposing that in England things happen differently. My question is: can he succeed?

Well, under this strategy he must suppose that in England (a) things can be seen through snow-covered mountains, (b) the sea can scale sheer cliffs, and (c) bullets can swerve without being physically deflected.

Furthermore, as clues accumulate, and as some occur containing "because," "since," "so," etc., the Betan must suppose that these strange things which can happen in England do regularly happen there in certain conditions: in England (a) an intervening snow-clad mountain *improves* one's view of dark objects beyond it, (b) waves *are drawn* up sheer cliffs by people at the top, (c) the availability of an alternative target *turns* a bullet in its tracks.

After dealing with variants of just three statements, the Betan already has a strange picture of English life; but there is worse to come. For one thing, each of his suppositions must be reconciled with the rest of what he reads about England, and this will force him into other, equally wild suppositions. (False factual beliefs, indeed, may not suffice. For our rules allow him to read such statements as "In England waves are *not* drawn up cliffs by the presence of people at the top," which would require him to make a semantic error about — of all words — "not.") And those three examples plus their progeny are only tiny fragment of all the contingent clues he can encounter. There will be others, involving thousands of familiar, fundamental aspects of the behavior of the macroscopic world; and each will require him to think that England is different in the relevant respect and in hosts of other respects which follow from that.

If the Betan executes even a small portion of this clue-canceling strategy, he will lose control of his picture of how things happen in England: it is humanly impossible to go any distance with this strategy. To take it all the way, however, is not just psychologically but logically impossible for the Betan as we have described him; for if we suppose him to adopt, remember, and retain all the beliefs about England demanded by his strategy, we must retract our original stipulation that he does, in the main, understand the English language. For example, we cannot say that he knows what "bullet" means if he has endless false beliefs about how the things properly called "bullets" behave, what they look like, what their structure is, and so on. Yet the proposed strategy, if applied to a suitable range of contingent clues, will indeed leave the Betan with hardly any true beliefs about bullets: when shown a real bullet he certainly won't classify it as an object of the sort called "bullet" in English, and the longer he studies it the less inclined he will be to classify it thus. But this is just to say that he doesn't know what "bullet" means — and the argument can be re-applied to virtually every English word.

So the proposed strategy is impossible. To save the Betan from correction by contingent clues we must try — as with logical clues — to credit him with matching semantic errors; and we have seen what that leads to. This result, like the one in Section VIII, is not peculiar to "between." Other pairs of spatial words certainly yield the same result, and I conjecture that the story would run in essentially the same way for any meaning-switch involving a pair of spatial expressions, just so long as it was not logically equivalent to the "left"/"right" switch. I shall give some evidence for this in Section XIII.

XI. CONTINGENT CLUES: THE ALPHAN

Here, perhaps, are some contingent clues for the Alphan:

(a) "Most clock-hands move downward while to the right of center and upward while to the left of center."

(b) "I, like most people, am stronger in my right hand than in my left."

These, on the Alphan's interpretations of them, may conflict with generalizations which he knows to hold true in Alpha, i.e., in that part of the world of which he has knowledge not gained through reading English. Can he disarm them by supposing that England differs from Alpha in the relevant respects? Yes, he can. This strategy is open to him as it was not to the Betan, for reasons which constitute the second big contrast between enantiomorphism and betweenness.

First, there are fewer contingent clues to the Alphan's error than to the Betan's. For every true generalization that becomes false under the "left"/"right" switch there are hundreds that become false under the transformation of "x is between y and z" into "y is between x and z."

Secondly, the beliefs about England which the Alphan must initially adopt under his clue-canceling strategy will include only such items as that the English are mostly lefthanded, that their hearts are on the right, that their clocks run counter-clockwise. None of these will ramify, demanding more and more suppositions about matters not directly concerning right and left.

Thirdly, each generalization which is challenged by a contingent clue to the Alphan's mistake concerns a relatively limited class of things. Where the Betan has to suppose the falsity (in England) of laws of elementary impact-mechanics which govern the behavior of all middle-sized objects, the Alphan has only to suppose the falsity (in England) of certain generalizations about (i) classes of artifacts and other upshots of human decisions and conventions, and (ii) certain biological species. With one exception from sub-atomic physics, which I shall discuss in Section XIV, the only generalizations I know of whose truth-value changes under the "left"/"right" switch are ones which quantify over classes of one of these two kinds.

So the Alphan can easily believe what his clue-canceling strategy requires him to believe. (i) Since the kinds of asymmetry in clock-movements, alphabets, rules of the road, positions of guests of honor, etc., are all matters of social choice, it is likely enough that in England "they order these things differently." (ii) Nor should the Alphan find it unbelievable that in England the relevant biological generalizations are false; for this is just to suppose that England differs from Alpha in its basic stock of biological material, like the supposition — which would be very believable if our planet weren't so well explored — that on some Pacific island there are green sparrows and white crows. It would be different if the Alphan had to suppose that England contains animals with the proportions of mice and the bulk of elephants: he would choke on this, because it involves a ratio of leg-thickness to body-weight which goes against certain elementary and basic physical generalizations that hold true in Alpha. But nothing like that is involved in supposing that Englishmen are mostly lefthanded, or in supposing, of a certain species of asymmetrical Alphan snail, that they do not occur in England though their incongruous counterparts do.

Another point worth noticing about these biological truths that become false under the "left"/"right" switch is that most of them give rather specialized information. The strength of human hands and the placing of human hearts are exceptions to this; but I can think of no other generalizations of this kind which would be known to everyone who led a full, normal, observant, intellectually active life. This is in striking contrast with the ones the Betan has to wrestle with. In Section XIV I shall revert to this point.

XII. THE AMBIDEXTROUS UNIVERSE

There are endless matters which might seem to give the Alphan contingent clues which he cannot easily cancel by the proposed strategy. For guidance on these, and for other pleasures, see Martin Gardner's exceptionally fine book *The Ambidextrous Universe*. I shall discuss a few "pseudo-clues" which I have found to be popular, showing that each fails in at least one of the three following ways: it is not a clue, because the generalization involved does not become false under the "left"/"right" switch; it is not a legitimate clue because it breaks the rule forbidding reference to independently known particulars; it is a clue which can easily be canceled by the proposed strategy.

Mechanical phenomena won't correct the Alphan's error, but it is not obvious that this is so. Given a layout of billiard balls on a billiards table, and a choice of two ways (differing only as right and left) of striking the cue ball, the choice may make a big difference to the final positions of the balls. Does not this supply a basis for contingent clues for the Alphan? It does not. If the initial layout is symmetrical, then the result of striking the cue ball one way will be an incongruous counterpart of the result of striking it the other way, and so for the Alphan all will go smoothly. If the initial layout is not symmetrical, then the Alphan – interpreting our description of it according to his semantic error – will begin not with our initial layout but with its incongruous counterpart; and *then* striking the cue ball one way will give him a final position which is an incongruous counterpart of the one we got by striking it the other way; so again he will have no grounds for suspecting error. This example fairly illustrates the situation with regard to the entire range of mechanical phenomena.

Nor is there any guidance for the Alphan in the common run of electrical phenomena. Rules of thumb relating current-flow to direction of magnetic field, etc., will simply lead him to switch "north" and "south" as applied to magnets; and, short of the recherché matter to be discussed in Section XIV, that switch would not ramify through causal laws or semantic links.

Of two enantiomorphic forms of a certain acid, only one reacts in a certain way with quinine. But that is a fact about the (asymmetric) form of quinine which happens to be the only one biologically available on our planet. Its enantiomorph is chemically and (given the right stock) biologically possible, and it would react in the given way with the other form of the acid in question. Like all other pseudo-clues involving organic molecules, this falls under the heading of generalizations over certain biological species.

As I implied in Section IX, the Alphan can get logical clues from the interrelations of "north," "east," etc., and so we must credit him with a meaning-switch in respect to these too: specifically, he must think that the orientation of any English map can be expressed in English by the pattern $\begin{smallmatrix} & N & \\ E & & W \\ & S & \end{smallmatrix}$, suitably rotated. He may arrive at this through reading "As one stands facing north, east is to one's right"; or, more elaborately, through reading how places in England relate to one another, these relations being expressed both in terms of "left" and "right" and in terms of compass-points. In the latter eventuality, he will find that $\begin{smallmatrix} & N & \\ E & & W \\ & S & \end{smallmatrix}$ works beautifully on the map of England which he gradually builds up. It will of course be a mirror-map of England, but it will give him no trouble unless he gets some independent knowledge of England — e.g., by trying to tour it with the aid of his map.

If he has a correct map of Alpha, can he comfortably impose $\begin{smallmatrix} & N & \\ E & & W \\ & S & \end{smallmatrix}$ upon it? He has no right to assume that it belongs on his map at all; but never mind that. If he does try to impose it on his map of Alpha — or on Alpha — will he encounter any positive obstacles which will serve as contingent clues? To do so, he will have to have (1) something dictating how $\begin{smallmatrix} & N & \\ E & & W \\ & S & \end{smallmatrix}$ should be rotated before being placed on the map of Alpha, and (2) something else casting doubt on that placement. That is, he needs *two* contingent correlates of compass-points — correlates which he is told are valid in England, and which concern matters in respect to which he cannot easily believe that England differs from Alpha. This might be an example:

> "North is the direction toward which compass-needles point. East is the direction from which the sun rises."

One point to notice about these double-correlates which are needed if the Alphan's $\begin{smallmatrix} & N & \\ E & & W \\ & S & \end{smallmatrix}$ is to yield contingent clues is that if they can generate contingent clues at all they can do so without reference to "north" etc., thus:

> "As one stands looking in the direction toward which compass needles point, the rising sun is toward one's right."

The vital point, though, is that the Alphan cannot have even *one*, let alone the needed two, of these correlates of "north," "east," etc. That is, he cannot have good reason to think that any such correlates which he knows to obtain in Alpha must also hold good in England.

Compass needles cannot help us to correct the Alphan, because they point south as well as north. That their ends are differently shaped, and how they are shaped, is a matter of convention.

Still, let us concede compass needles in order to get the sunrise to work. If the Alphan is to get a contingent clue from this, he must say: "Surely the compass-direction of the sunrise in England must be the same as in Alpha!" But why should he say this? Not because a particular star shines on a particular rotating planet which contains both England and Alpha. Of the items which the Alphan knows about in ways other than by reading about them in English, he must not identify any one as *the item* to which the Englishman refers as "the sun" or "the earth (Terra)," though he may recognize some as *items of the kind* the Englishman calls "sunlight" or "stars", "ground" or "planets." (I repeat that this niggardliness is not *ad hoc* or arbitrary. If the Alphan can read English statements about "the earth" and "the sun," and identify these with items known to him in other ways, then he might as well read about and independently identify the constellation Orion, or the box *A* in Section IV above, or his right shoulder. From the point of view of the Kantian Hypothesis, any such use of an independently known particular is on a par with our touching the Alphan's right shoulder while saying "I am now touching your right shoulder." This does not make the Hypothesis trivial: its rules for the Betan are just as stern, yet *he* is deluged with logical and contingent clues to his semantic error.)

To mention just one more popular pseudo-clue: since the Alphan may not identify a particular planet — let alone its Northern Hemisphere — as the one containing both England and Alpha, he cannot have any contingent clues involving the direction from which the cold winds blow, or the like.

I cannot anticipate and criticize every plausible pseudo-clue to the Alphan's mistake, but my treatment of the ones I have mentioned may help to show how others should be dealt with.

It is time to consider what the Alphan is to make of the samples of

English writing he receives. Clearly, he must not encounter English statements about these samples; and indeed if he encounters English statements about English writing in general — statements which become false under the "left"/"right" switch — then he must suppose that his English correspondent eccentrically writes mirror-English, or that his missives come through a censorship office which photographs them and forwards the negatives, or some such nonsense. These are trivial details. What is not trivial is the following question. Suppose that the Alphan is somehow deprived of samples of written English, but is sent — in Morse-code English, say — very full instructions for writing English letters and words and sentences and paragraphs: what will he write if he follows those instructions as he, with his "left"/"right" switch, understands them? The answer is that he will, without any hitch or hesitation, write perfect mirror-English letters and words and sentences and paragraphs. Sceptics should try it for themselves. There are no clues for the Alphan here.

Of course it would have saved trouble if, following Borel (§§33—36), I had at the outset explicitly placed the Alphan and the Betan on a cloud-covered planet at a great distance in an unknown direction from ourselves. This would have had us communicating in Morse-code from the start; it would also have automatically ruled out all the biological, geographical, meteorological, and sociological overlap between Alpha and England, as well as much of the astronomical overlap; and thus it would have reduced the number of tempting pseudo-clues for the Alphan. I did not adopt this course because, although it would have made things easier, it would not have made clear just what sorts of overlap were being excluded or why they were being excluded.

XIII. SOME OTHER SWITCHES

When I first worked on this topic I contrasted left/right with large/small, but was charged with unfairly exploiting the fact that large/small is metrical. So I re-worked the contrast using "between" instead. The latter, like anything else I might use, also has special features; but they have not essentially contributed to the contrasts I have drawn. To get *prima facie* evidence for this, consider how the story would go for certain alternatives to "between."

Had the switch involved "large" and "small" and their grammatical cognates, there would have been such logical clues as:

> "My house is smaller than Jones's — indeed it is the same size as Jones's largest room."

This and its like would require matching semantic errors involving "part" and "whole," "inside" and "outside," "contain," "surround," and hundreds more. And they would ramify: for example, "surround" would infect "grasp," "penetrate," "hole," etc. There would also be contingent clues:

> "I couldn't see the water because the house between myself and the shore was so large."
> "The rock wasn't small enough for a child of ten to lift it."

It is clear that a "large"/"small" switch would be Betan rather than Alphan.

Suppose we had tried a "round"/"square" switch. These words connect through logical clues with "angle," "smooth," "equidistant," "straight," "curve," "circle," "triangle," and so on. And there would also be many contingent clues involving the role in English life of wheels, building bricks, land surveys, tree trunks, and so on. For example,

> "Roundabouts are so-called because the path of someone riding on one is *round*. This is because roundabouts are built and operate as follows . . .",

with the blank filled by a correct account of how roundabouts work. Someone who had switched the meanings of "round" and "square" would have to adjust his semantics and/or his English physics in such a way that that account really would explain to him why the path of someone riding on a roundabout is *square*. Another Betan situation.

Perhaps I needn't offer details on "near" and "far," or "inside" and "outside," or "toward" and "away from." These will very quickly connect, causally and semantically, with "large" and "small" — and we have seen where that switch leads.

What about "head"/"feet" and "front"/"back"? Either of these switches would generate a Betan situation, though an uninteresting one — like a switch in the meanings of "teapot" and "breadboard," or of "nose" and "elbow." It could seem interesting only to someone who was still in thrall to the mistaken view, discussed in Section IV above, that the human body is essential to enantiomorphism rather than merely the basis for some convenient terminology for it.

The question "What would happen with an 'up'/'down' switch?", though it could be motivated by the same mistake, has more inherent interest. It is in fact hard to decide what a "switch in the meanings of 'up' and 'down'" would be. The word "down(ward)" can be defined as "the direction of normal fall" or "the direction toward the ground" or both; and analogously for "up(ward)." This suggests three possible interpretations of the switch, but I cannot control the details of any one of them. Part of the trouble is that the logical/contingent dichotomy, which has served well enough for the other switches, lets us down here. "The direction of normal fall" connects with "the direction toward the ground" at least to this extent: if those two directions were different we should have no objects left except ones that were fixed to the ground. Is that a merely contingent matter?

Still, without knowing just what an "up"/"down" switch would be, I think I can show that it would have to be Betan rather than Alphan.

Suppose the contrary. Suppose there is a Gamman who understands English except that he has switched "up"/"down" and certain related pairs such as "above"/"below." He reads many statements about England, interprets them according to his matching set of semantic errors, and believes them. If he is to be analogous to the Alphan, the Gamman's false beliefs about England must not be so far-reaching as to conflict with the postulate that he mainly understands English. So we must suppose that if he came to England with expectations based on what he had read, he would find it fairly much as he had expected, and yet unlike his expectations in some systematic respect.

What respect? The fact that we don't bounce around on our heads? The fact that unimpeded objects don't tend to shoot up into the sky? Anything along the lines of either of those mistakes would obviously generate a Betan state of affairs, not an Alphan one. The only other suggestion I have heard or can think of is this: "He expects unimpeded objects to shoot upward toward the ground — i.e., he has false beliefs about the direction of object-fall in England *and* about where the English ground is in relation to the English sky." But those two "false beliefs" cancel out. The Gamman, in this version of him, would get no surprises if he arrived in England; which is just to say that his English reading has not been infected by any semantic error.

What about a switch in the meanings of "before" and "after"? Nothing in my paper requires an answer to this question; and that is just as well, for even a sketchy answer — which is all I know how to give — would take up far too much space.

XIV. THE CONSERVATION OF PARITY

The principle of the Conservation of Parity says in effect that if in any general truth of physics we substitute "right" for "left" and vice versa, the result will also be a truth of physics. This principle, though long accepted, is now thought to be false: work which began in Princeton in 1956 has satisfied physicists that there is a basic physical law which becomes false under the "left"/"right" switch.

The point could be put as follows. Suppose that we have two experimental set-ups with initial states I_1 and I_2 and resultant states (arising from the initial ones in ways that can be wholly explained by basic physical laws) R_1 and R_2. The Parity principle implies that if I_1 is an enantiomorph of I_2 then R_1 is an enantiomorph of R_2; and it now turns out that this is sometimes false.

Our Alphan can now be exposed to contingent clues with far more force than any so far mentioned. Let us send him a description of a Parity-refuting experiment, with initial state I_1 and resultant state R_1, and suppose that he tries to reproduce the experiment in Alpha. He will in fact start not with I_1 but with its enantiomorph I_2, and he will end up with a resultant state R_2 which is *not* an enantiomorph of R_1. So the resultant state won't be the one the Englishman predicted, nor will it be the one that the Alphan, with his semantic error, thinks the Englishman predicted. This should lead the Alphan to suspect that he has misunderstood the English description of the experiment, and if he perseveringly tests that suspicion he will learn what the misunderstanding was.

If we are to prevent this, we must try one of two courses.

(i) We may try to credit the Alphan with further semantic errors such that he will understand the English description of R_1 as an accurate description of R_2. But this takes us back into Betan territory; for these semantic errors — concerning words which are not meaning-linked with "right" and "left" — would ramify into the rest of the language. I believe the Alphan would in fact have to switch the meanings of "more" and "fewer" or some equivalent pair of terms (see Gardner and Frisch), and that switch would obviously lead the whole Alphan story to collapse.

(ii) We may present the Alphan as supposing that if he had performed his experiment in England it would have come out differently. But if he thus distinguishes sub-atomic particles into "the ones we have here in Alpha" and "the ones they have there in England," he will be cutting a very poor figure. Anything which seems to be a fundamental physical law may turn out to be a relatively local accident, but a scientist in his right mind would not accept such a conclusion if he could rationally avoid it. The Alphan has an alternative starting him in the face, namely that he has switched the meanings of "right" and "left" and therefore constructed I_2 instead of I_1.

At long last, we have got him. (I here suppress more recent developments and discoveries in the physics of this matter, as they lie beyond the scope of my present concerns. For details, see Gardner, *op. cit.*)

This certainly refutes the Kantian Hypothesis as I formulated it: we can now *tell* the Alphan which is which as between right and left. But then we could have told him anyway, using "port" and "starboard." The fact is that the Kantian Hypothesis has served less as a sharp-edged proposition whose truth was to be tested than as a guide to the exploration of some contrasts between enantiomorphism and other spatial distinctions. And those contrasts have not been entirely lost.

Despite the failure of the Parity principle, it remains true that the left/right distinction constitutes, so far as meaning-relationships go, a self-contained unit with simple internal relations and no external relations — that is, no ramifications into the rest of the language.

What of the other contrast? Well, the generalization which we are now using to correct the Alphan is not one which he can easily suppose false in England — it is neither sociological nor biological but is a matter of fundamental physics. Still, there is a contrast between it and the kinds of causal law which were available for correcting the Betan. Quite recently, two Nobel Prizes were awarded for the discovery of a physical law which does not survive the left/right switch, and so knowledge of that law is a perfect paradigm of specialized, non-common knowledge. Someone who grasps all the underlying fact and theory will not find the law "easy to suppose false"; but we can lead busy, observant, intelligent lives without having the slightest need to think that the law is true. This situation could change, if our technology came increasingly to depend upon the law in question; but knowledge of that law and of any asymmetries depending upon it is, and will long continue to be, optional intellectual equipment. So, even with the failure of the Parity principle taken into account, the right/left distinction (by which, always, I mean the distinction between any enantiomorphic pair) still differs enormously from every other spatial distinction: it remains unique in its degree of isolation in the layman's language and the layman's *Weltanschauung*. Is it too ambitious to suggest that these simple facts help to explain physicists' surprise at the failure of the Parity principle?

It cannot be a coincidence that there are these two differences — logical and contingent — between enantiomorphism and other spatial distinctions. I am sure that the differences which show up semantically ("logical clues") are to be *explained* by the differences which show up in the extralinguistic world ("contingent clues") — the explanation depending upon the fact that our language is our reasonable attempt to cope with that sort of world.[2]

NOTES

[1] See the Bibliography at the end of the paper.

[2] A version of this paper was read to several philosophy departments in 1966 and 1967, and was much helped by the comments and criticisms of many people to whom, though I cannot separately name them, I here express my sincere thanks. In other ways the paper has benefited from the generous collaboration of Lewis White Beck, D. G. Brown, Martin Gardner, James G. Hopkins, and Douglas F. Wallace, to all of whom I am deeply grateful.

BIBLIOGRAPHY

Averroës, *Tahafut al-Tahafut* (trans. S. van den Bergh, London, 1954).

Borel, Émile *Space and Time* (New York, 1960).

Caird, Edward *A Critical Account of the Philosophy of Kant* (Glasgow, 1877).

Frisch, O. R. "Parity not Conserved: a New Twist to Physics?", *Universities Quarterly*, vol. 11, (1957).

Fritsch, Vilma *Left and Right in Science and Life* (London, 1968).

Gardner, Martin *The Ambidextrous Universe*, revised edition (New York, 1969).

Ghazali, Abu Hamid Muhammad *Tahafut al-Falasifa* (trans. S. A. Kamali, Lahore, 1958).

Jammer, Max *Concepts of Space* (Cambridge, Mass., 1954).

Kant, Immanuel "Von dem ersten Grunde des Unterschiedes der Gegenden im Raume" (1768), in *Gesammelte Werke* (Akademie Ausgabe), vol. 2. Translated as "On the First Ground of the Distinction of Regions in Space" in John Handyside (trans.), *Kant's Inaugural Dissertation and Early Writings on Space* (Chicago, 1929). Reprinted in this volume, pp. 27—33.

Kant, Immanuel, Inaugural lecture: "De mundi sensibilis et intelligibilis forma atque principiis" (1770), in *Gesammelte Werke* (Akademie Ausgabe), vol. 1. Translated as "Dissertation on the Form and Principles of the Sensible and Intelligible World" in Handyside (see preceding entry). A selection reprinted in this volume, pp. 35—36.

Kant, Immanuel *Prolegomena to any Future Metaphysic that willbe able to present itself as a Science* (trans. P. G. Lucas, Manchester, 1953). Reprinted in Smart.

Kemp Smith, Norman *A Commentary to Kant's Critique of Pure Reason* (New York, 1962).
Leibniz, G. W. *The Leibniz-Clarke Correspondence* (ed. H. G. Alexander, Manchester, 1956).
Mayo, Bernard "The Incongruity of Counterparts," *Philosophy of Science*, vol. 25 (1958), pp. 109—115. A reply to Pears.
Pears, D. F. "The Incongruity of Counterparts," *Mind* n.s. vol. 61 (1952), pp. 78—81.
Reichenbach, Hans *Philosophy of Space and Time* (New York, 1958).
Remnant, Peter "Incongruent Counterparts and Absolute Space," *Mind* n.s. vol. 72 (1963), pp. 393—399. This volume, pp. 51—59.
Russell, Bertrand *Essay on the Foundations of Geometry* (Cambridge, 1897).
Scott-Taggart, M. J. "Recent Work on the Philosophy of Kant," *American Philosophical Quarterly* vol. 3 (1966), especially pp. 178—180.
Smart, J. J. C. (ed.) *Problems of Space and Time* (New York and London, 1964).
Weyl, Hermann *Philosophy of Mathematics and Natural Science* (Princeton, 1949). See also the same author's *Symmetry* (Princeton, 1952), especially pp. 16—38.
Wittgenstein, Ludwig *Tractatus Logico-Philosophicus* (New York, 1961).

JOHN EARMAN

KANT, INCONGRUOUS COUNTERPARTS, AND THE NATURE OF SPACE AND SPACE-TIME*

The purpose of this paper is twofold. First, I want to examine some rather curious arguments of Kant's which purport to show that some alleged properties of space can be derived from some alleged facts about incongruous counterparts. Secondly, I want to give some preliminary answers to some important questions about the distinction between right and left and the nature of space and space-time which are raised by Kant's argument. As a byproduct, I hope that the discussion will provide an example of how science can illuminate philosophical issues and of how philosophy can at least lead one into interesting scientific issues.

I

Kant's views on the nature of space and time went through a number of major changes during the period between 1747 and 1770, and incongruous counterparts, e.g., right and left hands, played an important and shifting role in the development of Kant's thought. I will be primarily concerned with one stage of this development, but I will begin with some remarks about the flanking stages.

From 1747 to sometime before 1768 Kant believed that space was something objective and real. He rejected Newton's conception of space as something absolute, a boundless receptacle which is prior to the bodies it contains, and he accepted Leibniz's view of space as something relational, the mutual order of relations among existent things.[1] In 1770 Kant rejected both the Newtonian and the Leibnizian views of space. He now maintains that space is something subjective and ideal, 'and, as it were, a schema issuing by a constant law from the nature of the mind, for the coordinating of all outer sensa whatever'.[2] His main premise, as stated in the *Dissertation*, was that the difference between right and left cannot be stated 'in terms intelligible to the mind through verbal description . . .' (ibid., section 15C, p. 60; this volume, p. 35). His conclusion: 'It is therefore clear that in these cases the diversity, that is, the incongruence, cannot be apprehended except by pure intuition' (*Dissertation*, section 15C, p. 60; this volume, p. 35). Presumably, his final conclusion is supposed to follow from the additional premise that objects of intuition must be ideal.[3]

As Jonathan Bennett[4] has pointed out, it would be indeed surprising if Kant's main premise were not false. That it is false is shown by the fact that we do have expressions like 'right-handed' and 'left-handed'. Bennett suggests that what Kant meant to claim was that one cannot explain or tell which hand is which, which hand one wants to call right and which one wants to call left; rather one must point or show. However, I can find no evidence in Kant's writings that suggests Kant was concerned with the left–right distinction in the which-is-which sense. In any case, this second claim is also false; I can, for example, tell you that what I call left-handed is given by the hand which in most human beings is attached to the arm on that side of the body nearest the heart. Restrictions can be put on the ways in which I am allowed to explain which is which; for instance, I might be allowed to use only the most fundamental laws of nature. In this case, I will have a good deal of trouble explaining to you which is which unless you understand elementary particle physics, and until the 1950s it was thought to be impossible that the job of telling could be done by the laws of physics alone.[5] But this and other peculiarities in the left–right distinction do not seem to me to bear on the question of whether space is a pure intuition or the question of whether it is something subjective and ideal.

Just two years earlier, in 1768, Kant defended a quasi-Newtonian view of space, and he used the same example of incongruous counterparts to argue that '*absolute space has a reality of its own, independent of the existence of all matter, and indeed is the first ground of the possibility of the compositeness of matter.*'[6] His argument, as reconstructed by S. Körner,[7] was as follows. There is a difference between right and left since, e.g., a right-handed and a left-handed glove cannot, even at different times, occupy the same space. But these differences cannot lie in the internal spatial relations of their parts, for these relations are entirely similar. Therefore, the differences must lie in the relations of the gloves to the absolute space. Max Jammer's reconstruction is essentially the same except that he takes Kant as saying that 'the intrinsic relations among the individual parts of our left hand are the same as in our right hand'.[8] This reconstruction seems justified by Kant's assertion that so far as 'the mode of connection of the parts of the body [i.e., the right or left hand]' go 'everything may be completely identical in the two cases' (*Regions in Space*, p. 27; this volume, p. 32).

J. J. C. Smart[9] treats a slightly different version of the same argument. Although Smart refers to the *Prolegomena*, the version in question comes from Kant's earlier 1768 essay. Kant asks us to imagine that the first thing created by God is a hand. Then, since whether the hand is a left or a right hand 'there is no difference in the relations of its parts to one another', it would follow on the assumption of the relational theory of space that 'the hand would in respect of this character be absolutely indeterminate, i.e., would fit either side of the human body, which is impossible' (*Regions in Space*, p. 28; this volume, p. 32).

Kant's argument is almost universally rejected; purported refutations are offered by Smart, Jammer, H. Weyl,[10] C. B. Garnett,[11] and P. Remnant.[12] I find three things wrong with these criticisms of Kant. First, none of these authors gives an accurate and full reconstruction of Kant's argument, and they do not succeed in identifying its real weaknesses. Secondly, some of the above authors employ counter-arguments which in several respects are even stranger than Kant's own argument. Thirdly, all of Kant's critics grant him a premise which is crucial to his characterization of the distinction between right and left but whose underlying assumption stands in need of detailed critical examination. I believe that it is by seeing that part of this assumption can be false that one can see that there is something right about Kant's ideas on the left–right distinction.

I should also add that an air of unreality pervades the discussions of the above-mentioned authors, and this for two reasons. (1) If we are to believe modern science, space *is* absolute, i.e., non-relational in Leibniz's sense. (2) There is not one example of a relational theory which is worked out in any detail, and, prima facie, there are strong reasons for believing that such a theory cannot be successful without becoming very much like an absolutistic theory. I will have more to say on these points below.

Let me now turn to Kant's critics. Garnett swallows the major portion of Kant's argument but objects to Kant's final conclusion.

> Kant's argument here proves only that the nature of a body is determined partially in some way other than by the relations of the inner parts to each other. It in no way shows what the new or additional principles are (p. 113).

Garnett's reservation may well be justified, but in stating his reservation he concedes too much to Kant, as we shall see below.

Jammer objects that a right-handed object can be interchanged with a left-handed one by "moving" the object about in a four-dimensional space. 'It is clear, therefore', says Jammer, 'that mathematically no essential mark distinguishes one sense from the other,' (pp. 131–2). Jammer's claim about interchange is perfectly correct, but it is irrelevant to the point at issue — the nature of the distinction between right and left. There are all sorts of things true of $n + 1$-dimensional Euclidean space which are not true of n-dimensional Euclidean space; but the differences between four-dimensional and three-dimensional Euclidean spaces do not seem relevant to the distinction between three-dimensional right and left hands, and the differences between five-dimensional and four-dimensional Euclidean spaces do not seem relevant to the distinction between four-dimensional right and left hands, etc. One might as well argue that 'It is clear that mathematically no essential mark distinguishes a set of vectors which is maximal with respect to the property of being linearly independent and a set which is not, for a set which is maximal in n-dimensional space is not maximal in $n + 1$-dimensional space.'

Remnant attempts to show that Kant's argument is incoherent. He maintains that if there is no difference between the internal spatial relations of the parts of right and left hands, then

> In a universe which contained nothing but a single hand, it would not just be empirically undecidable whether the hand were right or left; it would be strictly indeterminate . . . so, if I were creating a universe, I could choose whether the hand with which I had begun was to become left or right by deciding next to create a standard body or its incongruous counterpart . . . (p. 399; this volume, p. 58).

It seems to me that Remnant's attempt fails. If space were absolute then a hand standing alone in an otherwise empty universe would not be indeterminate with respect to its handedness; it would have a definite handedness — the parity operation for, say, the (approximately) Euclidean three-space of the room I am now sitting in would not collapse if all the objects in it save one hand were to vanish, at least if this space is as Newton conceived it. How one could come to know whether the hand in the Kant–Remnant hypothetical universe is what we call a right hand or whether it is what we call a left hand is a difficult question, and it is part of the larger question of how identity across possible worlds is to be established; but these questions are not crucial to the present issue. Thus, if Remnant's claim were correct, Kant's argument could be turned completely round and could be used to refute absolute space; but such an attempted refutation of absolute space seems to me to have no more plausibility than Kant's attempted demonstration of absolute space.

Smart also rejects Kant's premise that it is absurd to suppose that if the first thing created by God is a hand, then its handedness would be indeterminate; but unlike Remnant, he does not maintain that Kant's argument is incoherent, for his rejection of Kant's premise is conditional on the assumption of a relational theory of space. Smart says that a proponent of the relational theory could reply to Kant by saying that 'if there were only one thing, a hand, in the universe, it would be meaningless to call it either a left hand or a right hand,' ((b), p. 217). (A similar move is advocated by Weyl, pp. 21—22.) Smart immediately adds one qualification:

> Or at least the relationist would say that this is what would be said on the assumption that the laws of nature are symmetrical with respect to mirror reflections. The recent discovery of the non-conservation of parity suggests that, provided that we knew whether the hand consisted of matter or anti-matter, there would be a meaning to saying whether it was a left hand or a right hand. But this once more would consist of a *relation* between the 'handedness' of the elementary particles of which the hand is composed and the 'handedness' of the hand as a whole. Once more, there would be no need for absolute space. Moreover, there would be some difficulty about the meaningfulness of saying whether the hand consisted of matter or anti-matter if there were only one thing in the universe ((b), pp. 217—18).

Several comments seem called for. First, we are either being treated to a curious mixture of epistemological and ontological questions or a curious theory of meaning: the meaningfulness of the statement 'The hand is left-handed' somehow depends upon our knowledge of whether the hand is composed of matter or anti-matter. Secondly, there may be some difficulty about the meaningfulness of saying whether the hand consisted of (what we would call) matter or (what we would call) anti-matter, but the only source of such a difficulty would seem to be the difficulty in making trans-possible world identifications. So exactly the same difficulty would apply to any assertion attributing an empirical property to the isolated hand, and the (alleged) difficulty does not, therefore, show anything peculiar about the particle—anti-particle or the left—right distinction. Thirdly, there must presumably be some distinction between right and left if one can meaningfully ask whether or not the distinction is one which is codified in the laws of nature; and if a distinction exists, be it *de facto* or nomological, then it seems meaningful to ask of some object, even if it is the only object in the universe, 'Is it right- or left-handed?' Next, it seems doubtful that the handedness of a hand need be or can be accounted for by the handedness of the particles of which the hand is composed. It seems to make perfectly good sense to speak of a right-handed object which is composed of particles which have no definite handedness at all; indeed, the handedness or lack of handedness of the composing particles seems irrelevant to the handedness of the particles — what is important is the spatial arrangement of the particles. Nor is there any need to become involved in the question of the relation of the handedness of the object and the handedness of the particles composing the object — if Kant's argument as reconstructed by Smart is going to work at all it will work when applied to elementary particles standing alone, and macroscopic objects need not be brought into the discussion. Finally, as we shall see later on in detail, parity non-conservation is relevant to the distinction between right and left, but in a way very much different from what Smart imagines.

It is not really a matter of great importance whether Smart's defense of the relationist succeeds or fails; Smart's defense is against a Kant armed with one of the discoveries of modern science, but modern science rejects the relational view of space. Thus even if Smart's defense succeeds, it will fail at a later stage when more science is brought to bear. Moreover, what we want to know is whether Kant's argument by itself forces a conception of space as absolute. Smart's defense is relevant to this question. The relationist could accept Kant's premise that the internal spatial relations of the parts of right and left hands are the same and could argue that, therefore, a hand standing alone in an otherwise empty universe would be neither a right hand nor a left hand. He would then be left with the task of explaining congruity and incongruity in a universe with more than one hand. There is no example of a relational theory which is worked out in great enough detail to say whether such an explanation is possible, but in any case Kant's argument does not seem to rule out the possibility of such an explanation. (I will have more to say on this point below.)

I believe not only that Kant's argument is ineffective but also that there is no coherent argument at all. Let us return to the argument for a closer look. It seems true that, as Körner puts it, the internal relations of the parts of left and right hands are 'entirely similar' — at least if this is taken to mean that the relations are different but resemble each other in a systematic way; the relations for one hand mirror, in an obvious and literal way, the relations for the other hand. But on this interpretation, it is false to say that the difference between right and left cannot lie in the differences between the internal spatial relations of the parts of left- and right-handed objects. But isn't this incorrect? Isn't Kant correct in claiming that the internal relations of the parts of a right hand are the *same* as for a left hand? The answer obviously depends upon how one construes the notion of internal relations and how internal relations are supposed to contrast with external relations. Let us assume that Kant's conclusion is correct and that space is absolute in the sense that it is a kind of container in which bodies reside. The most obvious move would be to construe the external relations of the body to the encompassing space as involving, say, the position and orientation of the hand in the space. The distinction between the internal spatial relations of the parts of the hand and the external relations of the hand to the encompassing space would be that the latter but not the former would change as the body is moved about in the space. But it would follow from Kant's own view that the handedness of the object arises from the internal relations of the parts of the object since, according to Kant, the handedness of an object does not change as it is transported about in space. If we limit the internal spatial relations of the parts of the hand to those which can be defined purely in terms of the distances between points of the hand and angles between lines lying in the hand, then obviously Kant's claim is correct since mirror image reflection preserves these distances and angles. But if one continues to limit oneself to these kinds of relations, then introducing points and lines of the space external to the hand will not help to distinguish between right and left since whatever the relation in this sense a right hand has to an external point or line, these relations can be mirrored by a left hand. Parroting Kant, one can then argue that the difference between right and left cannot lie in the relation of the hands to the encompassing space since these relations are the same; therefore, the difference must lie in something external to absolute space.

At this juncture, I must admit to being grossly unfair to Kant in one respect; Kant recognized the above points. He says quite explicitly that the difference between right and left cannot lie in the relation of the hands to the external space in the sense of the points of space lying outside the hand:

the region towards which the ordering of the parts is directed [i.e., the handedness of the object] involves reference to the space outside the thing; not, indeed, to the points of the wider space — for this would be nothing but the position of the parts in an outer relation . . . (*Regions in Space*, p. 20; this volume, p. 27).

Rather the difference between right and left depends upon the relation of the hands 'to universal space as a unity of which every extension is a part' (ibid., p. 20; this volume, p. 27). The notion of 'universal space as a unity of which every extension is a part' seems to refer to Kant's doctrine of the unity of space — the idea that space is singular in that no place can be apprehended except as part of one and the same Space since 'Space' is a proper name and 'x is a space' means that x is part of Space.[13] But it is hard to see how Kant's appeal to the unity of space helps to make his argument coherent. I think that it is a reasonable conjecture that Kant recognized that he had not supplied an explanation of the difference between right and left and that this was part of the cause of the shift in his views which took place between 1768 and 1770. Reflecting on the unity of space led Kant only indirectly to a new explanation of the difference between right and left: since space is a singular concept it is a pure intuition, and the difference between right and left can be apprehended only by pure intuition and cannot be made intelligible by any verbal description, in particular through a verbal description referring to the points of the space external to the hand.

There is, as a matter of empirical fact, a kernel of truth in Kant's views as expressed in his 1768 essay. Space is absolute, and it is at least locally orientable. The sense of a hand, 'the quarter towards which it lies', depends upon the relation of the hand to the external space — external in the sense of being prior to the hand and the other bodies which it contains but not necessarily external in the sense of lying outside the hand — and the local orientation of this space. The fact that the handedness of an object is not in at least one sense a purely intrinsic feature of the object but depends upon its relation to the local orientation of space would be brought home if one could give examples of three-dimensional spaces in which this relation could be inverted by continuous rigid transport of the hand — this kind of transport can be taken by definition to guarantee that in the sense intended the intrinsic relations of the parts of the hand to each other are preserved — so that a hand which was originally right-handed would be changed into a left-handed one or vice versa.[14] But such an example would contradict the assumption underlying Kant's characterization of the incongruity of right and left hands. However, Kant's characterization can be retained if it is weakened: the interchange of right and left hands cannot be accomplished by local twisting and turning; non-local transport is required to take advantage of the non-simple connectedness of the non-orientable space. These points will become clear in the following section.

In summary, there is, as a matter of empirical fact, something right about Kant's notion that the handedness of an object does not arise purely from the internal spatial relations of its parts; but Kant's arguments do not reveal what this something is and they do not show that space is absolute rather than relational. In particular, Kant does not give any reason why the relationist should not legitimately include among the list of primitive relations among bodies to which all other spatial relations are (allegedly) reducible the relation of standing in a right-handed (or left-handed) configuration with respect to each other (curiously, none of Kant's critics mentions this fact); so although Kant's treatment of incongruous counterparts is effective against a version of the relational theory which takes the primitive relations to be, say, distance and angle – and apparently Kant took Leibniz to be putting forward such a version – Kant does not show why Leibniz could not produce a version which is immune to his criticisms. Perhaps the added consideration of non-orientable spaces would be effective against even the modified version suggested above; for it is hard to see how the relationist could explain how and why rigid transport changes the relation of standing in a right-handed configuration into that of standing in a left-handed configuration and vice versa, but I will not pursue this matter here.

Though Kant does not mention any of them, there are a number of other criticisms which he could have brought to bear against Leibniz's theory. I will consider one example here. In his correspondence with Clarke, Leibniz was forced to admit a distinction between absolute and relative motion in an attempt to account for what Newton would characterize as cases of absolute acceleration:

> I grant that there is a difference between an absolute true motion of a body, and a mere change of situation with respect to another body. For when the immediate cause of the change is in the body, that body is truly in motion; and then the situation of other bodies, with respect to it, will be changed consequently, though the cause of that change be not in them.[15]

Leibniz does not explain what it is for the cause of the change to be in the body itself; and, furthermore, this notion would seem to be in conflict with Newtonian dynamics and the principles of the conservation of energy and the conservation of momentum. More importantly, Leibniz makes absolute motion a species of relative motion: *X* is in absolute motion just in case *X* is in motion relative to some system of bodies *Y* and the cause of the change of the situation of *X* with respect to *Y* is in *X*. But in certain cases, e.g., the case of Newton's rotating globes, there is no change of situation of bodies with respect to other bodies, and therefore on Leibniz's theory, no motion, absolute or relative, at all. Newton's theory predicts and explains why it is that in some cases but not in others there is a change of situation of the globes with respect to each other if the cord connecting them is cut. Leibniz apparently has no explanation of this effect; and in general he has no explanation of effects typically associated with accelerations when there is no change of situation of bodies with respect to each other. Thus, one can hardly agree with Leibniz that he has 'left nothing unanswered, of what has been alleged for absolute space'.[16]

Finally, I want to correct the impression given by some philosophers that Leibniz's views on space and time have been vindicated by modern science. According to the orthodox interpretation of relativity theory, space-time is an entity whose existence is prior to the existence of any physical process which performs in the space-time arena; *a fortiori*, space is a container whose existence is prior to the existence of the bodies it contains. And there are non-orthodox interpretations, e.g., Wheeler's,[17] which stand the relational theory on its head: curved empty space-time is seen not as an arena but as everything; everything is built out of curved empty space-time. To be sure, relativity theory shows that space is less absolute in a number of senses than Newton would have had us believe. For example, special relativity rejects Newton's assumption that there is a unique projection of space-time onto space, and general relativity rejects Newton's further assumption that space is fixed and immobile, uninfluenced by its material (and energetic) content. However, these latter senses in which our conception of space has become less absolute since Newton, do not bear on the issue of absolute versus relational theories of space as Newton, Leibniz, and Kant conceived the issue.

II

Kant held that it is an *a priori* truth that space is Euclidean. Today this position is not tenable. Moreover, it has been asserted that there are empirical reasons for believing that as a matter of fact space is neither flat nor infinite. This assertion is misleading. As stated in Section I, there is no unique projection of space-time onto a physical three–space; and one can construct examples of space-time such that relative to one projection space is flat, but curved relative to another projection, and one can also construct examples where the space of one projection has the topology of a hyper-plane whereas the space of another projection has the topology of a hypersphere.[18] But at least one can say that the space associated with certain natural frames of reference in cosmological models which are in accord with known data are curved and/or finite (compact).

What seems obviously true (if not *a priori*) in Kant's discussion of incongruous counterparts is that left- and right-handed objects cannot be substituted for each other; they cannot, even at different times, occupy the same space. And in one respect it is true: if what was originally a left hand came to occupy the space just vacated by a right hand, we would cease to call the former a left hand. However, Kant's premise was intended to convey the non-trivial but obvious (and *a priori?*) truth that such a thing cannot be made to happen by rigid continuous transport of the hands through space; in a more mathematical vein, this truth is expressed by saying that mirror image reflection is not a continuous operation. But this obvious truth has gone the way of so many obvious truths; today it is not obviously true, and it is even conceivable that it is not true — there are mathematical examples of spaces in which this obvious truth is a falsehood, the Möbius band being the simplest example.

In order to treat this question in more generality, let us assume that space-time is a four-dimensional, differentiable manifold equipped with a Lorentz signature metric. Roughly, a manifold is a Hausdorff topological space which is locally Euclidean.[19] The Lorentz signature allows us to define in the tangent space T_x at every point $x \in M$ an object called the null (or light) cone. A tangent vector at $x \in M$ is said to be timelike (respectively, null, spacelike) if it lies inside (on, outside) the light cone at x.

A space-time M is said to be temporally orientable if, intuitively, one can label in a continuous way one lobe of every light cone the future lobe and the other lobe the past lobe. There are two equivalent ways of making this concept more precise. First, one can say that M is temporally orientable if and only if there is a continuous, non-vanishing, timelike vector field on M. Secondly, an equivalent characterization can be achieved by means of the holonomy group of M. Consider a point $x \in M$ and the collection of all closed paths beginning and ending at x. The transport of vectors around one such path, keeping each vector everywhere parallel to itself, defines a linear mapping of the tangent space T_x onto itself, and the collection of all such mappings forms a group. One can speak of *the* holonomy group $H(M)$ of M since the groups defined for any two points $x, y \in M$ are isomorphic. For a Lorentz-signature manifold, $H(M)$ will be a subgroup of the Lorentz group. Now we can say that M is temporally orientable if and only if $H(M)$ does not contain any time inversions, i.e., $H(M)$ lies in the component $L\uparrow$ of the Lorentz group for which the matrices (in any four-dimensional representation) all have 44 components which are positive (the coordinate system being so chosen that x_1, x_2, x_3 represent space and x_4 represents time).

Parallel characterizations of spatial orientation can also be given. We can say that M is spatially orientable if and only if there is a continuous field of orthonormal triads of spacelike vectors on M. Equivalently, we can say that M is spatially orientable if and only if $H(M)$ does not contain any space inversions in the sense that $H(M)$ lies in $(L\uparrow \cap L_+) \cup (L\downarrow \cup L_-)$, where $L\downarrow$, L_+, L_- are respectively the components of the Lorentz group for which the corresponding matrices possess the property that the 44 components are all negative, the determinant is positive, the determinant is negative.

If $H(M)$ lies in L_+, M is said to be orientable. If M is spatially and temporally orientable, then obviously M is orientable; but the converse is not necessarily true. A connected M which is orientable (spatially orientable, temporally orientable), possesses two opposite orientations (spatial orientations, temporal orientations).

Space-times which are not spatially or temporally orientable are not mere mathematical curiosities, for there are examples of such space-times which satisfy the field equations of general relativity.[20] Very little is known about the conditions under which a space-time M can fail to be spatially or temporally orientable. It is known that M must have an interesting topology; for example, M must not be simply connected, i.e., it must not be possible continuously to contract all of the closed paths of M down to a point. Markus (ibid.) has shown that any space-time which is of constant curvature and which is geodesically complete (i.e., the geodesics of M can be extended to any arbitrary value of some affine parameter) must be spatially orientable. But, so far as I know, there are no published results for the case of non-constant curvature.

I believe that the above considerations show that Reichenbach was wrong in his assertion that

> Time therefore seems much less problematic since it has none of the difficulties resulting from multidimensionality. Time does not have the problem of *mirror image congruence*, i.e., the problem of equal and similarly shaped figures that cannot be superimposed, a problem which has played some role in Kant's philosophy.[21]

The temporal analogue of a spatially extended figure would be a temporally extended figure, e.g., a timelike vector. Consider then two vectors at $x \in M$, one pointing into the future lobe of the light cone, the other pointing into the past lobe. These vectors may be 'temporal mirror images' of each other in the sense that they can be obtained from each other by time inversion. If M is temporally orientable then these vectors are incongruous in that they cannot be made to coincide with each other by means of any transport in space-time which is continuous and which keeps timelike vectors timelike. Similarly, if M is spatially orientable then a right-handed triad of spacelike vectors cannot be superimposed on its 'mirror image', and left-handed triad by any transport in space-time which is continuous and which keeps the spacelike vectors spacelike and linearly independent. It is not clear what Reichenbach thought was problematic about spatial incongruous counterparts, but it does seem that if there is a problem for space due to the existence of spatial incongruous counterparts, then there is a parallel problem for time.

III

Granted that the assumption that the transformations interchanging right and left and earlier and later may be continuous does not stand in contradiction to mathematics or the general theory of relativity, is there any evidence that in our universe these transformations are discrete rather than continuous?

One possible source of evidence for the existence of the spatial and temporal orientability of our space-time has been discussed recently. It is known that the existence of a 'spinor structure' for a space-time M implies that M is both spatially and temporally orientable.[22] This is easily seen in the case of a non-finite (i.e., non-compact) *M* since in this case a necessary and sufficient condition for a spinor structure is the existence of a continuous field of orthonormal tetrads.[23] For our purposes, the important feature of a spinor is that under a rotation through 360° it changes sign. Utilizing this fact, Aharonov and Susskind[24] have shown how it is in principle possible to construct what I shall call an *A—S* box. The box consists of two parts, the *A* part and the *S* part. When these parts are coupled together in a certain way, an electric current is observed to flow from one part to the other, say from the *A* part to the *S* part. The following operations are then performed: the parts are separated; one part is rotated through 360° relative to the other; and then they are recoupled in the same way as before. After these operations, the current is observed to flow the opposite way, from the *S* part to the *A* part! Penrose (op. cit.) has suggested that if it could be shown that the parts of the *A—S* box could retain their memory of the number of relative rotations even when separated by large space-time distances, we would have a fairly conclusive argument for the existence of a spinor structure of space-time and, therefore, for the existence of spatial and temporal orientations.

It has also been claimed that evidence for the *P* and *CP* non-invariance of the laws governing the interactions of elementary particles counts as evidence for the spatial orientability of our universe. (*C* represents the operation of interchange of particle and anti-particle and *P* represents the operation of space inversion.) Ya. B. Zel'dovich and I. D. Novikov[25] argue that

> In a non-orientable space there exists a contour such that a circuit over it transforms a right-hand system into a left-hand one, i.e., a circuit over such a contour is equivalent to the operation of space reflection (*P*) . . . In 1956, the discovery of parity non-conservation in weak interactions made it possible to define uniquely the concepts of a 'right' and a 'left' system, and therefore a real physical space cannot be non-orientable (p. 237).

However, they continue by adding that

> This statement cannot be made if it assumed that a circuit over a non-oriented contour transforms particle into anti-particle, i.e., the circuit corresponds to combined parity (*CP*), since, according to Landau's hypothesis, it is precisely such combined inversion which conserves the left—right symmetry of empty space (p. 237).

They then add that recent experiments indicate that the laws of nature are not *CP* invariant and that *CP*-non-invariance excludes the possibility of a non-orientable space.

The first point which needs mention is that the experiments referred to above are open to various interpretations, not all of which imply *CP* non-invariance. However, for sake of discussion I will assume that this implication can be drawn. Even on this assumption there are still some bothersome things about Zel'dovich and Novikov's argument. First, some principle of global significance seems to be lacking. Here and now the laws of elementary particle physics may give us a distinction between right and left, but what about distant regions of space-time? To fill this hiatus, let us assume what Dicke[26] calls the strong principle of equivalence which says roughly that for every $x \in M$ there exists a sufficiently small region containing x and a coordinate system on this region in terms of which all of the laws of nature take on some standard form and have some standard numerical content. Secondly, there are two readings of what Zel'dovich and Novikov call Landau's hypothesis. On one reading *CP* invariance would preserve a symmetry between left and right only in an epistemological sense: *P* non-invariance would imply the existence of a lawlike difference between left and right, but *CP* invariance would mean that one could not explain by means of laws alone the difference in a which-is-which sense to anyone who did not know what you were calling matter and what you were calling anti-matter since such a person would not know whether you were talking about a left-handed system composed of matter or a right-handed system composed of anti-matter.[27] This interpretation is suggested by Zel'dovich and Novikov's stated concern with the conditions which would make it possible to 'uniquely define' the concepts of left- and right-handed systems. But on this interpretation their argument does not even get off the ground since even *CP* non-invariance would not make it possible to give an experimental definition of left- and right-handed systems. For one thing, it would be necessary to assume in addition *C* non-invariance, but even this additional assumption (which is experimentally justified) is not enough. Zel'dovich and Novikov talk about three-space; but the basic entity we have to work with is space-time, and this brings in the operation *T* of time inversion. Every local Lorentz invariant field theory of elementary particles is known to be invariant under the combined operation *CPT*; therefore within the context of such a theory one can give an experimental definition of right and left (of particle and anti-particle, or of earlier and later for nearby events) only *relative* to some choice of a system vs. its *CPT* image.[28]

The second possible interpretation of Landau's hypothesis is suggested by what I shall call Wigner's proposal.[29] Wigner suggested that we reinterpret the parity operation so that the new operation *P′* has the effect of combined *CP* reflection; the mirror image, in this new sense, of a particle would be an anti-particle. Wigner's proposal plus the assumption of *CP* invariance allowed physicists to believe in the left—right symmetry of laws of nature despite the experiments of the 1950s; the observed asymmetries were blamed on the difference between matter and anti-matter so that no need was seen for postulating a lawlike difference between left and right. The discovery of *CP* non-invariance means that Wigner's proposal cannot save left—right symmetry; but even so, I fail to see how we can conclude the existence of a globally consistent left—right distinction.

This situation has been clarified by Geroch (op. cit.); he argues that *CPT* invariance, *C, P,* and *CP* non-invariance, and the strong principle of equivalence together imply that space-time is orientable, though not necessarily spatially and temporally orientable. Let us begin with the simpler case in which no combination of *C, P,* and *T*, not even *CPT*, is a symmetry; in this case space-time is spatially and temporally orientable. For in this case it is possible to give an experimental definition of right and left and of earlier and later (for nearby events). Now suppose, for example, that $H(M)$ does contain space inversions. Then there must exist two points $x, y \in M$ and a path connecting x and y such that the definitions of right and left at x and y do not agree when compared by parallel transport along the given path. But since parallel transport is continuous, this means that there is a discontinuity at some point in the definitions of right and left, and such a discontinuity contradicts the sameness of the laws of nature at every point as formulated in the strong principle of equivalence.

CPT invariance plus *C, P,*and *CP* non-invariance imply that *CPT* is the only combination of *C, P,* and *T* which represents a symmetry. This means that an experimental definition of, say, right and left can be given only up to some choice of a system vs. its *CPT* image. Thus, there is no need for the agreement under parallel transport of the definitions of right and left since different choices of a system vs. its *CPT* image may be made at different points; but since *CPT* is the only symmetry which obtains, it follows that if there is disagreement there will also be disagreement on the definitions of earlier and later and of particle and anti-particle. Thus, appropriate transport of a system will give total *CPT* reflection if it gives any reflection at all so that $H(M)$ lies in L_+ and M is at least orientable.

Thus it seems that particle physics in its current state takes us only part of the way towards the spatial and temporal orientability of our space-time; but the conclusion that space-time is orientable is an important one, for it follows from this that if space-time is spatially (temporally) orientable, then it is also temporally (spatially) orientable.

The reader may have guessed that I have purposely avoided one important line of inquiry: what sense do certain kinds of laws of nature make in a space-time which is not spatially or temporally orientable? My reason for avoiding this issue is that I simply have nothing to say about it. For some interesting but, I think, not definitive remarks on this subject see Schrödinger (op. cit., pp. 13—14).

IV

In summary, the criticism of Kant offered by Jammer, Smart, *et al.*, seems to me to fail in four ways. First, these authors fail to locate the major fault in Kant's argument — namely, its incoherence. Secondly, their counter-arguments are at some places irrelevant to the issue which Kant raises and at other places rely on highly questionable premises. Thirdly, they fail to realize that Kant's fundamental assumption about the distinction between right and left is open to question. Finally, they fail to present an adequate alternative characterization of the left—right distinction.

A number of the considerations raised by Jammer and Smart are relevant to the distinction between right and left, but not in the ways in which these authors take them to be relevant. As Jammer would have it, the behavior of objects when 'moved about' in four-dimensional space is relevant; but the relevant four-dimensional space is space-time which has an indefinite metric in contrast to the positive definite metric of four-dimensional Euclidean space. The result is that the relevant four-dimensional space is not isotropic but has certain distinguished directions due to the null-cone structure, and this anisotropy must be taken into account when objects are 'moved about' — we want to keep timelike vectors timelike and spacelike vectors spacelike. As Smart would have it, the nature of the symmetry group of the laws of nature is also relevant to the distinction between right and left; certain non-invariances will give a local, lawlike distinction between right and left, and if these non-invariances are coupled with the global invariance postulated by the srong principle of equivalence, then a global chirality must exist. But contrary to Smart qua-relationist, *P* non-invariance is not a necessary condition either for a local or a globally consistent distinction between right and left though, of course, *P* non-invariance is a necessary condition for being able to explain, by means of lawlike features of the world, the distinction in the which-is-which sense.

There are several issues about the distinction between right and left and several issues about absolute and relational theories of space which were not discussed here but which seem to me to merit philosophical examination, but it is my belief that these two sets of issues do not overlap in any significant way.[30]

NOTES

* I was somewhat bemused upon rereading these youthful reflections. I agreed to have the paper reprinted because it may help the reader to understand the development of the recent philosophical discussion of these issues.

[1] It is not clear to me in what sense Leibniz maintained a relational theory of space. For some discussion on this point see C. D. Broad, 'Leibniz's last controversy with the Newtonians', *Theoria*, vol. 12 (1946), pp. 143—68, and H. Ishiguro, 'Leibniz's denial of the reality of space and time', *Annals of the Japan Association for the Philosophy of Science*, vol. 3 (1967), pp. 33—6.

[2] *Dissertation on the Form and Principles of the Sensible and Intelligible World*, translated by J. Handyside, *Kant's Inaugural Dissertation and Early Writings on Space*, Open Court, Chicago, 1929, section 15D, p. 61.

[3] See Kemp Smith, *Commentary on Kant's Critique of Pure Reason*, New York, Humanities Press, 1962, pp. 163—5. Reprinted in this volume, pp. 43—48.

[4] 'Incongruous counterparts', read at Princeton University, 1966. A version of Bennett's paper has been published under the title 'The difference between right and left,' *American Philosophical Quarterly*, 7 (1970), 175—91. This volume, pp. 97—130.

[5] And even today it may not be possible; see section III of this paper.

[6] 'On the first ground of the distinction of regions in space', in Handyside, op. cit., p. 20; this volume, p. 28. The italics are Kant's.

[7] *Kant*, Penguin books, Baltimore, 1966, pp. 33—4.

[8] *Concepts of Space*, Harper, New York, 1960, pp. 129—32.

[9] (a) *Problems of Space and Time*, Macmillan, New York, 1964, pp. 6—7; and (b) *Between Philosophy and Science*, Random House, New York, 1968, pp. 217—18.

[10] *Symmetry*, Princeton, 1952, ch. 1.

[11] *The Kantian Philosophy of Space*, Columbia University Press, 1939.

[12] 'Incongruous counterparts and absolute space', *Mind*, vol. 72 (1963), pp. 393—9. This volume, pp. 51—59.

[13] See J. Bennett, *Kant's Analytic*, Cambridge, 1966, section 16.
[14] In Riemannian spaces, rigid transport of geometrical figures of any finite size is possible only if the space is of constant curvature. Thus, if one does not want to limit himself to spaces of constant curvature — and I do not — then one must either talk about transport which is only approximately rigid but rigid enough so as not to distort the handedness of the figure (e.g., transport which preserves the linear independence of vectors), or else one must limit himself to sufficiently small figures. In the discussion below it is to be understood that I am assuming one of these alternatives.
[15] A. H. Alexander, *The Leibniz—Clarke Correspondence*, Manchester University Press, 1965, p. 74.
[16] Ibid., p. 74. Additional criticisms of Leibniz's views on space and time are given in my paper 'Who's afraid of absolute space' unpublished, 1969. In 1748 Leonhard Euler (Refexions sur L'Space Et Le Tems', *Historie de l'Academie Royale des Sciences et Belles Lettres* (*Berlin*), vol. 4 (1748), pp. 324—33) argued that Newton's laws of motion do not make sense if a relational theory of space is adopted. Euler's paper seems to have provided the main impetus for Kant's conversion in 1768 to an absolutistic view of space. After his conversion, Kant discovered in incongruous counterparts what he thought to be another demonstration of the absoluteness of space.
[17] See J. A. Wheeler, *Geometrodynamics*, Academic Press, New York, 1962.
[18] See E. Schrödinger, *Expanding Universes*, Cambridge, 1967.
[19] A topology consists of a space T and a collection of subsets of T called open sets. A topology is Hausdorff if distinct points of T are contained in disjoint open sets.
[20] See Schrödinger, op. cit., and L. Markus and E. Calabi, *Annals of Mathematics*, vol. 75 (1962), pp. 63—76.
[21] H. Reichenbach, *Space and Time*, Dover, New York, 1958, p. 109.
[22] See R. Penrose, in *Battelle Rencontres* (edited by C. M. De Witt and J. A. Wheeler), W. A. Benjamin, New York, 1968.
[23] R. Geroch, *Journal of Mathematical Physics*, vol. 9 (1968), pp. 1739—43.
[24] Y. Aharonov and D. Susskind, *Physical Review*, vol. 158 (1967), pp. 1237—8. For some further discussion and some criticism of Aharonov and Susskind's paper, see G. C. Hegerfeldt and K. Draus, *Physical Review*, vol. 170 (1968), pp. 1185—6; and J. S. Dowaker, *Journal of Physics A* (*Proceedings of the Physical Society*), vol. 2 (1969), pp. 267—73.
[25] *JETP* (*Journal of Experimental and Theoretical Physics*) *Letters*, vol. 6 (1967), pp. 236—8.
[26] R. H. Dicke, *Experimental Relativity*, Gordon & Breach, New York, 1964, pp. 4—5.
[27] See K. Nishijima, *Fundamental Particles*, W. A. Benjamin, New York, 1964, pp. 339—40.
[28] We still might be able to explain the difference in the which-is-which sense to a distant observer with whom we are in radio contact: it might be argued that every human-like observer carries with him a time sense and that agreement of these time senses is a necessary condition for any meaningful communication between two such observers. However, I do not see how to make this argument precise, and I will not pursue the matter here.
[29] E. P. Wigner, *Reviews of Modern Physics*, vol. 29 (1957), p. 238.
[30] For further discussion of the former, see my paper 'The anisotropy of time', in *The Australasian Journal of Philosophy*.

GRAHAM NERLICH

HANDS, KNEES, AND ABSOLUTE SPACE

1. COUNTERPARTS AND ENANTIOMORPHS

My left hand is profoundly like but also profoundly unlike my right hand. There are some trifling differences between them, of course, but let us forget these. Suppose my left hand is an exact mirror-image replica of my right. The idea of reflection deftly captures how very much alike they might be, while retaining their profound difference. We can make this difference graphic by reminding ourselves that we cannot fit left gloves on right hands. This makes the point that one hand can never occupy the same spatial region as the other fills exactly, though its reflection can. Two objects, so much alike yet so different, are called 'incongruent counterparts'.

In my usage that phrase expresses a relation among objects, just as the word 'twin' does. Thus, no one is a twin unless there is (or was) someone to whom he is related in a certain way. Call this relation 'being born in the same birth'. Then a man is (and has) a twin if and only if he is born in the same birth as another. Let us call the relation between a thing and its incongruent counterpart 'being a reflected replica'. Then, again, a thing *is* an incongruent counterpart if and only if it *has* one. My right hand has my left hand, and the left hands of others, as its incongruent counterparts. If people were all one-armed and everyone's hand a congruent counterpart of every other, then my hand would not be (and would not have) an incongruent counterpart.

But incongruent counterparthood gets at something further and deeper than twinhood does. No actual property belongs to me because I could have been born in the same birth as another. But all right hands do share a property, just because incongruent counterparts of them are possible. Though there are incongruent counterparts of my left hand, it is not their existence that makes it left. It appears to be enough that they could exist for my hand to have this property. Let us express this new further and deeper idea by calling hands 'enantiomorphs' and by saying that they have handedness. Then that is not an idea that depends on relations of hands to other things in space, as incongruent counterparthood does.

We can show this in various ways. A well-worn method is to invent a possible world, say, one where all hands are right, as I did a moment ago. A more vivid possible world is one that contains a single hand and nothing else whatever. This solitary hand must be determinate as to being either left or right. It could not be indeterminate, else it would not be determinate on which wrist of a human body it would fit correctly, that is, so that the thumb points upward when the palm touches against the chest. Enantiomorphism, then, is not a relation between objects that are in space.

Let us not make too much mystery out of this. We can say how hands, or anything else, come to pose a problem: they have no centre, axis or plane of symmetry. The sphere has all three: many common shapes have at least a plane of symmetry, the human body being an example. But the failure of symmetry yields no difference between left and right hands, since each is asymmetric. It gives just a neccessary condition of enantiomorphism.

2. KANT'S PRE-CRITICAL ARGUMENT

These ideas originate in Kant, of course. Though enantiomorphs crop up in the *Prolegomena,* they feature in and around the *Critique* as illustrating that space is the form of outer sense. But in an early paper of 1768, 'Concerning the Ultimate Foundations of the Differentiation of Regions in Space', he used them to argue for the ontological conclusion that there is absolute space.[1] It is this earlier argument that this chapter is all about. Kant rightly regarded his paper as a pioneering essay in *Analysis situs*, or topology, though an essay equally motivated by metaphysical interests in the nature of space. He recognised only Leibniz and Euler as his predecessors in this geometrical field. The argument has found little favour among geometers and philosophers since Kant first produced it. I guess no one thought he was even half right as to what enantiomorphism reveals about space. Kant himself had second and even third thoughts about his argument. My aim is to show that his first ideas were almost entirely correct about the whole of the issue.

This is not to deny that Kant's presentation is open to criticism. He was ignorant of several important and relevant facts which emerged later.[2] But knowledge of these leads me, at least, only to revise the detail of his argument, not to abandon its main structure and content.

In the remainder of this section I will set out a version of Kant's argument. I try to follow what I think were his intentions closely and clearly, but in such a way as to indicate where further comment and exposition are needed. Hence this first statement of the argument is rather bald.

A_1: Any hand must fit on one wrist of a handless human body, but cannot fit on both.
A_2: Even if a hand were the only thing in existence it would be either left or right (from A_1).
A_3: Any hand must be either left or right (enantiomorphic) (from A_2).
B_1: Left hand and right hand are reflections of each other.
B_2: All intrinsic properties are preserved under reflection.
B_3: Leftness and rightness are not intrinsic properties of hands (from B_1 and B_2).
B_4: All internal relational properties (of distance and angle among parts of the hand) are preserved in reflection.
B_5: Leftness and rightness are not internal properties of hands (from B_1 and B_4).
C_1: A hand retains its handedness however it is moved.
C_2: Leftness and rightness are not external relations of a hand to parts of space (from C_1).
D: If a thing has a non-intrinsic character, then it has it because of a relation it stands in to an entity in respect of some property of the entity (from A_3, B_3, B_5, C_2, D).
E: The hand is left or right because of its relation to space in respect of some property of space.

Kant places several glosses on *E*. He claims that it is a hand's connexion 'purely with absolute and original space'[3] that is at issue. That is, space cannot be The Void, a nonentity, since only something existing with a nature of its own can bestow a property on the hand. Again, to have made a solitary right hand rather than a solitary left would have required 'a different action of the creative cause',[4] relating the hand differently 'to space in general as a unity, of which each extension must be regarded as a part'.[5]

The premise *D* is not explicit in Kant. It is, quite obviously, a highly suspect and rather obscure claim. But I think Kant needed something of this very general nature, for he did not really know what it was about space as a unity that worked the trick of enantiomorphism. It will be possible to avoid this metaphysical quicksand if we can arrive at some clearer account of just what explains this intriguing feature.

3. HANDS AND BODIES: RELATIONS AMONG OBJECTS

A number of relationists have replied to Kant's argument by claiming flatly that a solitary hand must be indeterminate as to its handedness. But it is very far from clear how a hand could possibly be neither left nor right. The flat claim just begs the question against Kant's lemma *A*, which argues that a hand could not be indeterminate as to which wrist of a human body it would fit correctly. Nevertheless, there is an influential argument to the effect that this lemma is itself a blunder. It is instructive to see what is wrong with this relationist contention.

Kant's reason for claiming that it cannot be indeterminate whether the lone hand is right is perhaps maladroit because it suggests that the determinacy is constituted by a counterfactual relation to a human body. The suggestion is hardly consistent with Kant's view that it is constituted by a relation between the hand and space. However, it provides Remnant with a pretext for foisting on Kant an essentially relationist view of the matter.[6] He takes Kant to be offering the human body as a recipe for telling whether the lone hand was left or right. He shows quite convincingly that even the introduction of an actual body into space will not let us tell whether the hand is left or right. For suppose a handless body is introduced. The hand will fit on only one of the two wrists correctly, i.e., so that when the arm is thrown across the chest with the palm touching it, the thumb points upward. But this does not settle whether the hand is left or right unless we can also tell whether it is on the left or right wrist that the hand fits. But we can be in no better position to settle this than to settle the original question about the single hand. Possibly we are in a worse position because the human body (handless or not) is not enantiomorphic (except internally, with respect to heart position, etc.). Even if every normal human body had a green right arm and a red left (so that it became enantiomorphic after a fashion), that would not help. For incongruent counterparts of it, with red right and green left arms are possible. News that the hand fits the green wrist enables us to settle nothing about its rightness unless we know whether the body is normal. Settling this, however, is just settling handedness for a different sort of object. Thus, it is concluded, a solitary hand is quite indeterminate in respect of handedness.

What Remnant shows, in these arguments, is that no description in terms of relations among material things or their material parts ever distinguishes the handedness of an enantiomorph. This is an objection to Kant only if his maladroit reason is construed as a lapse into the view that what makes the hand left is its relationship to a body. Since this casts his reason as a contrary of the conclusion he draws from it, the construal is improbable. Remnant takes Kant to have been guilty of the blunder of supposing that, though it is impossible to tell, of the single hand, whether it is left or right, it is quite possible to tell, of a lone body, which wrist is left. It would have been a crass blunder indeed. But Kant never says that we can tell any of these things. In fact, he denies it. Insofar as Remnant does show that relations among the parts of the hand and the body leave its handedness unsettled, he confirms Kant's view. His attack on Kant succeeds only against an implausible perversion of the actual argument.

Kant's introduction of the body is aimed, not at showing the hand to be a right hand or a left hand, but at showing that it is an enantiomorph. It shows this perfectly clearly. For the hand would certainly fit on one of the wrists correctly. It seems equally certain that it could not fit correctly on both wrists. Whether it fits a left wrist or a right is beside any point the illustration aims to make. Though this expository device may suggest a relationist view of enantiomorphism, it does not entail it. It is simply a graphic, but avoidable way of making the hand's enantiomorphism clear to us. The idea that a hand cannot be moved into the space that its reflection would occupy is certainly effective, too. But it is less striking and less easily understood (I shall say more about it later). Kant was quite well aware that this method was also available and, indeed, used a form of it in his 1768 paper.

Thus the objection to the determinacy of handedness in solitary objects fails.

4. HANDS AND PARTS OF SPACE

John Earman charges Kant with incoherence.[7] No relation of a hand to space can settle handedness. Kant appeals to a court that is incompetent to decide his case. It is useful to look into Earman's objection.

As I understand Earman, he argues as follows: we can plausibly exhaust all relevant spatial relations for hands under the headings of internal and external relations. Kant takes the internal relations to be solely those of distance and angle which hold among material parts of the hand. But these do not fix handedness, since they are invariant under reflection, but handedness is not (B_5). What makes Kant's argument incoherent is that the external relations (to the containing space) cannot fix it either. For the external relations can only be those of position and orientation to points, lines, etc., of space outside the hand. (Incidentally, whether these parts of the container space are materially filled or not appears to make no relevant difference.) But for every external relation a right hand has to a part of space, a left hand has just that relation to a similar part, as reflection makes quite clear. In any case, changes in these relations occur only if the hand moves through the space. However, although movement through the space can alter these relations, it cannot alter handedness (C_2). Evidently none of these relations settles the problem. This presents us with an unwelcome parody of Kant's argument: since handedness is not settled by either internal or external relations in space, it must consist in the hand's relation to some further well-defined entity beyond space.[8]

That is an unlucky conclusion. We will have to think again.

Earman offers us a way out of the wood. The spatial relations we have been looking at do not exhaust all that are available. He suggests that we say, quite simply, that *being in a right configuration* is a primitive internal relation among parts of the hand. If we could say that it would certainly solve the problem in a very direct fashion. It would be a disappointing solution since we cannot explain the difference between left and right by appeal to a primitive relation. But, in fact, there is no such relation, as I will try to show in Sections 6 and 7.

However, my interest in the present section lies in other things that Earman has to say. He frankly concedes that Kant was well aware of the kind of objection (as against the kind of solution) that he offers, so that, in this respect, his criticism is 'grossly unfair'. For Kant did insist that it is 'to space in general as a unity' that his argument appeals. Earman says that he does not see how this helps.[9] But that merely invites us to take a closer look.

We can get some grip on the idea of 'space in general as a unity' by taking a quick trip into geometry. We need not make heavy weather of the rigour of our journey. A rigid motion[10] of a hand is a mapping of the space it fills which is some combination of translations and rotations. Such mappings make up an important part of metrical geometry. A reflection is also a mapping of a space, and, like a rigid motion, it preserves metrical features. This last fact, so important for Kant, is easily seen by supposing any system of Cartesian coordinate axes, x, y and z. A reflection maps by changing just the sign of the x (or the y or the z) coordinate of each material point of the hand. In short, it uses the y—z (or the x—z or the x—y) plane as a mirror. Thus it preserves all relations of distance and angle of points in the hand to each other, since only change of sign is involved. Thus the lemma B_1, B_4, B_5 is sound.

Now we can express the idea of enantiomorphism in a new way which has nothing to do with possible worlds or with the relation of one hand to another (actual or possible) hand or body. It does, however, *quantify over all mappings* of certain sorts. We can assert the following: each reflective mapping of a hand differs in its outcome from every rigid motion of it. That is a matter of space in general and as a unity. (Thus lemma C can be gained virtually by definition.) This quantification over the mappings seems to have nothing to do with any object in the space; not even, really, with the hand that defines the space to be mapped. Though this terminology is much more recent than Kant's, the ideas are old enough. I see no reason to doubt that they are just what he intended. Space in general as a unity is exactly what is at issue.

5. KNEES AND SPACE: ENANTIOMORPHISM AND TOPOLOGY

Kant was right. The enantiomorphism of a hand consists in a relation between it and its containing space considered as a unity. This is more easily understood if we drop down a dimension to look at the problem for surfaces and figures contained in them. We need to see how hands could fail to be enantiomorphs.

Imagine counterpart angled shapes cut out of paper. They are like but not the same as L's, since L's are directed. Let me call them knees, for short (but also, of course, for the legitimacy of my title). They lie on a large table. As I look down on two of them, the thick bar of each knee points away from me, but the thin bar of one points to my left, the thin bar of the other to my right. Though the knees are counterparts, it is obvious that no rigid motion of the first knee, which confines it to the table's surface, can map it into its counterpart, the second knee. Clearly,

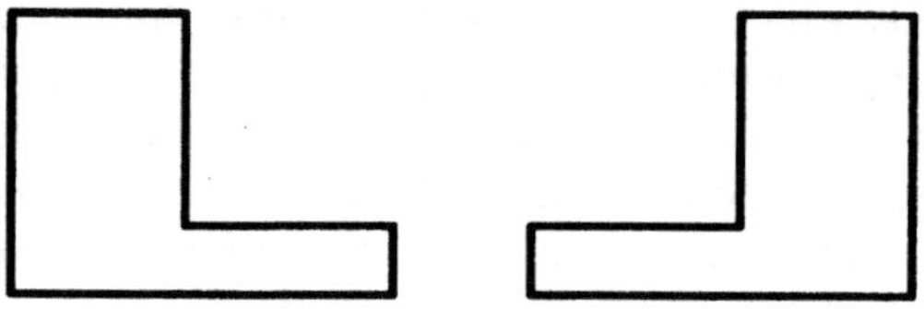

Fig. 1.

this is independent of the size of the table. The counterparts are incongruent. The first knee I dub left. The second is then a right knee.

Their being enantiomorphs clearly depends on confining the rigid motions to the space of the table top, or the Euclidean plane. Lift a knee up and turn it over, through a rigid motion in three-space, and it returns to the table as a congruent counterpart of its mate. That the knees were incongruent depended on how they were put on the table or how they were in the space to which we confined their rigid motions.

A different picture is more revealing. Suppose there is a thin vertical glass sheet in which the knees are luminous angular colour patches that move rigidly about. They are *in* the sheet not *on* it. Seen from one side of the sheet, a knee will be, say, a left knee. But move to the other side of the sheet and it will be a right knee. That is, although the knees are indeed enantiomorphs in being confined to a plane of rigid motion, any knee is nevertheless quite indeterminate as to being a left rather than a right knee, even granted the restrictions on its motion. It could hardly be clearer, then, that nothing intrinsic to an object makes it left or right, even if it is an enantiomorph. Our orientation in a higher dimensional space toward some side of the manifold to which we have confined the knee prompts our inclination to call it left, in this case. It is an entirely fortuitous piece of dubbing. Hence, if the knees were in a surface of just one side (and thus in a *non-orientable manifold*) they would cease to be enantiomorphic even though confined to rigid motions in that surface. (Thus B_3 is correct independently of B_1 and B_2 and despite Earman's objection.)

There is a familiar two-dimensional surface with only one side: the Möbius strip (see Figure 2). If we now consider knees embedded in the strip, they are never enantiomorphic. A rigid motion round the circuit of the strip (which is twisted with respect to the three-dimensional containing space) turns the knee over, even though it never leaves the surface. However, the Möbius strip might be considered anomalous as a

Fig. 2. Möbius strip.

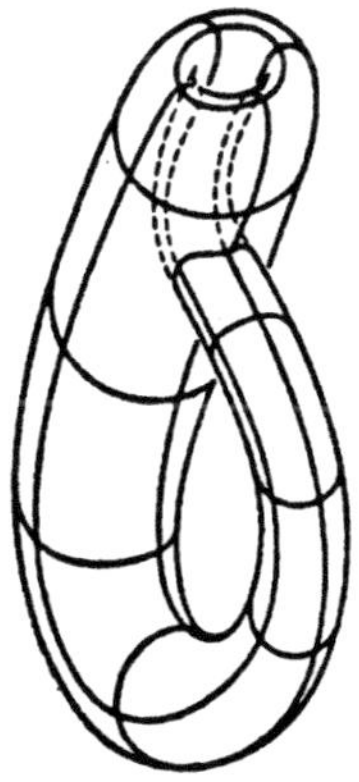

Fig. 3. Klein bottle.

space. It is bounded by an edge. We need the space of Klein's bottle, a closed finite continuous two-space (see Figure 3).[11] Rigid motion of a knee round the whole space of Klein's bottle maps it onto its reflection in the space. Knees are not enantiomorphic in this one-sided, non-orientable manifold. They are indifferent as to left or right. Let us say that here they are *homomorphic*.

These general results for knees as two-dimensional things have parallels for hands as three-dimensional things. It seems to be pretty clearly the case that, as a matter of fact, there is no fourth spatial dimension that could be used to turn hands 'over' so that they become homomorphs. No evidence known to me suggests that actual space is a non-orientable manifold. But it cannot be claimed beyond all conceivable question that hands are enantiomorphic, and it is not too hard a lesson to learn how they could be homomorphic. So Kant's conclusion A_3, a principal lemma, is false. But this has nothing to do with whether there is one hand or many. It has nothing to do with an obscure indeterminacy that overtakes a hand if there are no other material things about. It is false because spaces are more various than Kant thought.

Thus, whether a hand or a knee is enantiomorphic or homomorphic depends on the nature of the space it is in. In particular, it depends on the dimensionality or the orientability, but in any case on some aspect of the overall connectedness or topology of the space. Whether the thing is left depends also on how it is entered in the space and on how the convention for what is to be left has been fixed. Kant certainly did not see all this. Nevertheless, it should be obvious now how penetrating his insight was.

6. A DEEPER PREMISE: OBJECTS ARE SPATIAL

Clearly enough, Kant's claim that a hand must be either left or right springs from his assumption that space must have Euclidean topology, being infinite and three-dimensional. The assumption is false, and so is the claim. This suggests an advantageous retreat to a more general disjunctive premise for the argument, to replace A_3. Rather than insist that the hand be determinately either left or right, we insist rather that it be determinately either enantiomorphic or homomorphic. Thus, if there were a handless human body in the space, then either there would be a rigid motion mapping the hand correctly onto one wrist but no rigid motion mapping it correctly onto the other; or, there would be rigid motions some of which map it correctly onto one wrist and others which correctly map it onto the other wrist. Which of these new determinate characters the hand bears depends, still, on the nature of the space it inhabits, not on other objects. The nature of this space, whether it is orientable, how many dimensions it has, is absolute and primitive.

What underlies this revision of Kant's lemma *A* is the following train of ideas. We can dream up a world in which there is a body of water, without needing to dream up a vessel to contain it. But we can never dream up a hand without the space in which it is extended and in which its parts are related. To describe a thing as a hand is to describe it as a spatial object. We saw the range of spaces a knee might inhabit to be wide; the same goes for spaces in which a hand might find itself. So dreaming up a hand does not determine which space accompanies it, though Kant thought it did. But it does not follow that there could be a hand in a space that is indeterminate (with respect to its global connectivity, for example). We can describe a hand, leaving it indeterminate (unspecified) whether it is white or black. But there could not be a hand indeterminate in respect of visual properties. Like air in a jar, even an invisible hand can be seen to be invisible, so long as we know where to look. (I am here shuffling under a prod, itself invisible here, from David Armstrong.) No considerations mentioned yet admit of a hand that could be neither enantio- nor homo-morphic. There seems no glimmer of sense to that expression.

But I spoke just now of there being a wide range of spaces that hands or knees might occupy. What does this mean, and does it offer a route to the relationist between the alternatives I am pushing? What it means is that we can describe a knee, for example, as a mass of paper molecules (or continuous paper stuff) which is extended in a certain metrical two-space. We can regard this two-space as bounded by extremal elements that make up edges (of surfaces for the three-space of a hand). These elements define the shape the mass of matter is extended in by limiting it. A wide variety of global spaces can embed subspaces isomorphic (perhaps dilated) with our knee- and handspaces. In short, a kneespace is a bit of our ordinary space. Nothing mooted here is meant to suggest that a hand might somehow be taken from one space to another while being spaceless in the interim. Let us, for a moment, toy with the idea that a kneespace need not be a subspace at all, but that it could just end at its material extremities without benefit of a further containing space. Does the hand or knee become indeterminate as to enantiomorphism if we consider it just in its own handspace or kneespace?

Kant evidently feared that it would, and it might seem that he was right. Can't we argue that, for the hand or knee to be either enantiomorphic or homomorphic, there must be enough unified space to permit both the reflection and some class of rigid motions to be defined in it? Otherwise the question whether any rigid motion maps the hand onto its reflection does not have the right kind of answer to yield either result. But this is wrong. What counts is not whether the particular object has a reflection or whether it can be rigidly moved in the kneespace, but whether suitable objects in general do. It depends, first, on the space, not on the object. Both handspaces and kneespaces are orientable spaces. This is easy to see by imagining a much smaller hand or knee in the space and considering its reflections and the class of its rigid motions in that limited space. Clearly a hand in a handspace is enantiomorphic. This does not mean that hands are intrinsically enantiomorphic. It means that handspaces are certain kinds of spaces. Any hand must always lie in a handspace at least as a subspace, but there is no necessity about which space it is a subset of. Nor does there seem to be anything to say about the hand as material which could determine even so small a thing as whether the space extends beyond the matter of the hand or is confined to it. Dreaming up a hand means dreaming up a space to contain it, of which the being and nature are independent and primitive. It is introduced in its own right as a well-defined topological entity.

An oddity here is that, though a hand filling its handspace can be a dilated incongruent counterpart of a small hand in its space, we could never compare or contrast one hand in its own handspace with another in its own handspace. That requires mutual embedding in a common space. I conclude that leftness is not a primitive relation with respect to which hand parts are configured. I touch on this topic again in Section 7.

But handspaces or kneespaces as anything other than subspaces have, so far, been mere toys of our imagination. The thought that space could just come to an end is one at which the mind rebels. Let us express our distaste for such spaces by calling them pathological.[12] I want to do more now than toy with the idea of pathological spaces. It was an important conviction in Kant's mind, I think, that pathological spaces cannot be the complete spaces of possible worlds. Though I can think of no strong defence for this deep-lying conviction, which most of us share, I can think of an interesting one. At least, it interests me. Why might one think that space cannot have boundaries?

The thought that space cannot simply come to an end is ancient. The argument was that if, though impossible, you did come to the end you could cast your spear yet further. This challenge is quite ineffective against the hypothesis that there is just nowhere for the spear to go. But what the challenge does capture, rather adroitly, is the fact that we cannot envisage any kind of mechanics for a world at the point at which moving objects just run out of places to go. What would it be like to push or throw such an object? We can't envisage. It would be unlucky if this boggling of the mind tempted us to regard pathological spaces as contradictory. To go Kant's way on this is to go the way of synthetic necessary truth. In the 1970s, the prospects for following that road are dim. But there are prospects, more tangible and, perhaps, a bit brighter, since Kripke's semantics for modal logic. It is tempting to pursue this defence of Kant's inference from pathological spaces to global ones which properly contain them. It would be a protracted and rather tangential undertaking, however. The best defence for the disjunction enantiomorphic/homomorphic is to argue, as I did earlier, that hands and knees must be either homomorphic or enantiomorphic whether their containing spaces are pathological or not. This rests the disjunction on the deeper premise that hands are spatial objects and there can be no hand without a space in which it is extended.

Nevertheless, I am inclined to offer the suggestion that, when it is our task to conceive how things might be, as a whole, we should ask for what I will call an *unbounded-mobility* mechanics. (I intend the phrase to recall Helmholtz's 'free mobility', which he used to express the possibilities of motion in spaces of constant curvature.) That would rule out pathological spaces for possible worlds.

7. DIFFERENT ACTIONS OF THE CREATIVE CAUSE

So far I have not let the relationist get a word in edgeways. But his general strategy for undermining our argument is pretty obvious. He must claim that there cannot be a space that is a definite topological entity unless there are objects that define and constitute it. Of course, he has to do more than simply to assert his claim; he has to make it stick. That needs at least two things. First, he needs to show how bringing objects into the picture can settle topological features of a space in some way — for example, settle the feature that it is an orientable manifold. Then he has to show, next, why only objects can give it. The second of these tasks has the virtue of familiarity, at least, but I know of no discussion whatever of the first.

Let us pick, for our example, the case that still lies before us, of distinguishing orientable from non-orientable manifolds. Suppose we begin with hands or knees in their pathological or limited spaces. These are orientable spaces which can be subsets of non-orientable spaces. How could relations among these hands or knees make some wider space orientable? Knees are simplest. I will talk about them.

Suppose there are two knees, each in its own pathological space, neither being primitively taken as part of a wider, mutually inclusive manifold. The supposition is expressed more accurately, perhaps, in this sentence:

$(\exists x)(\exists y)(x$ is a knee in a kneespace & y is a knee in a kneespace & $x \neq y)$

According to this hypothesis there is no primitive spatial relation between the knees. Let us suppose further that some change in one knee causes a change in the other. If cause is the transmission of an effect, then this suggestion of a causal connexion presupposes that there is, after all, a more primitive spatial relation between the things. But if cause is not the transmission of an effect, then how does the supposition of a causal connexion even begin to relate the things in some derivative spatial way? I see no glimmer of light in this sort of approach. We can add knees to knees in this fashion till the cows come home and be no nearer constituting a wider non-orientable manifold, or indeed any wider manifold at all. The trouble is that it is not just our kneespaces that are pathological; so is the method of adding just described. What seems always to have been taken for granted is that adding objects is putting them together in the same unified space. I do not see how that can fail to mean that we posit paths (continua of points) across which they are related (e.g. across which they are at some distance). The reason for always taking that for granted, I guess, is the belief, shared with Kant, that pathological spaces are not possible.

Consider Melissus' paradox: by virtue of there being nothing between two distinct objects, there is empty space between them; that is, there is something between them. The remarks just made about pathological adding and pathological spaces should make it clear that I rest no weight on this intriguing argument. Nevertheless, I do claim that in all possible worlds, as we ordinarily envisage them in philosophy, there always is a path between two objects that do not touch each other.

Pathological adding cannot help the relationist. Let us press on with standard adding. A model will help us focus our ideas. Suppose there are two strips of paper, one white and the other red. We cut several knees out of the red paper and we intend to embed these in the white strip by cutting appropriate shapes out of the white paper and fitting the knees into them. Clearly, once we have made a cut in the white strip to admit the thick bar of a knee, there are two directions in which we can cut to admit the thin bar.

Well then, a space is orientable if it can[13] be covered by an array of directed entities in such a way that all neighbouring entities are like-directed. In our case, the question is whether we can make the white strip orientable or non-orientable by entering the knees in some ways rather than in others. The answer is that these entries have no bearing on the matter at all. Suppose our white strip is joined at its ends to form a paper cylinder. Then we can cover it with knees in such a way as to illustrate its orientability. But clearly this illustrates without constituting its orientability. That is obvious once we see that we can cut across the strip, give it a twist and rejoin it without changing the way any knee is embedded in it. What we now have is a Möbius strip which is non-orientable. Some pair of neighbouring knees will be oppositely directed. That is solely a matter of how the space (the white strip) is pathwise connected globally. It is quite irrelevant how many knees are embedded in the strip and how their shapes have been cut out for them.

This makes it look, more than ever, as if the space as a definite topological entity can only be a primitive absolute entity; that its nature bestows a character of homomorphism, leftness or whatever it might be, on suitable objects. My conviction of the profundity of Kant's argument rests on my being quite unable to see what the relationist can urge against this, except further relationist dogma. As always, of course, that might mean just that I still have lessons to learn.

But so do relationists. The difficulty of our going to school with open minds on the matter is strikingly shown in Jonathan Bennett's paper 'The Difference between Right and Left'.[14] His paper is devoted largely to the question whether and how an English speaker whose grasp of the language was perfect, save for the interchange of the words 'left' and 'right', could discover his mistake. He has to learn it, not ostensively, but from various descriptions in general terms relating objects to objects. Bennett concludes that the speaker could not discover his error that way, though he could discover a similar error in the use of other spatial words, such as 'between'. It is a long, careful, ingenious discussion. But it is an utterly pointless one. Bennett states on pp. 103—104 and again on p. 107 that in certain possible spaces (some of which we know) there may not be a difference between left and right. So how could one possibly discover 'the difference' between left and right in terms of sentences that must leave it entirely open whether there is any difference to be discovered? That cannot be settled short of some statement about the over-all connectivity of the space in which the things live. The same may be said, moreover, of Bennett's discussion of the case of 'between'. For, given geodesical paths on a sphere (and Bennett is discussing air trips on earth more or less), the examples he cites do not yield the results he wants. The familiarity of relationist approaches appears to fixate him and prevent the imaginative leap to grasping the relevance of those known global spatial results which clarify the issue so completely. No doubt relationism has some familiar hard-headed advantages. But it can also blinker the imagination of ingenious men. It can make them persist, despite better knowledge, in digging over ground that can yield no treasure. Empiricist prejudice is prejudice.

Let us get back to my paper strip and the knees embedded in it. It is a model, of course. It models space by appeal to an object, one which also defines what is essential to the space. That is, the subspaces, paths, and mappings that constitute the space are modelled by the freedoms and limitations provided by the constraint of keeping objects in contact with the surface of the modelling body (or fitted into it). This modelling body is embedded in a wider true space, which enables the visual imagination to grasp the space as a whole. Realists do treat space just like a physical object in the sense of such analogies. A space is just the union of pathwise connected regions.

The model also shows us a distinction already mentioned, in a clearer light. Whether an object is enantiomorphic or not may depend, in part, on its shape; spheres are homomorphs, even in Euclidean space. It also depends on the connectivity of the space, standardly, in global terms. But, what differentiates a thing which is an enantiomorph from one of its incongruent counterparts is a matter of how it is *entered into the space* — how we cut the hole for it in the white strip. Whether we call a knee in an orientable strip a left knee or a right is wholly conventional; it does not really differentiate the knees themselves at all, but simply marks a difference in how they are entered.

The idea of entry is a metaphor, clearly. It springs from our ability to manipulate the knees in three-space, and turn them there so that they fit now one way, now another into the model space. Once in the model space they lose that freedom or mobility and are left only that which determines enantiomorphy. It is not easy to find a way of speaking about this which is not metaphorical. But a very penetrating yet not too painfully explicit way of putting the matter is Kant's own, though I believe it to be still a metaphor. The difference between right and left lies in different actions of the creative cause.

8. CAN GEOMETRY EXPLAIN THINGS?

There is a style of relationist argument which we have not yet said anything about, though we promised to do so in Chapter 1 at the end of Section 2.[15] Furthermore, it may look rather as if there is nothing we can say about it without sawing off a branch we need to sit on. In explaining how handedness is tied up with space in general, as a unity, we said roughly, this: a hand is homomorphic if it can be moved into its own reflection and enantiomorphic if it cannot. Now this statement is already halfway towards what some relationists see as a perfectly adequate way for them to deal with the problems of left and right. Once we begin to speak of how things could move and of how they would be related to things that could have existed, don't we give the relationists all the licence they need?

This is what Lawrence Sklar thinks.[16] If we allow relationists the freedom, which they standardly assume, to talk of possible objects and possible continuous rigid motions, then handedness and orientability can be explained perfectly well. The '*real* substantivalist objection to relationism' is twofold: either (i) it is nonsense to talk about possible things and possible motions, or (ii) to talk of what is possible is to speak indirectly of 'some actual substantival entity and its actual properties'.[17] These objections have not yet been considered. Remote questions like those of orientability and handedness have no special relevance to the 'real' objections. The crucial battle is over much broader issues, like the relationist's ability to deal with empty space. Hands and global properties of space are just beside the point.

I agree that the relationist sees his talk of what is possible and what not as indispensable for his enterprise. I agree that the standard absolutist objections are to this talk and they take a form something like the one Sklar says they take. But they have certainly not been seen as conclusive objections.

Nevertheless, though the main thrust of my paper was directed elsewhere, we added something to the standard debate. It was this. I might have been born in the same birth as another. To put it in Sklar's style, I have a possible relation to a possible twin. But no property related to twinhood accrues to me for that reason — I am not called a twon, or whatever. Yet all right hands do seem to share a property just if incongruent counterparts of them are possible. They are enantiomorphic, handed. The relationist needs not only to assure us that certain spatial relations to certain things are possible. He needs to tell us why we ignore possibilities in general, yet take note of these possibilities in just the way he says we do.

But, to come back to the traditional debate, I do not really think that we can deny relationists their use of modal statements. In particular, the operator 'it is possible that' is useful, to say the least; indeed we can hardly get on without it.[18] So one line of argument, which is open to

Sklar's substantivalist, is closed to me. But the second line of argument, as Sklar sketches it anyhow, looks quite implausible too. I believe that it is certainly false that 'possibly' generally relates to some 'actual substantival entity and its actual properties'. The two uses of modal expressions discussed in the paragraph after next have nothing at all to do with actual things and their properties.

Nevertheless, this second style of argument against relationism can be developed, much as Sklar says, into a powerful objection, provided that one applies it to a property like orientability. Sklar's way of using modal expressions somewhat obscures this, as I will try to show. He speaks throughout of possibilia: possible objects and possible continuous rigid motions. Now, unlike Sklar, I do not think that a relationist can afford to quantify over motions (i.e. refer to them), possible or not. Any absolutist whose blood is even faintly red, will insist that continuous rigid motions are a kind of spatial entity. They are paths, continua of points, and space itself is nothing but their sum. But, quite apart from running this risk of a warm welcome to the absolutist camp, we ought not to use the adjective 'possible' when we might use the sentence operators 'possibly' or 'it is possible that'. The operators make issues very much clearer. So a relationist should analyse (or reduce) the expression 'orientable' in some such style as this:

> *O*: It is possible that there are two hands *a* and *b* such that *a* is beside *b*, but it is not possible that *a* and *b* move rigidly with respect to *c* and *d* etc. so that *b* will stand to *c*, *d* etc. as *a* now does.

Here, we do not quantify over motions and we avoid 'possible' as an adjective in favour of operators on sentences, open or closed.

It is obvious that the relationist still has the job ahead of him. If these modal expressions mean what they ordinarily mean in philosophical contexts, vague though that is, *O* is just flatly false. I take the ordinary meaning, among philosophers, to be also expressed by 'It is logically possible' or 'it is conceivable'. In that case, the negative clause of *O* must be false just because non-orientable spaces are logically possible or conceivable. *O* is false, even if we construe the modals in the common sense of 'For all I know'. For all I know, our space is *not* orientable and I guess that this is so for all Sklar knows, too. So any relationist who wants to stay in the field owes us some account of what his modal expressions mean. There is no straightforward way of taking them.

Now, we cannot argue like this against the relationist's reduction of empty space — or so I said, by implication, earlier. A careful relationist will probably offer us something along these lines.

E: It is possible that there are two objects *a* and *b* and *a R b*

where *R* dummies for an expression which puts *a* or *b*, or both, in empty places (or whatever is required). Now *E*, unlike *O*, is true in both the familiar senses just mentioned, I believe. It is certainly true in the first. As I argued before, we boggle at the idea that space might come to an end, whether or not its boundaries coincide with the limits of a body. Any world we conceive of is a world with a space having no boundaries. So the relationist can claim *E* in one perfectly straightforward use of its modal expression. This has nothing to do with the actual properties of actual things.

Let us return to *O*. It is not even enough for the orientability of space that *O* be true. It might be true for any of a hundred reasons that are consistent with a non-orientable space. It might be true because it is impossible to move objects relative to others, or because it is impossible to move them rigidly. We can look for less perverse reasons than these. It might really be the case that *O* is true because, although our spacetime yields only a finite non-orientable space for every projection out into a space and a time, still, for all of these projections, the universe has a finite time-span, too short for circumnavigating the space in the limits imposed by the speed of light. So even if we grant *O* true in some sense, the relationist must make sure that this is the sense he means.

Sklar allows reference to and quantification over possible continuous rigid motions as 'perfectly acceptable relationist talk'. I think this obscures at least this last objection. For, if we speak of possible motions this fixes our gaze on the motions (the paths) as what we must look to. So we forget about the speed of light, in my earlier case, because there are the relevant paths in spacetime. This takes some of the heat off 'possible', over which the charge of gross circularity looms very black indeed in the preceding paragraph. This might even be proper, if we take motions as entities. But it's as plain as the nose on my face, that this commits us to absolutism.

It is not hard to make the charge of circularity stick if the relationist talks as I say he should. One way of doing it is to see modal ideas as Kripke does. We assume a range of worlds. Some are accessible to others in virtue of resemblances among them, expressed in the true necessary sentences of the modal language. For example, granted the standard philosophical sense of modals mentioned before, all the worlds accessible to ours resemble each other in having non-pathological spaces. Different senses of modal expressions correspond to different relations of access, that is, different respects of resemblance among worlds. Which worlds are in the range of a modal operator taken in a given sense depends on respects of resemblance among worlds. What we want the relationist to tell us is which range of worlds his operators cover; that is, how his worlds resemble each other. It is clear what he has to reply. He must say that the worlds at issue are just those alike in having orientable spaces. That is circular analysis and no reduction. I think the relationist account of empty space is no less circular and what I have said makes it obvious enough, no doubt, how I would try to pin that down. But it is not open to the same lines of attack which make the account of orientability so blatantly faulty. There can be little question that the relationist analysis of empty space is much more ingratiating, if not more correct, than his account of handedness. However, rightly or wrongly, in the earlier sections, my fire was directed another way. I should like to make it clear which way that was.

I see the argument just given as directed against the relationist's account of what it means to say that hands are handed and space is orientable. Where the account is faulty and why, seem not very abstruse questions. But the relationist might still save the day if he can show that any actual arrangement of things in space could determine its orientability and limit or empower how things might be moved globally. I threw out this challenge in the last section and offered suggestions as to why (as I think) it cannot be met. Postulating objects never strictly determines even whether they are spatially related or not. Postulating relations among them (specifying how to enter them in a space) never determines global properties of the space. This remains so, even in the context of a theory like General Relativity, as Sklar himself has lucidly observed.[19]

By contrast, the orientability of space does determine the handedness of hands, for it determines which paths there are in a space which a hand might take. It is a genuinely explanatory idea. This aspect of my argument was not sufficiently clear, perhaps because it is not easy to take orientability seriously as an explanatory concept. It does look rather abstruse, especially when we see it as intended to apply to a weird thing like space as a whole unitary entity. But recourse to the concrete models I used make it clearly a simple, concrete property when ascribed to material things. Among two-spaces, I spoke of table tops, cylindrical and Möbius strips of paper, Klein's (glass?) bottle and so on. It is easy to see how these things differ in the relevant ways. They differ in shape. The Möbius strip is like the paper cylindrical strip, except that it has a twist in it. Orientable spaces, like the two-spaces of the plane and the sphere, clearly differ in shape in many ways.

So, clearly, do non-orientable spaces, as Klein's bottle and elliptic two-space show. In each case we see, from the shape of the space, whether asymmetrical things in it will be handed or not. It is the shape which carries the primary load of explanation. Orientability is a disjunctive property, like being coloured (red or green or . . .). It is a very general topological aspect of shape, like having a hole. Nevertheless, it is a concrete, explanatory idea. So I claim this advantage: I can explain handedness simply by relating the shape of hands to the shape of the space a hand is in. It strikes me as a very simple and direct explanation, given that we can speak of the shapes of spaces.

The kind of explanation which the shape of space offers us is not really causal. Perhaps that is a main reason for dissatisfaction with it on the part of relationists. But, at this point, the offer of a geometric explanation rather than a causal, mechanical one looks too vulnerable to the charges of being strange and, essentially, empty. The idea of the shape of space is nebulous, remote, abstract. We must give it substance and familiarity. Explanations from the geometry of space are too different from our common paradigms of explaining to draw any strength from work they might do in a variety of other connexions. What else *can* the shape of space explain? Unless we find it other roles besides its part in handedness it will simply perish from isolation and remoteness from the familiar intellectual civilisation. Let us begin the long task of showing just what a shape for space means, in detail, and of showing its power to explain the world to us.

NOTES

[1] Kant, this volume, pp. 27—33.

[2] A lively account of many of the facts can be found in Martin Gardner, *The Ambidextrous Universe*. Pelican, 1964.

[3] Kant, this volume, p. 32.

[4] Kant, this volume, p. 32.
[5] Kant, this volume, p. 27.
[6] Peter Remnant, 'Incongruous Counterparts and Absolute Space', this volume, pp. 51—59. The paper is cited as definitive by Gardner (1964) and Bennett 'The Difference Between Left and Right', this volume, pp. 97—130.
[7] John Earman, 'Kant, Incongruous Counterparts, and the Nature of Space and Space-Time', this volume, pp. 131—149.
[8] Earman, this volume, p. 137.
[9] Earman, this volume, p. 138.
[10] Naively, a rigid motion is a movement of a thing which neither bends nor stretches it. As a mapping, it is just a function which takes us from the space of the thing to a space into which it might be rigidly moved. Reflection is a mapping but it is not a motion.
[11] This surface always intersects itself when modelled in Euclidean three-space. But this does not rule it out as a properly self-subsistent two-space. It can be properly modelled in four-space, for example.
[12] More technically, I believe that we find a space pathological when it can be deformed over itself to a point, or to a space of lower dimension. Topologists call such spaces contractible. (See E. M. Patterson, *Topology.* Oliver and Boyd; Interscience, (1956), p. 74). The sphere as a three-space is deformable through its own volume to a point, but the surface of the sphere, as a two-space, cannot be deformed over its own area to a point; so it is not pathological or contractible. The Möbius strip can be deformed across its own width to a closed curve (of lower dimension). It was for this reason that I moved, earlier, to Klein's bottle as a non-contractible proper space.
[13] Notice that a space's *non*-orientability is never constituted by how it *is* covered.
[14] Jonathan Bennett, 'The Difference Between Left and Right', this volume, pp. 97—130.
[15] We pointed out, in Chapter I, that the two sorts of existential relationism spill over into one another. It is interesting to see Lawrence Sklar speaking of "the lawlike features of collections of such relations", in Sklar, pp. 180, 183.
[16] Lawrence Sklar, 'Incongruous Counterparts, Intrinsic Features and the Substantiviality of Space', this volume, pp. 173—186.
[17] Sklar, this volume, pp. 183—184.
[18] Graham Nerlich, 'Pragmatically Necessary Statements', *Nous* **7**, 3 (September 1973), 247—268. See especially pp. 262—268.
[19] Sklar, this volume, p. 184.

LAWRENCE SKLAR

INCONGRUOUS COUNTERPARTS, INTRINSIC FEATURES AND THE SUBSTANTIVIALITY OF SPACE

Kant argued, as part of his argument that space is an a priori intuition, from the existence of incongruous counterparts (such as right- and left-handed gloves otherwise alike) to the existence of space as an entity over and above the material objects in it and their spatial relations to one another. Peter Remnant and John Earman have argued that Kant's argument is incoherent.[1] Graham Nerlich has recently invoked the dependence of facts about handedness on global features of space to attempt to revindicate Kant's argument.[2] I will argue here that, even taking account of the dependence of facts about handedness on global features of space, noticed by Earman and utilized by Nerlich, there is no good argument against relationism founded on facts about handedness. Or, more precisely, there is no good argument against relationism based on handedness which goes beyond the best standard arguments against relationism which invoke no facts about handedness at all.

I

Remnant and Earman believe Kant's argument incoherent for the following reason: Kant says that, since the internal relations of parts of left- and right-handed objects are the same, nothing about the structure of these objects could differentiate them. Therefore, what makes them different must be a difference in the relationship they bear to space itself. But if we invoke "space itself," the only account we could plausibly offer to explain the difference in handedness of the objects would be their differential congruity with parts of space itself. But a right- and a left-handed glove would each coincide with some "piece" of space itself; so what makes one left- and the other right-handed? The only answer could be the handedness of the piece of space with which they coincide. But if there is a feature of a piece of space, its handedness, which determines whether an object coincident with that piece of space is right- or left-handed, then why can't the same feature hold of objects themselves and serve to differentiate them? And if there is no such "internal" feature of the objects, then how does the invocation of space itself help us?

Earman goes on (this volume, p. 137) to maintain that there is such an internal feature of objects, the orientation of their parts. Right- and left-handed objects differ in their internal structure in that their parts have a different orientation with respect to each other. A failure on Kant's part to realize that orientation of parts is just as much an "intrinsic" feature of the objects as, say, size of parts and magnitude of angles between them is the fundamental mistake which led him into thinking that any argument from handedness to belief in space itself as an entity over and above the things in it was either needed or plausible. There is no more puzzle about handedness than there is about any other internal feature of objects which differentiates them. And if there is such a puzzle the invocation of space as an autonomous entity in no way resolves it.

But Earman retrenches a bit (this volume, pp. 138—139). Suppose our space is globally nonorientable. Then there will be a continuous rigid motion (crm) that brings a handed object into congruence with its incongruous counterpart, even though the two objects cannot be brought into congruence by any local crm.[3] But how could even a global crm change an intrinsic internal feature of an object? So the "intrinsicness" of handedness seems questionable.

It is just this theme that Nerlich takes up in his attempt to argue from facts about handedness to the substantiviality of space. At least I take it that Nerlich is arguing for substantivalism. He calls his opponent a "relativist," but never makes it fully clear what constitutes a relativist. His actual claims are:

> Which of these . . . determinate characters [being enantiomorphic, i.e., a member of a possible pair of incongruous counterparts or, instead, homomorphic, i.e., being bringable into congruence with any object of the same size and with the same magnitude of angles between parts] the hand bears depends, still, on the nature of the space it inhabits, not on other objects. The nature of this space, whether it is orientable, how many dimensions it has, is absolute and primitive (this volume, p. 160).

> . . . space [is] a definite topological entity [and] can only be a primitive absolute entity; . . . *its* nature bestows a character of homomorphism, leftness or whatever it might be, on suitable objects. My conviction of the profundity of Kant's argument rests on my being quite unable to see what the relativist can urge against this, except further relativist dogma (this volume, p. 165).

> . . . what differentiates a thing which is an enantiomorph from one of its incongruous counterparts is a matter of how it is *entered into the space*. . . .

> The idea of entry is only a metaphor, clearly. . . . It is not easy to find a way of speaking about this which is not metaphorical. But a very penetrating but not so painfully explicit way of putting the matter is Kant's own, though I believe it still to be a metaphor. The difference between right and left lies in different actions of the creative cause (this volume, p. 166).

I am not sure just what it means to say that space is "absolute and primitive." I believe that the notion of handedness being the result of a "creative cause" is irrelevant to the debate between substantivalists and relationists. I take it that a relativist is one who espouses a relationist theory of space. I think that Nerlich may be right in saying that handedness is a matter of how an object is "entered into space," but I believe that the existence or nature of handed objects is irrelevant to deciding between a relationist and a substantivalist theory of space.

II

Consider two objects. Are they incongruous counterparts, or can they be made congruent by a crm? One might think that there was some "intrinsic feature" of the objects that decided this question. In one sense this turns out to be true, but we must proceed with some caution.

Suppose that the objects are two-dimensional and are both on the same plane. It may be possible to bring them into congruence by a crm that takes at least one of them out of the plane, but not possible by any crm that keeps both confined to the plane. Similarly for objects in our apparently three-dimensional space. If our space really is three-dimensional and if it is orientable, then the objects may be such that no crm can bring them into congruence. Yet *had*, contrary to fact, space been four-dimensional, or if it *had*, contrary to fact, been nonorientable, then a crm *would* have existed which *would* have made them congruent, if the objects are typical "incongruous counterparts."

So, bringability into congruence or its impossibility depends not only upon "the structure of the objects themselves," but upon the structure of space "as a whole," its dimensionality and its global orientability.

Now if space is such that the objects are incongruous counterparts, then we can talk about some "intrinsic feature" of them, that is, some feature of them preserved under all crm's of them, which differentiates them – their handedness. But we must realize that (1) the very existence of such a feature depends upon the overall structure of the space, and that (2) which object has which "handedness" depends upon "how it is situated" in space as a whole.

I think this is true, as far as it goes; but the consequences of this truth must be examined with caution. Notice first that a similar argument can be constructed about a feature of an object that has nothing to do with its handedness.

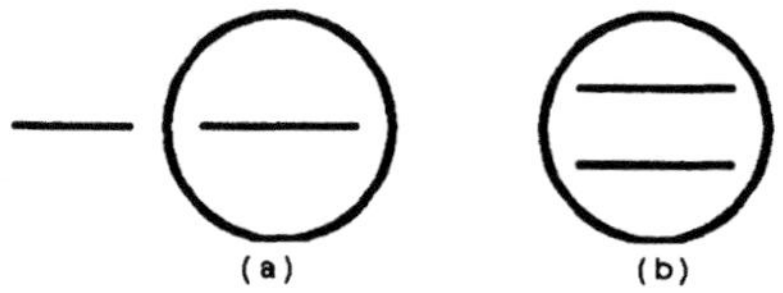

Consider the objects in diagram (a). Is there a motion that transforms them into the objects in diagram (b) which (1) is continuous and which (2) never brings a bar into coincidence with the circle? Well, that depends. What is the space like in which the objects are situated? If it is two-dimensional, the answer to our question is "No." If it is three-dimensional, the answer is "Yes."

Call the arrangement in (a) "opposite-sided." Call that of (b) "same-sided." Is there an "intrinsic feature" of the object arrangements which is opposite- (same-) sidedness, that is, a feature of the arrangement preserved under all continuous transformations that never bring a bar into coincidence with the circle? That depends upon the space in which the objects are placed. If there is such a feature, what determines which object arrangement has it? The answer is: "How the objects are placed in the space." Perhaps there is some crucial disanalogy here with the left-right case, but I fail to see it. There is nothing mysterious about incongruous counterparts. Many features of a given set of objects, with a specified set of internal relations of its parts to one another, depend both for their existence, and, if they exist, for their nature, on the nature of the space in which they exist and on how the parts of the object are situated in the space.

If that is what Nerlich (and Kant?) mean, then they are right. But Nerlich (and Kant) also think that such facts refute the relationist theory of space. They do no such thing.

What is the relationist view of space? It is, at least in the version familiar from Leibniz to Reichenbach, that space is nothing but the collection of actual and possible spatial relations among actual and possible material objects. There are, or may be, some material objects. And there are, or may be, some spatial relations among them. And spatially speaking, that is all there is. There is no such thing as "space itself considered as an entity" which "exists over and above" the material objects and their spatial relations.

The invocation of *possible* objects and *possible* spatial relations among them is crucial here, just as "permanent possibilities of sensation" are crucial for the phenomenalist. If we wish to be phenomenalists and yet talk about unobserved material objects, then we must, if we are to translate all material-object talk into sense-datum talk without loss of content, tolerate subjunctive as well as indicative sense-datum assertions. Just so, if we wish to talk about places in the world at which no material objects exist, and even more if we wish to be able to talk about spaces totally devoid of contained matter, then, if we are going to translate all talk about "space itself" into talk about the spatial relations among material things, we had better allow talk about possible objects and their possible spatial relations as well as talk about actual objects and their actual spatial relations.

Now the reader might not like the invocation of possibilia, their possible relations to one another, subjunctive or counterfactual assertions, etc. If he finds these totally abhorrent, then he will probably reject the relationist theory of space. Since versions of relationism that eschew such notions are pretty implausible, he may opt for substantivalism immediately.

Alternatively, he may argue like this: Talk about the possible relations among possible material objects is all right, so long as one understands that it is "grounded" in belief in the *actual* nature of *actually* existing substantival space. Just so, we can understand the language of "possible sense-data of possible observers in possible perceptual situations" only because of our belief in actual material objects.

Each of these positions rests on a deep philosophical objection to relationism. The arguments may even constitute devastating objections to the relationist account. All that I wish to claim here is that the following assertions are correct:

(a) Given the full relationist resources, including possible objects and possible spatial relations among them, we can account for all the interesting features of left and right, etc.
(b) Or, more correctly, we can account for these features just as well

on the relationist account as we could on any substantivalist account. If there are any "mysteries" about left and right unsolvable on the relationist account, the invocation of space as a substantival entity will be of no use in solving these puzzles.

(c) If there are uneliminable difficulties with the relationist account of space, they have nothing to do with features of the left-right distinction, the enantiomorph-homomorph distinction, dimensionality, or orientability.

(d) The notion of the "creative cause" of the spatial features of objects is of no relevance to the dispute between the relationist and the substantivalist accounts of space.

The notion of space as "absolute" is ambiguous, but, I will argue, notions of orientation are irrelevant to "absoluteness" in any of its senses. It is not clear what sense is to be given to the notion of space as "primitive." Whatever sense we can give to it is such that there seems to be no good argument from the facts about oriented objects to space being primitive in any way that would disturb a relationist.

(a) Suppose that there is a three-dimensional hand. What, from the point of view of the relationist, makes it an enantiomorph, or, alternatively, a homomorph? If it is an enantiomorph, what makes it a left hand or a right hand? Is leftness (rightness) an intrinsic feature of the hand in this case? Suppose that the hand is not an enantiomorph, but is, instead, a homomorph. How can this be the case? If it is the case, is there still some sense in which the hand is still left (or right)? And does leftness (rightness) in this sense now constitute an intrinsic feature of the hand?

According to the relationist, the hand is an enantiomorph if and only if there is a *possible* incongruous counterpart for it. That is, if and only if there is a possible object such that (1) its parts have the same lengths as the parts of the original hand and the same absolute values of the magnitudes of the angles between them, but (2) there is no crm that will bring the hands into congruence. All this is perfectly acceptable relationist talk. No reference is made to "the entity, space, itself," but only to possible objects and possible spatial relations among them.

Now if the hand is an enantiomorph it will be either right- or left-handed. That is, in a world in which there are enantiomorphs the members of a pair of incongruous counterparts are possessed of an intrinsic feature, that is, a feature preserved under all crm's, which is their handedness.

Now Nerlich claims that the handedness of an object is dependent upon "the way it is entered into the space." Is this correct? That depends upon what you mean. If this means that the handedness of an object is dependent upon the spatial relation of its parts to one another, then the claim is certainly true. But then, the triangularity of a triangle depends upon the spatial interrelation of its parts, and so triangularity would also depend upon "how the object is entered into the space."

Is there any interesting way in which handedness differs from triangularity? Well, yes. Handedness, in the full sense, exists only if there is an orientation property of the object which is preserved under all crm's. That is what we mean by the object's handedness in the full sense. But, as we know, the existence of such a feature depends upon topological features of the space as a whole – its dimensionality and its orientability. This is not surprising, for the dimensionality of the space and its orientability determine the class of all crm's and, hence, what is preserved under them. There is no analogous dependence of the very existence of triangularity on the over-all topology of the space, and in this sense handedness differs from triangularity. If that is what it means for the handedness of an object to depend "on the way the object is entered into the space," then handedness is so dependent. But that does not mean that handedness is not an intrinsic property of the object in a space in which handedness exists, and there is no good argument against relationism in these interesting topological facts.

But is handedness *really* an intrinsic feature of the object? If you mean by 'intrinsic' a feature of the object which is preserved by all transformations of a specified kind, then handedness may well be an intrinsic feature of an object. For example, if space is an orientable three-space and the object a three-dimensional hand, then if by 'intrinsic' you mean "preserved under all continuous rigid motions," than handedness will be an intrinsic feature of the hand.

If by 'intrinsic' you mean, however, that the feature is one that any object similar in construction specifiable in only local terms (lengths of parts, magnitudes of angles between parts and what we will soon call 'local handedness', for example) will have, irrespective of the nature of the space in which the object is embedded, then handedness is not intrinsic. For an object of a given construction so specified may not even be "handed" in the full sense at all – if it is in a nonorientable space, for example; whereas another possible object, describable in the same terms in the local way, but now taken as embedded in an orientable space, will indeed have full handedness. If 'intrinsic' means "independent of the topology of the space as a whole," then handedness is not intrinsic.

What does it mean, according to the relationist, for the hand to be a homomorph? Only that, given any possible counterpart to the hand, that is to say any possible object whose parts have the same lengths and the same magnitude of angles between them, then there is a possible crm that brings them into congruence.

Suppose that there are no enantiomorphs. In what sense can the hand still be said to be left-handed or right-handed? Well, suppose, there is a possible crm that takes the hand into its counterpart because although the hand is in an oriented three-space, the three-space is embedded in a four-space. Note that all this talk about "being in an oriented three-space" and "there being an embedding four-space" is all perfectly intelligible from a relationist standpoint. The assertions are, of course, explicated in terms of possible spatial relations among possible objects, and the lawlike features of the collection of such relations.

Now although any two three-dimensional hands which are counterparts and which are in the three-space will be bringable into congruence by a crm that takes at least one of the hands out of the three-space into other parts of the embedding four-space, it will still be the case that there are pairs of possible incongruous counterpart hands in the three-space, in the sense that (1) the members of the pair are counterparts, and (2) no crm that keeps both hands in the three-subspace will bring the hands into congruence.

In this case, one of the hands will be "left in the three-space" and the other "right in the three-space." And each of the hands will be characterized by an "intrinsic feature": "being three-left" or "being three-right." The feature is intrinsic in the sense that an object that has it continues to have it no matter how many crm's it undergoes, *so long as the motions keep it confined to its original three-space*. What we see here is an illustration of the following general truth: What we mean by an intrinsic feature of an object is relative to some particular kind of transformation of the object we have in mind. A feature may be intrinsic relative to transformations of one kind, but not so relative to transformations of a different kind. There is nothing special about handedness here. For, as we saw, a structure can be intrinsically same-sided relative to one transformation (keeping the bars in the diagram in the plane) yet neither intrinsically same-sided nor intrinsically opposite-sided relative to some other transformation (allowing motions of the bars through all of three-space). That is just what we mean by 'intrinsic'.

Suppose that the hand is a homomorph despite the fact that it is in a three-space not contained in any embedding four-space. Once again, this statement is perfectly intelligible from a relationist point of view. Why is the hand a homomorph despite the absence of an embedding four-space? Perhaps because the three-space is globally nonorientable. Then all hands can be brought into congruence with their counterparts by global crm's; although there will still be possible pairs of hands such that (1) they are counterparts and (2) no *local* crm can bring them into congruence.

Suppose this is so. The hand is now obviously neither a left hand nor a right hand in the full sense. Kant thought this absurd, but Nerlich reasonably asserts that this just shows that Kant never considered the real possibility of higher-dimensional spaces or nonorientable spaces being the case. It is clear that our hand is now not "three-left" or "three-right" either. Is there any intrinsic feature of the hands which distinguishes the members of a locally incongruent pair of counterparts even though they are bringable into congruence by a global crm?

By this time the answer should be evident. In such a non-globally-orientable three-space each hand is still either *locally-left* or *locally-right*. Of course there is no sense in asking whether a hand locally-left at point p is locally-left or locally-right when moved to point q. On the other hand, if p, q and the path taken by the object between them are all contained in a region of the space over which the local orientations can be extended in a globally consistent way, then partial global extensions of the purely local notion are possible.[4] Nor, of course, will it generally be true that an object locally-left at p at one time, then continuously rigidly moved about in the space and then returned to p will still be locally-left at p upon its return to the point of origin of its travels. This is just what it means to say that the space is globally nonorientable.

Notice also the following: Suppose we have a three-dimensional hand. Consider its "local-three-handedness." This will be an intrinsic property of the hand in the sense that it will be invariant under any local crm's that keep the hand confined to the original three-space in which it was located.

Even if that three-dimensional space is embedded in a four-dimensional superspace and even if the three-space is globally nonorientable, the local-three-handedness of the object is still well defined. It is, in fact, independent of the embedding of the three-space in any higher-dimensional space and independent of the global connectivity and orientability of the three-space. So in this sense of 'intrinsic', which connotes independence from questions of embedding of a space in higher-dimensional spaces and of global orientability of a space, local-three-handedness is a truly intrinsic feature of an object. I think that it is this fact that makes us want to say that a hand in our world is either left-handed or right-handed irrespective of the existence of any four-dimensional embedding space unknown to us or of the fact that the three-space of our world might, in fact, be globally nonorientable. The kind of handedness we normally have in mind is local-three-handedness, and this is independent of these possibilities about our physical space.

More fully and correctly: (1) we have very good reason to believe that there are locally-three-handed objects, since there clearly exist possible counterparts incongruous under any local crm that keeps them confined to three-dimensional space; (2) we have very good reason to believe that this local-three-handedness can be extended to a regional-three-handedness over the presently observable spatial universe, since the three-space we are presently aware of seems quite globally orientable; (3) in so far as we have good reason to believe that our three-space is not embedded in a four-space, we have good reason to believe that there are regionally-handed objects; and (4) in so far as we have good reason to believe that our three-space is a globally orientable three-space, we have good reason to believe that there are handed objects in the world, *simpliciter*.

So all the crucial notions: being an enantiomorph, being a homomorph, having a specified number of dimensions, being orientable or nonorientable, being left and right — in the full, local or "subspace" senses, are all completely intelligible from a relationist point of view.

(b) If someone objects to any of the notions I have invoked above, or rather, to the particular definitions of particular geometric spaces or objects — such as enantiomorphism, local-handedness, etc. — let him ask himself whether postulating space as a substantival entity is going to leave him any better off.

If he wants to talk about the dimensionality of his "space" or its orientability, he will need just the same characterizations for this entity that we needed for our lawlike-governed collection of possible relations among possible material objects. Treating space as an object, rather than as a collection of possible relations among possible material objects, solves none of the difficulties in defining the various geometric notions necessary to characterize the "structure of space".

And if he objects to my postulation of relativized intrinsic features of objects, like enantiomorphism, local-enantiomorphism, or enantiomorphism-in-a-subspace, he should consider this: The only good that the postulation of substantival space is going to do for him in defining the relevant notions of the features of objects is that he will be able to predicate these features of the "space" in which an object is contained instead of predicating them of the object itself. He can then "explain" the nature of the object as being its relation to its containing space.

But this explanation seems no explanation at all, and the relationist objection is familiar. If all these features are well definable of the containing spaces of possible objects, why not just predicate them of the objects themselves and be done with it? The relationist argument, as always, consists partly in the claim that postulating substantival space provides no explanations and no understanding over and above postulating possible objects and their possible spatial relations. Worse yet, such postulation confuses the issue by making it look as though there were additional features of the world (for example, locations in substantival space) which really don't exist at all.

Notice that substantivalism is equally useless in answering such traditional positivist questions as: "How do we know which objects are left and which are right?" and "Mustn't there be some independent feature lawlike-connected with handedness in order for it to make sense to say that an object is left- or right-handed?" For if there are any real puzzles here, and I think that there are not but only confusions and pseudo-problems, then these real puzzles are just as much puzzles about handed pieces of space as they are about handed objects in space.

(c) What is the *real* substantivalist objection to relationism? The best philosophical argument I know is the claim either that (i) it makes no sense at all to talk about actual objects and possible spatial relations, but only about actual objects and their actual relations, and on this ground spatial talk is not wholly translatable into relationist talk without loss of content, or (ii) that talk about possibilia makes sense only because of the underlying assumption of some actual substantival entity and its actual features, and in the case in question this can only be substantival space and its actual geometrical structure.

Now there are deep philosophical issues here. But one thing is clear, and that is the fundamental irrelevance of particular facts about enantiomorphism, homomorphism, or handedness. For if the relationist must invoke possibilia in order to explicate these notions, he must invoke them to explicate far more basic spatial concepts — for example, there being an empty spatial location in the actual world or there being a possible world of totally empty space. It was to account for these notions relationally that the idea of possible objects and their possible spatial relations was originally invoked. If the substantivalist wants to refute relationism on these grounds, he need not go to such *recherché* lengths as the invocation of questions about orientation and orientability. His quarrel with the relationist lies on much broader issues.

The only other arguments that I know for substantivalism which have any persuasiveness are those from particular aspects of physics, say from the "absoluteness" of absolute acceleration in Newtonian mechanics, special relativity and, perhaps, general relativity as well. It is these arguments which, I take it, Earman finds persuasive. I am not sure that they are at all convincing, but in any case they hardly rest upon the possibility or nature of enantiomorphic, homomorphic, or handed objects.[5]

(d) Is handedness the result of a creative cause? Perhaps so. For there is neither more nor less reason to believe that the handedness of an object is the result of some causal factors than there is to believe any feature of the object to be the result of causes. But features of an object which consist in the relation of its parts to one another, or of the actual and possible relations of the object to other actual and possible objects, can surely be the result of causes. So even if handedness is "the result of the action of a creative cause," this in no way indicates anything inadequate whatever in the relationist account of what handedness *is*.

Is space "absolute?" Well, if that means "Is space a substance", we have already seen the irrelevance of the consideration of notions of orientation to that issue. If is means "Do absolute motions in Newton's sense exist?", then surely nothing could be more irrelevant than the existence or nature of enantiomorphic objects.

Is space "primitive?" If that means that the existence of space temporally precedes there being objects in it, then all arguments from orientation are irrelevant. If it means that we can imagine space without material objects, but no spatial objects without space, then this is conceded by the relationist. There can be a collection of possible relations among possible objects without there being any objects or any actual relations among them; but there can't be an actual object with actual spatial features unless there are some possible objects with possible spatial features. Once again, questions of orientation are completely irrelevant.

If "Space is primitive" means "The relationist claim that all spatial talk can be translated into talk about the spatial relations among material objects is wrong," then, as we have seen, the allegation of primitiveness is just the allegation of the substantivalist about the incorrectness of the relationist account, and, as we have seen, the existence or the nature of features of orientation are completely irrelevant to the issue.

Is the relationist or the substantivalist account of space (or space-time) correct? I don't know. What is clear, however, is that the existence and nature of incongruous counterparts are irrelevant to the issue.

If Kant and Nerlich are claiming that questions of orientation can be understood fully in the context of an understanding of the topology of space as a whole, then they are right. This is an important topological fact, and Kant may, indeed, have been anticipating this in his discussion of the handedness of objects. But if they are claiming that any facts about the existence or nature of orientation features add any weight to the other well-known arguments of the substantivalist against relationism, then they are plainly in error.

NOTES

1 Remnant, "Incongruent Counterparts and Absolute Space," *Mind*, n.s., LXXII, 287 (July 1973): 393—399; this volume, pp. 51—59. Earman, "Kant, Incongruous Counterparts, and the Nature of Space and Space-time," *Ratio*, XIII, 1 (June 1971): 1—18; this volume, pp. 131—149; parenthetical page references to Earman are to this volume.

2 "Hands, Knees, and Absolute Space," Journal of Philosophy, LXX, 12 (June 21, 1973): 337—351; this volume, pp. 151—172; parenthetical page references to Nerlich are to this volume. A similiar claim that the dependence of orientation properties on global features of space provides a refutation of relationism, is made, without argument, in Ted B. Humphrey, "The Historical and Conceptual Relations between Kant's Metaphysics of Space and Philosophy of Geometry," *Journal of the History of Philosophy*, XI, 4 (October 1973): 483—512, p. 488, n. 11.

[3] Throughout this article I assume, without loss of philosophical generality, that the space is one of constant curvature.

[4] In more detail the situation is like this: Even if a space is globally nonorientable there may be subregions of it such that we can divide all the counterpart objects in the region into two classes of opposite handedness. An object of given handedness in the region cannot be brought into congruence with its counterpart of opposite handedness in the region by any crm that keeps the object in the region. If this is so we can talk about "handedness with respect to the region." Of course in a nonorientable space the following situation can arise: (a) there is a region, *A*, such that there are pairs of objects in the region which are counterparts and such that no crm of the objects confined to *A* can bring them into congruence; (b) there is another region, *B*, partially overlapping *A*, which is, like *A*, regionally orientable; but, (c) the region that is the union of *A* and *B* is such that any two counterparts in the united region can be brought into congruence by a crm in the united region.

[5] For a detailed discussion of the philosophical debate between the substantivalist and the relationist, and of the relevance or irrelevance of the results of physics to the philosophical debate, see my *Space, Time, and Spacetime*, chap. III, "Absolute Motion and Substantival Spacetime," University of California Press, forthcoming.

RALPH WALKER

INCONGRUENT COUNTERPARTS

Kant maintains that a right- and a left-hand glove are enough to shatter Leibniz's theory of space. For there is a difference between them that the relational theory can never capture. The difference in orientation, in handedness, cannot be reduced to any difference in the relations between things or between the parts of things; the two gloves share all their relational properties and all their non-spatial properties as well, yet they are quite clearly and obviously different. We can see the difference between them, but we cannot specify what it amounts to in any way that would be acceptable to Leibniz. There must be more to space, therefore, than Leibniz will allow.

The argument is a little hard to assess, and Kant himself seems uncertain just how much it shows. The reason is partly that there are two arguments here, not one. The first makes a perfectly good and straightforward point against Leibniz, though not against everyone who has held that space is no more than a system of relations. According to Leibniz objects must be identifiable independently of their spatial (and temporal) relations: we first identify the objects and then observe the relationships that hold between them. Now Strawson, for instance, differs from him on this, agreeing indeed that to determine spatial relationships we must rely on the identification of things, but holding also that the identification of things depends upon knowledge of their spatial relationships – the two come together and the dependence is mutual.[1] Against Leibniz, though not against Strawson, Kant can offer his gloves. For we find no difficulty in telling the gloves apart, though what distinguishes them is their orientation in space and not any feature which can be identified before spatial relations are discovered. The case of the incongruent counterparts is one in which it is strikingly clear that we distinguish the objects by means of a spatial property, though other examples can be used to make the same point, and we have already seen Kant elsewhere citing two drops of water for the same purpose: they can be qualitatively indistinguishable, but we individuate them without difficulty through their different positions. The advantage of the more complicated example of incongruent counterparts is perhaps that it helps circumvent a reply which Leibniz would make. For he argues that there must always be some hidden internal difference between the drops if we are to be capable of distinguishing them, a difference we may not be consciously aware of but which subconsciously we must apprehend. This contention was made more plausible by the microscope, which revealed endless minute differences between water-drops; and Leibniz took great pride in his triumph over M. d'Alvanslebe, a Hanoverian courtier who offered to find him two qualitatively indistinguishable leaves, and spent a morning vainly trying to do so.[2] But with the two gloves, or a single glove and its mirror image, it seems particularly obvious that the salient feature which differentiates them for us is their orientation in space, not any qualitative distinction consciously or subconsciously perceived.

This first point is a valid objection to Leibniz, though not to Strawson, because Strawson does not hold that things can be identified on the basis of their non-spatial (and non-temporal) properties alone. But Kant also has a second and more far-reaching point to make. For just as the difference between the gloves is patently a spatial one, so it seems equally evident to him that it is not a difference in spatial *relations*: it is internal to the gloves, and they therefore have spatial properties not susceptible of any relational analysis. They differ in absolute orientation.

An obvious counter to this is to point out that any two actual gloves clearly will differ in their external relations to other objects. The internal relations of their parts will be the same, but if they lie side by side face down on my dressing table the thumb of one will be closer to certain objects (my hairbrush, or the Eiffel Tower) than are its fingers, while for the other the reverse will be true. So long as there is some asymmetry in the arrangement of objects in the universe, there will be differences of this sort to appeal to.

Kant would reply that it is not this asymmetry in external relations which we rely on in telling the gloves apart. We can see a difference just by looking at them, without taking external relations into account. This would be parallel to his earlier reply to Leibniz on the d'Alvanslebe question, for there also he was concerned less to deny that there might be hidden qualitative differences between the two drops or leaves than to dismiss the point as irrelevant, since it is not these differences that we go on in making the distinction. But he can strengthen his case by asking us to consider a more impoverished universe than our own, in which there exist only the two gloves side by side. Everything is then symmetrical about a plane mid-way between the gloves, one side being as it were a mirror image of the other. The gloves do not differ in respect of the external or internal relations of their parts, yet they remain distinct and the difference in handedness remains. Leibniz must say that this could not be so; there could not be two gloves any longer, but one glove only, and that neither left- nor right-handed — for these terms become meaningful only when there are other objects to relate to. But this is unconvincing, for the situation envisaged is easy to imagine; one needs some better reason to dismiss it as incoherent than the dogmatic commitment to an analysis of space which has that consequence.

Kant himself does not consider a universe of two gloves symmetrically arranged, but in his original paper of 1768 he does raise the possibility that there might have existed only a hand and nothing more:[3]

> Let it be imagined that the first created thing were a human hand, then it must necessarily be either a right hand or a left hand. In order to produce the one a different action of the creative cause is necessary from that by means of which its counterpart could be produced.

This does seem plausible; the difference between a right and a left hand is such that one can hardly imagine a hand neutral as to which it is. In a similar way one can say of our own (presumably) asymmetrical universe that there might have existed in its stead a counterpart, exactly similar save for being orientated the other way, as its mirror image would be. But I think these examples carry less conviction than does the case of the two-glove universe; it is easier to suspect an incoherence. Remnant has recently taken Leibniz's side here, objecting that 'in a universe which contained nothing but a single hand, it would not just be empirically undecidable whether that hand were left or right; it would be strictly indeterminate.'[4] But Kant would say that Remnant had followed Leibniz in failing to see that there is an irreducibly ostensive element in the meanings of 'left' and 'right'. A left-handed and a right-handed glove look different regardless of their relations to other things, and a left-handed and a right-handed universe would look different too.

Someone might reply that in imagining these situations one cheats by surreptitiously introducing a relation to one's own asymmetrical body. It is true that I first learnt the words 'left' and 'right' by reference to my body, and there may still be circumstances in which I appeal to it to remind myself which is called 'left' and which is called 'right'. But the difference between the gloves is immediately obvious without reference to my body or to anything else; I should receive it in just the same way if my body were itself symmetrical about the plane that forms the axis of symmetry between the gloves.

A more satisfactory manoeuvre for Leibniz would be to accept all this, but to say it does not show space is more than a set of spatial relations. What it shows is that among the spatial relations we must include orientational ones, which are not reducible to any other sort: relations like '*x* is to the left of *y* as seen from *z*' (assuming, of course, that the hypothetical observer is standing on his feet not his head, but that is a detail). Leibniz himself would not have been very happy with this for metaphysical reasons which are not relevant here,[5] but the move appears open to those who do not have his metaphysical preoccupations. One could then distinguish between the gloves: from a point on the plane of symmetry between the gloves and palmwards of them, in glove A the thumb is to the left of the fingers and in glove B the fingers are to the left of the thumb.

The natural reaction to this suggestion seems to be that it is over-simple: one cannot treat relations like '*x* is to the left of *y* as seen from *z*' as unanalysable, and as soon as one tries to provide an analysis one has to appeal to features of space which cannot be regarded as mere relationships between objects. But it is not very clear why one should not treat such relations as unanalysable. The only attempt I know to argue that they cannot be handled in that way is due to Nerlich.[6]

Nerlich points out that amongst the topological spaces we can imagine there are some which are orientable, but also some which are non-orientable. An orientable space is one within which counterparts like the gloves can never be made congruent with one another by any combination of translations and rotations. To take a two-dimensional example, he considers a pair of 'knees' cut out of paper, as in Figure 1.

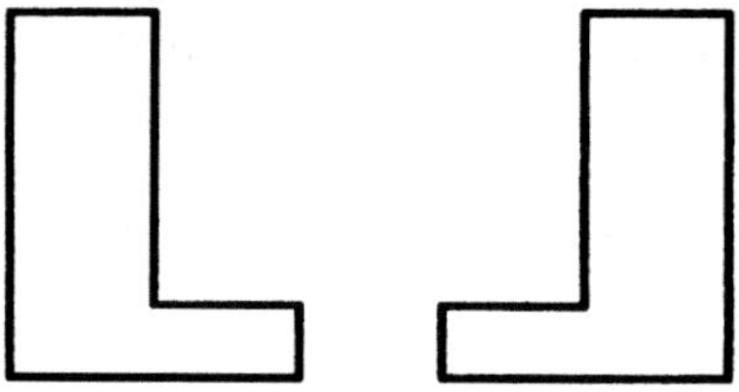

Fig. 1. Nerlich's knees

Confined in their movement to the surface of a table these knees are incongruent counterparts, and they show that the two-dimensional space of that surface is orientable. (Of course they can be made congruent by turning one of them over, but then one has to lift it off the table top.) But one can also imagine a two-dimensional space in which it is possible to move one knee round until it becomes congruent with the other; the surface of a Möbius band constitutes such a space, as would the surface of a Klein bottle (see Figures 2 and 3).[7] Philosophy offers few chances for experiments, but it is entertaining to make a Möbius band of sticky paper and cut holes in it for knees, to investigate these possibilities. Non-orientable spaces are equally possible in three dimensions, and similar things can be done in them with gloves, but unfortunately they are not so easily modelled.

The possibility of moving a right-handed knee or glove around in a non-orientable space until it becomes congruent with a left-handed one means that we can hardly claim its right-handedness is intrinsic to it; for it depends on how we move it. And it also seems to depend upon how we move. If Alice follows a glove round a non-orientable, three dimensional space and comes back to where she started — or if a two-dimensional Alice follows a knee round a Möbius band — it will look the same to her all the way, and when she gets back again it will still look to her as it did before she started out. Only everything else in that region will now appear to her right-left reversed; the exactly similar glove she left behind will have apparently changed its handedness; she will seem to have come back to a mirror image of the place she started from. To those at home she will appear to have undergone mirror-image reversal and to use 'right/left' vocabulary in a switched way; she will hold her fork with her right hand, her knife with her left, and so on.

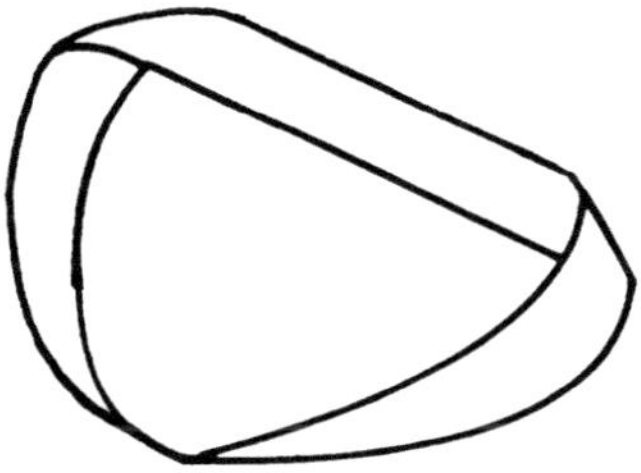

Fig. 2. Möbius band

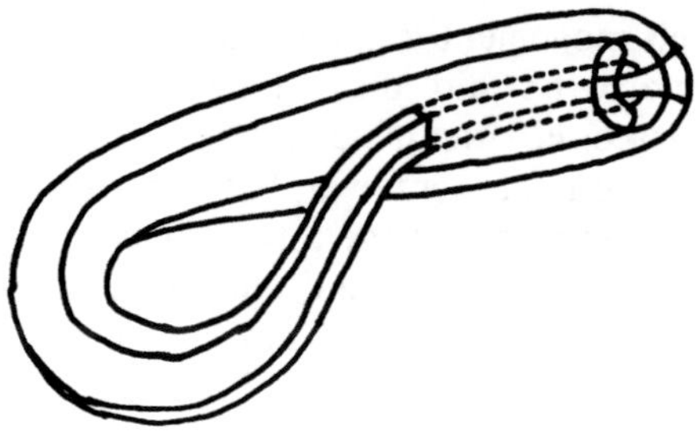

Fig. 3. Klein bottle

The stay-at-homes will naturally say that Alice has got 'right' and 'left' mixed up. But their only support is in numbers. If half of them had travelled round too there would be no settling who was right; if, from the same position, one group said that *A* was to the left of *B*, then the other group would say that *B* was to the left of *A*. How things look to you at home will depend on how many times you have travelled round the space. So it cannot be that '*x* is to the left of *y* as seen from *z*' is an unanalysable relation; and this is Nerlich's conclusion. There could be an objective answer to the question whether *A* was to the left of *B* if they were to define 'to the left of' in terms of asymmetrical fixed objects in the neighbourhood, but then this is not what 'to the left of' means.

Nevertheless there is something that Alice and the stay-at-homes can agree upon without appeal to the asymmetrical objects, namely that there is a difference between the two gloves — and a difference in handedness. All they cannot decide on is which to call the right-hand glove, but this does not really matter very much. They agree that the adventuresome glove used to have the same handedness as its twin that stayed behind, and now no longer does. '*x* is to the left of *y* as seen from *z*' will not do as an unanalysable spatial relation, but perhaps 'as seen from *z*, *x* lies to *y* in the same orientation as *v* lies to *w*' will. There is no dispute between Alice and the rest as to which gloves are similarly handed; the only dispute is over which orientation to call right and which to call left.

This has to be complicated slightly, for Alice might see stretched out before her a whole series of gloves going all the way round the (finite) space and back to where she is standing. Then she would see herself in the distance, right-left reversed, with the gloves beside her there looking reversed correspondingly. (Assuming, of course, that in this strange world the light rays follow geodesics.) The unanalysable relation therefore does not hold between the gloves, but only between them as they appear from a given position at such-and-such a distance and in such-and-such a direction. So it is a slightly complex relation to admit as primitive, but if one is willing to then we must conclude that one can handle the difference between the gloves within a Leibnizian theory of space.

But if (like the historical Leibniz) one is unwilling to treat orientational relations as primitive one will then have to take Kant's argument seriously. He is not quite right, though, when he says that to distinguish the counterparts we must refer to 'space in general as a unity'.[8] All that is necessary is that we should be able to refer directly to spatial points – points which cannot be adequately characterized in terms of the (non-orientational) relations that hold between objects. Once we can do this we can distinguish between the two gloves, using demonstratively specified points to play the part of the hairbrush or the Eiffel Tower: the thumb of one glove is closer to here than its fingers are, whereas for the other it is not so. Some other point may have the same relationship to the second glove, but even if the gloves are symmetrically arranged and alone in the universe that point is not this one. Leibniz thinks we can use a demonstrative only when it could be replaced, at least in principle, by some purely descriptive expression capable of individuating the item referred to, but in this case that is not possible. Kant, and common sense, see Leibniz as making a gratuitous stipulation, and recognize that we can pick out places and times directly, regardless of whether they differ from one another in any way expressible without a demonstrative.

It is this ability to make direct and irreducible reference to spatial points which is crucial for Kant's position. And it is plausible to hold that he is right about it, though the idea of direct reference does suggest a second way of avoiding the thrust of the argument from incongruent counterparts. For one could admit such reference not to points in space but to parts of objects – the fingers of the gloves, for example – and one would again be able to distinguish the left-handed from the right-handed glove, in a similar way. Kant, however, overlooks this quite attractive possibility, and concludes that we must be directly aware of space as a framework of demonstrable points; which is what he means by calling it an intuition. In the same way he thinks time is a framework of demonstrable moments, and though he has no parallel argument for this it is very natural to hold that we can make direct reference to the present moment. When he calls space and time 'forms' of intuition his point is still essentially the same: when we perceive objects we perceive them as laid out in space, and the spatio-temporal framework provides us with the matrix for identifying them.

He does also go further, though. He takes the additional step of claiming that we are directly aware of certain properties the spatial framework has, and to a lesser extent certain properties of time as well. We know space to be unlimited in extent (an 'infinite given magnitude') although the objects we experience cover only a limited area, and we know its geometry to be Euclidean. It is therefore, amongst other things, an orientable, three-dimensional space. These extra features of space are supposed to reveal themselves in our inability to imagine it in any other way. But nothing in the argument from incongruent counterparts entitles him to these further claims — not even to the claim that space is orientable. For we can distinguish between the orientations of the gloves even in a space which is not orientable, just as we can distinguish between the paper knees lying on the table. We just cannot use our terms 'left' and 'right' to mark the distinction without getting into the difficulties that faced Alice and her friends, since things can change their handedness by moving around.

NOTES

[1] Strawson, *Individuals*, p. 37.

[2] Leibniz to Clarke IV: 4, Gerhardt VII: 372; Leibniz to the Electress Sophia, 31 October 1705, Gerhardt VII: 563.

[3] *Von dem ersten Grunde des Unterschiedes der Gegenden im Raume*, Ak. II: 382f.

[4] P. Remnant, 'Incongruent Counterparts and Absolute Space', *Mind* N. S. 72 (1963), p. 399. This volume, p. 58.

[5] Ultimately relations are unreal, for Leibniz; a relation would be 'an accident in two subjects, with one leg in one, and the other in the other; which is contrary to the notion of accidents' (Leibniz to Clarke V: 47, Gerhardt VII: 401; the translation is Clarke's, p. 71 in H. G. Alexander's edition). Hence it must be possible to analyse all true statements into non-relational form. Statements about spatial relationships are to be analysed into statements ascribing to individual substances (monads) various different degrees of clarity and distinctness in perception; and although it is difficult at the best of times to see how this analysis can be adequate, it would be especially unpromising to attempt such a treatment of '*x* is to the left of *y* as seen from *z*'.

[6] G. Nerlich, *The Shape of Space*. This volume, pp. 151—172. The suggestion that they should be so treated was made by J. Earman, 'Kant, Incongruous Counterparts, and the Nature of Space and Space-Time', *Ratio* 13 (1971), pp. 1—18. This volume, pp. 131—149, esp. p. 139.

[7] Nerlich, pp. 157—160.

[8] Ak. II: 378.

MARTIN CURD

SHOWING AND TELLING: CAN THE DIFFERENCE BETWEEN RIGHT AND LEFT BE EXPLAINED IN WORDS?

Jonathan Bennett claims to find in Kant's writings a thesis concerning the meanings of 'right' and 'left' which he calls the 'Kantian Hypothesis'.[1] Bennett suggests that this thesis is refuted by messages based on the failure of parity conservation in weak interactions such as beta decay.[2] Whether or not Kant actually held the thesis that Bennett attributes to him I am not concerned to dispute but I do wish to argue that, when properly understood, the Kantian Hypothesis is *not* refuted by signals describing parity-violating experiments.

I shall begin with a explication of the Kantian Hypothesis, consistent with Bennett's analysis based on the distinction between two different types of signal. Bennett's partial defense of the hypothesis is then examined by exploring the extent to which so-called 'chiral terms' are semantically isolated from the rest of the language. Finally the nature of the signals describing parity-violating experiments and their relevance to the Kantian Hypothesis is examined. It is concluded that Kant's thesis that explaining the difference between right and left requires an ineliminable element of showing is consistent with the results of modern physics.

I

Bennett defines the Kantian Hypothesis as the claim that 'an explanation of the meanings of 'right' and 'left' requires showing, i.e., demands an appeal to sensorily presented examples' (this volume, p. 100). Elsewhere in his paper Bennett describes the thesis as the claim that 'one could explain the meanings of these words ['right' and 'left'] only by a kind of showing — one could not do it by telling' (this volume, p. 98) and he adds for clarification: 'The Kantian Hypothesis that I want to discuss says that we must use sensorily presented instances — must resort to showing — if we are to explain the *direction* of the left/right distinction, i.e., to explain which is which' (this volume, p. 110).

In order to understand the Kantian Hypothesis more precisely, imagine that there is an intelligent, English-speaking person on a distant planet, Alpha, with whom we can communicate only by means of a two-way radio. Suppose that the Alphan has available to him a set of enantiomorphic objects such as right hands and left hands. The problem is to send a signal, S, such that it ensures that the Alphan is able to correctly identify right hands as right hands and left hands as left hands based solely on the information in the signal. 'Being able to correctly identify' in this context means 'being able to label these objects the same way they are labelled on Earth' since it is having this kind of ability that is required in order to understand the meanings of 'right' and 'left'. The problem, then, is to tell the Alphan how to distinguish between right and left hands in the 'which is which' sense.[3]

In explicating the phrase 'based solely on the information in the signal' it is important to distinguish between two types of signal which are defined as follows:

Type 1: A signal, S, is a Type 1 signal if and only if the recipient of the signal, P, would be able to correctly label X's as X's at time t if P understands the meanings of all the terms apart from 'X' in S, even if P has not seen a correctly labelled X prior to t nor is seeing a correctly labelled X at t.

Type 2: A signal, S, is a Type 2 signal if and only if the recipient of the signal, P, would be able to correctly label X's as X's at time t if P understands the meanings of all the terms apart from 'X' in S only if P has seen a correctly labelled X prior to t or is seeing a correctly labelled X at t.[4]

The intuitive idea behind this distinction is that a Type 1 signal is a pure case of 'telling' whereas a Type 2 signal involves an essential element of 'showing' if it is to achieve its purpose. A simple example will illustrate the difference. Imagine that the Alphan knows what horses are but wants us to explain to him the difference between stallions and mares. He knows that the difference is a sexual one but he is unable to distinguish between stallions and mares in the which is which sense. Assume that we know that the only large animals in the fields of the Alphan's neighbors, Delta and Gamma, are a typical stallion and a typical mare respectively. Now compare the following two signals:

S_1: Stallions are male horses; mares are female horses.

S_2: The large animal in Delta's field is a stallion; the large animal in Gamma's field is a mare.

S_1 is clearly a Type 1 signal. Assumming that the Alphan has an appropriate knowledge of mammalian biology, as soon as he understands the signal he will be able to distinguish between stallions and mares despite the fact that he has not yet seen a correctly labelled example of either type of horse. S_2, on the other hand, is a Type 2 signal since unless the Alphan sees the animals in the fields of Delta and Gamma or has seen them and remembers what they look like, he will not be able to correctly distinguish between stallions and mares. In the first case we will have *told* the Alphan the difference between stallions and mares *simpliciter* whereas in the second case we will have explained how to identify stallions and mares only by *showing* him correctly labelled examples. The distinguishing characteristic of a Type 2 signal is that understanding the signal is necessary but not sufficient for being able to correctly identify objects as X's. A Type 2 signal is successful only if it directs the attention of the recipient to a correctly labelled example of the term to be defined, and thus it involves a kind of ostension-at-a-distance.

With this distinction in hand we can now define the Kantian Hypothesis (KH) as follows:

KH: The meanings of 'right' and 'left' cannot be explained by means of any Type 1 signal, unless P already correctly understands at least one member of the chiral group.

The significance of the qualification referring to the 'chiral group' will become clear by considering Bennett's partial defense of KH.

II

In his paper, Bennett discusses how two speakers of English, the Alphan and the Betan, whose understanding of the language is slightly defective, could come to discover their respective mistakes. The Alphan has a perfect grasp of English except that he gives to 'right' the meaning of 'left' and vice versa. The Betan's understanding of English is perfect except that he gives to the form 'x is between y and z' the meaning that we give to 'y is between x and z'. If we define the 3-place predicate 'botween' so that x is botween y and z just in case y is between x and z, then we can say that the Betan has switched the meanings of 'between' and 'botween'. Now, as Bennett convincingly argues, when we talk with the Betan over a two-way radio, the Betan can only retain his idiosyncratic understanding of 'between' by changing the meanings of practically every other word in the English language. The sheer number of changes that the Betan will have to make will soon lead him to suspect that perhaps the problem lies in his single initial mistake. But what about the Alphan? Can he detect his semantic error as easily? Bennett thinks not. All the Alphan has to do in order to protect *his* error from detection is to systematically switch the meanings of pairs of terms in a small subset of the words in the English language. Let us call this subset the 'chiral group'. The chiral group includes such pairs as 'left/right', 'right-handed/left-handed', 'clockwise/anticlockwise', 'port/starboard', 'S magnetic pole/N magnetic pole' and so on. The chiral group has two important features. First, its members can all be explicitly defined in terms of just one member of the group. Thus, for example, if one correctly understands the meaning of 'right-handed' then through explicit definitions (that is, by means of Type 1 signals) one will also come to correctly understand the meanings of 'clockwise/anticlockwise', 'S magnetic pole/N magnetic pole', and so on. Second, the chiral group is semantically isolated from the rest of the language. If the Alphan is consistent and systematically switches the meanings of each pair of terms in the chiral group in response to our statements involving chiral terms, then his error will remain undetected. Thus, on the basis of these two features of the chiral group, it seems as if the Kantian Hypothesis is true. We will now consider the relevance of parity violation to this conclusion.

III

For simplicity of exposition we shall confine our attention to the classic experiment performed by Miss Wu and her associates in 1956 in order to test the conjecture by Lee and Yang that parity conservation is violated in weak interactions.[5] In this experiment, a sample of radioactive cobalt was placed at the center of a ring of current. The purpose of the ring of current is to create a magnetic field perpendicular to the plane of the current which orients the spins of the nuclei of the radioactive cobalt atoms. During the disintegration of radioactive cobalt a neutron in the atom's nucleus decays into a proton emitting an electron and an e-antineutrino. If parity were conserved in this reaction the emitted electrons and antineutrinos should emerge with equal probability, upward and downward, with respect to the direction defined by the magnetic field. In the experiment it was found that the electrons are emitted asymmetrically. The vast majority are ejected in the direction of the magnetic field. This result can now be used to send a signal to the Alphan that will enable him unambiguously to distinguish right from left since the mirror image reflection of the Wu experiment does not correspond to a physically possible set of events. We phrase the signal as follows:

S_3: Place a sample of radioactive cobalt in the center of a ring of electric current. Note the direction in which the majority of the electrons are emitted. Take a hand and curl the forefinger in the shape and direction of the electron flow in the ring of current. If the thumb points in the direction of maximum electron emission, then the hand is a left hand; if the thumb points in the opposite direction then it is a right hand.

The beauty of S_3 is that even if the Alphan systematically misunderstands all the chiral terms in our vocabulary (so that what we call a S magnetic pole, he calls a N magnetic pole and what we call a clockwise flow of electrons, he calls an anticlockwise flow of electrons, and so on) he cannot fail to correctly identify a right hand as a right hand on the basis of our instructions. This is evident from the fact that S_3 does not contain any chiral terms apart from those which are being defined. It is assumed, of course, that the Alphan has an impeccable knowledge of physics. More importantly, it is assumed that (a) the Alphan and his laboratory are made of matter and not of antimatter, and (b) the Alphan has not switched the meanings of 'electron' and 'positron'. Subject to these assumptions, the only way that the Alphan can now avoid discovering his mistake is if he makes changes in the non-chiral terms in his language and starts switching the meanings of words like 'maximum' and 'minimum'. But he will then be in the same hopeless bind as the Betan. That way lies only semantic chaos and, sooner or later, he will discover his mistake.

IV

Does the success of a signal like S_3 in explaining the meanings of 'right' and 'left' refute the Kantian Hypothesis? Clearly S_3 does not require that its recipient, P, already correctly understand any chiral term. But is S_3 a Type 1 signal? I think not. Like the Type 2 signal S_2, S_3 essentially involves ostension-at-a-distance. Unless the instructions in S_3 *are actually carried out* and the Wu experiment performed, P does not acquire the ability to distinguish right from left hands in the 'which is which' sense. Merely contemplating the instructions is insufficient. It is necessary that P actually build the apparatus and *see* which way most of the electrons are emitted. When the instructions are carried out, P acquires the ability to distinguish right from left at precisely the moment when P has before him a correctly labelled hand (or some other correctly labelled asymmetric object to which a hand may be spatially related).

Admittedly signals S_2 and S_3 differ with respect to the degree of physical contingency of the facts they presuppose. It is usual to think that the sex of particular horses is a mere contingent fact whereas the behavior of the electrons in the Wu experiment is a lawlike fact. And, given our usual views about the isotropy of space, it is surprising to learn that a distinction as seemingly arbitrary as that between left and right should reveal itself in a law of nature. Nonetheless, both S_2 and S_3 are Type 2 signals requiring actual experience of the relevant spatial objects to which our attention has been directed. Thus the Kantian Hypothesis survives unimpugned. It has not been established that explaining the meanings of 'right' and 'left' does not require a kind of showing and that it could be done by telling alone.

NOTES

[1] Jonathan Bennett, 'The Difference Between Right and Left', *American Philosophical Quarterly*, 7 (1970), pp. 175—191. This volume, pp. 97—130. For a comprehensive discussion of Kant's arguments from incongruous counterparts and their modern critics, see Jill V. Buroker, *Space and Incongruence* (D. Reidel Publishing Company, Boston, 1981).

[2] After discussing the significance of parity violation, Bennett writes: 'This certainly refutes the Kantian Hypothesis as I formulated it: we can now *tell* the Alphan which is which as between right and left. But then we could have told him anyway, using 'port' and 'starboard' ' (this volume, p. 128). As I explain below, any signal using words like 'port' and 'starboard' should not be regarded as providing a counterexample to the Kantian Hypothesis since these words are what I call 'chiral' terms. If one does allow the explicit definition of one chiral term like 'right' in terms of another like 'starboard' to serve as a counterexample to the Kantian Hypothesis, then the Hypothesis is

obviously false. The distinctive feature of signals based on parity-violating experiments is that they do not essentially involve any chiral terms and hence the threat they pose to the Kantian Hypothesis is nontrivial.

[3] This problem is essentially the same as Martin Gardner's 'Ozma' Problem: 'Is there any way to communicate the meaning of 'left' by a language transmitted in the form of pulsating signals? By the terms of the problem we may say anything we please to our listeners, ask them to perform any experiment whatever, with one proviso: *There is to be no asymmetric object or structure that we and they can observe in common*'. Martin Gardner, *The Ambidextrous Universe*, 2nd edition (Charles Scribner's Sons, New York, 1979), p. 155. This volume, p. 77.

[4] 'Seeing a correctly labelled X and t' is intended to include not only the case of P seeing an X at t and knowing at t that it is an X, but also the case in which P remembers at t what an X looks like and knows at t that it is an X. 'Seeing a correctly labelled X prior to t' means simply that P has seen an X prior to t and knew *at that time* that it was an X.

[5] For a lucid account of this and other relevant experiments, see Eugene Wigner, 'Violations of Symmetry in Physics', *Scientific American* 213 (1965), pp. 28—36 and Martin Gardner, *The Ambidextrous Universe, op. cit.*

JAMES VAN CLEVE

RIGHT, LEFT, AND THE FOURTH DIMENSION

Mollusk shells, narwhal tusks, twining plants, and human hands — all of these may come in pairs that are examples of what Kant called *incongruent counterparts*. The members of such pairs are counterparts in that one is a perfect mirror image of the other, yet incongruent in that one could never be made to occupy the region of space just vacated by the other. In 1768 Kant believed that the existence of such objects furnished proof of a Newtonian or absolutist as against a Leibnizian or relationist view of the reality of space: space is a thing in its own right, not just a construction out of material bodies and the relations among them. ("Absolute space has a reality of its own, independent of the existence of all matter.")[1] Kant's argument is worth examining both for its own interest and for its connection with other issues, such as the logical status of relations and the possibility of a fourth spatial dimension.

I. TWO SENSES OF 'RIGHT' AND 'LEFT'

Kant's favorite examples of incongruent counterparts were right and left hands.[2] To understand his argument, we must begin by distinguishing two different senses of the terms 'right' and 'left'. These terms can be labels either for *shapes* (as in 'Bring me a right glove') or for *directions* (as in 'Turn right at the corner'). Thus, 'right hand' can mean either a hand of a certain shape or a hand attached to a certain side of the body.[3] The shape sense is often overlooked. Dictionaries tend to define the direction sense to the exclusion of the shape sense, and children typically learn to distinguish right from left by direction before they realize that their hands have distinctive shapes.[4] But hands do have distinctive shapes; that is the point of Kant's observation that a right glove will not fit a left hand.[5]

We shall in fact have occasion later to reconsider whether 'left' and 'right' are ever names for shapes; the crucial thing to realize for now is that these terms do have a sense in which they mark *something* more, whatever exactly it may be, than a mere difference in direction from one's body. If you are not convinced of this, imagine that you lose your right hand in an accident and go to the hospital to get it sewn back on. You awaken later and find to your chagrin that the doctors have given you a left hand instead. It is fastened to your right wrist, of course, but that does not make it a right hand. The plain fact of the matter is that although in one sense you have a right hand and a left hand just like everyone else, in another sense you have two left hands. This further sense is what I am provisionally calling the shape sense, and it is with right and left in this sense that Kant's argument is concerned.

II. KANT'S ARGUMENT OF 1768

With this understanding, we may now proceed to set forth Kant's argument.

(1) A hand is left or right (as the case may be) either (a) solely in virtue of the *internal* relations among the parts of the hand, or (b) at least partly in virtue of the *external* relations of the hand to something outside it — if not other material objects, then Absolute Space.[6]

(2) But a hand is not right or left solely in virtue of its internal relations, since these are the *same* for right and left.[7]

(3) Nor is a hand right or left even partly in virtue of its relations to other material objects, since a hand that was alone in the universe would still be right or left.[8]

Therefore,

(4) A hand is right or left (as the case may be) at least partly in virtue of its relation to Absolute Space.[9]

Once (4) is established, one need only be granted the existence of left and right hands in order to infer the existence of Absolute Space.

Since the argument is plainly valid, there are four possible positions one can take in response to it. One can accept its conclusion, agreeing with Kant that right and left consist in external relations to Absolute Space; this position I shall call *absolutism*. Or one can reject its third premise, maintaining that right and left consist in external relations to material objects; this position I shall call *externalism*. Or one can reject its second premise, maintaining that right and left consist in internal relations among the parts of a thing; this position I shall call *internalism*. Finally, one can reject the first premise, maintaining that right and left do not consist in relations at all, but are irreducible monadic properties of hands as wholes; this position I shall call *holism*.[10]

Internalism and holism may be classified together as varieties of "intrinsicism," since according to them the leftness or rightness of a hand does not depend on anything outside it. Externalism and absolutism, by contrast, are varieties of "extrinsicism," since according to them the leftness or rightness of a hand *does* depend on something outside it. They differ only on the question as to what this something is.[11]

Let us now proceed to examine each of the positions.

III. HOLISM

According to holism, a hand's being right or left is a thoroughly nonrelational affair — neither a relation of the hand to something outside it, nor a result of the relations of the parts of the hand to each other. Rightness and leftness are monadic properties of hands and are not reducible to any properties of or relations among the parts. No one I know of has held this view, but it deserves to be discussed if only for the sake of completeness.

Holism collides head-on with the doctrine of *mereological reductionism* (MR). According to this doctrine, a whole's having a monadic property is *always* a matter of its parts having such-and-such properties and standing in such-and-such relations. Schematically, any truth of the form

$$Qw$$

(where Q is monadic) is equivalent to (or at least entailed by) some truth of the form

$$P_1x_1 \,\&\ldots\&\, P_nx_n \,\&\, R(x_1\ldots x_n),$$

where $x_1 \ldots x_n$ are parts of w none of which possess Q.[12] Thus something is a ladder if and only if it consists of rails and rungs related in a certain way, a hydrogen atom if and only if it consists of an electron orbiting around a proton, and so on.[13]

That holism conflicts with MR is not necessarily an objection to it, since MR is itself a controversial thesis. *Color*, as common sense conceives of it, is an exception to MR, since every part of a colored thing is itself colored. Moreover, even if we limit the application of MR to properties that are in some sense *spatial*, it will remain controversial. For it will imply that extended objects are composed of extensionless points — thus conflicting with the conviction of some that points are at best logical constructions and with the conviction of others that extensionless entities, even if they existed, could never "fill space through their mere aggregation."[14]

If holism is to be rejected, then, it cannot be on the basis of its conflict with any uncontroversial general doctrine of mereological reducibility.[15] We shall see presently, however, that if any version of intrinsicism is to be accepted, there is reason to favor internalism over holism. So I shall say no more about holism.

IV. EXTERNALISM

Externalism maintains that a hand's being left or right depends on how it is related to other material objects — in particular, on how it is related to other asymmetrical objects like human bodies. This is the position Kant sought to refute by the famous thought-experiment of the solitary hand. Imagine a hand all alone in the universe; would it not be either left or right? If so, handedness cannot depend on relations to other material objects, for *ex hypothesi* there are none.

Some of Kant's critics, notably Peter Remnant and Martin Gardner, have responded to this thought-experiment by simply denying that a solitary hand could be left or right.[16] According to them, a solitary hand would simply be indeterminate or neuter. Their position is bound to appear at first sight as a desperate denial of the obvious. How, after all, could a hand be neither left nor right — unless it were a *freak* hand with (say) its thumb protruding from the middle of the palm?

To scotch the idea that a normal hand could be indeterminate, Kant notes that in that case "it would fit either side of the human body, which is impossible."[17] On the strength of this remark, Gardner takes Kant to be arguing as follows: Consider a universe whose only material object is a human hand. If a handless human body were introduced alongside the original hand, the hand would have to fit either the right or the left wrist of the body, and would therefore be either right or left. But the introduction of the body would not change the hand in any way; therefore, it must have been right or left all along.

If this were indeed Kant's reasoning, it would be open to the following parody: Consider a universe that contains nothing but a single ball. If another object, larger or smaller than the ball, were introduced into the universe, the ball would (in relation to it) be either large or small. But the introduction of the second object would not change the ball in any way. Therefore, the ball must have been large or small all along.

Here the conclusion is absurd and the fallacy leading to it not hard to spot. Nothing is large or small *simpliciter*, but only in relation to something else, as the parenthetical addition acknowledges. Consequently, the addition of the second object (call it O), though it makes no change in the nonrelational properties of the ball, endows it for the first time with the property of being large or small in relation to O.

Now can Kant's argument really be judged fallacious in a parallel way? If so, it would presumably be on the ground that 'right' and 'left', like 'large' and 'small', are implicitly relational predicates. 'Left hand' would have to be regarded as elliptical for 'left hand in relation to x'. It is not easy to see how this could be the case. If 'left hand' were short for 'left hand in relation to x', it would have to be possible for the same hand to be left in relation to one thing, x, and to be right in relation to another thing, y. For what choices of x and y could this be true?

Some of Remnant's and Gardner's remarks suggest an answer to this challenge. According to them, a hand is a right (left) hand in relation to body B if and only if it properly fits the wrist on the side of B opposite (nearest) the heart. To fit a wrist properly is to be attachable to it when the palm is placed on the chest with the thumb pointing up. (This assumes a limit to the flexibility of the wrist joint, but let that pass.) Given these stipulations, it is easy to imagine the needed x and y. A hand that fits the wrist opposite the heart on a normal human body (with heart on the left) will fit the wrist nearest the heart on an abnormal human body (with heart on the right). The same hand will therefore be right in relation to a normal human body, left in relation to an abnormal one.

It should now be clear that something has gone wrong. Remnant and Gardner call a hand left or right not according to its shape, but according to its relation to the heart. If a surgeon moves my heart from one side of my chest to the other, he will not affect the shapes of my hands; but he will affect their designation as left or right according to Remnant's and Gardner's stipulations. This shows that their stipulations do not govern 'right' and 'left' in the shape sense at all. But our question was whether a hand can be right-*shaped* in relation to one body and left-*shaped* in relation to another. Remnant's and Gardner's observations are thus entirely beside the point.

I must hasten to add that I have not yet done justice to Gardner's critique of Kant. His criticisms are based in part on facts about the fourth dimension, about which I have so far said nothing, and also on a certain doctrine about relations, about which I have also said nothing. So I will return to Gardner's defense of externalism once I have said something about these further topics.

V. INTERNALISM

If the rightness or leftness of a hand is not constituted by its relations to other material objects, perhaps it is constituted by the relations among its own parts. This is the answer given by internalism, which we shall now examine.

Kant's argument against internalism was that internal relations are the *same* for a right and a left hand, so cannot be the basis of the distinction between them. Evidently he assumed that internal relations are exhausted by relations of distance and angle – for example, the distance from middle of wrist to tip of middle finger, the angle between inside edges of thumb and forefinger. It is mathematically demonstrable that these relations are indeed the same for right and left; they are preserved under any mapping that takes a hand into its mirror image.

But why must internal relations be limited to relations of distance and angle? Why may they not also include relations of *direction*? Then we could give the following simple-minded analysis:

A hand is a right hand if and only if: when its fingers point up for an observer facing its palm, its thumb points to the observer's right.

There is an appearance of circularity here, but it is merely verbal; what we are doing is defining 'right' in the shape sense in terms of 'right' in the direction sense.

The reference to an observer in the definition just given is not essential, for (as Bennett has pointed out) it could be replaced by reference to any object sufficiently asymmetrical to define directions in three dimensions. For instance, we could give the following definition instead:

A hand is a right hand if and only if: were it placed palm down against the face of a standard clock with its heel at 6 and the tip of its middle finger at 12, the direction from thumb to little finger would be the same as the direction from 9 to 3.

Kant was aware that incongruent counterparts could be distinguished by "the definite direction in which the arrangement of their parts is turned."[18] So what is wrong with the definitions just given? Kant would have said: nothing, except that they are not *internalist* definitions. He thought that in appealing to direction one was tacitly appealing to Absolute Space: "the region towards which this ordering of the parts is directed involves reference to . . . universal space as a unity."[19] To appeal to direction would thus be to give in to absolutism. This contention of Kant's is by no means obvious, however; I shall return to it in Section VIII.

Others will object to the definition of right and left in terms of direction for a different reason. To specify direction we must refer to clocks, observers, or other objects outside the hand; are we not thus reverting to externalism?[20]

This objection may be met by distinguishing two senses in which a property may involve an external relation. Let us say:

A property P is *existentially relational* if and only if: it is impossible for an object x to have P unless there also exists some concrete object y lying wholly outside x.

Examples of such properties are *being the father of Socrates* and *being six feet from an oak tree*. And let us also say:

A property P is *referentially relational* if and only if: it is impossible to think of an object x as having P unless one also refers in thought to some thing, stuff, or kind of thing in addition to x.

Examples of such properties are *being water-soluble* (since you cannot think of sugar as being water-soluble without thinking of water) and *thinking of a unicorn* (since you cannot think of anyone as thinking of a unicorn unless you think of a unicorn yourself).

This distinction is meant to recall Brentano's distinction between the *Relativ* and the *Relativlich*.[21] The important point about it for our purposes is that a property can be referentially relational without being existentially relational: if a cube of sugar were the only thing in the universe, it could still be water-soluble.

The objection that to define right and left in terms of direction is to revert to externalism may now be answered. Having some parts lying in certain directions from other parts, as specified by reference to clocks, observers, or what have you, involves existentially relational properties of the parts, but only referentially relational properties of the whole. This rescues the internalist, for it shows that the relations he uses to define rightness and leftness do not depend on the *existence* of objects outside the hand.

Even if I am wrong about this, however, the day is not yet lost for the internalist. John Earman has suggested that the internalist should recognize *standing in a left-configuration* and *standing in a right-configuration* as primitive internal relations, not reducible to any other relations and not supervenient on relations of distance and angle.[22] If this move is allowable, the internalist has another way of evading Kant's argument against him.

It should not be thought that positing primitive internal relations is simply *ad hoc*, for the following argument makes it plausible that there must be *some* internal relations that constitute the basis of left and right: "Take a right hand, remove its thumb, and re-attach it on the other side of the palm; the result will be a left hand. (It may be attached to the right side of the body, but its shape will be that of a left hand.)[23] Thus, by rearranging the parts of a hand — that is to say, by altering its internal relations — you can change it from right to left. Conversely, rearranging the parts of a hand seems to be the *only* way to change it from right to left; how else would you do it? So an appropriate change in internal relations brings about a change in handedness, and a change in handedness is possible only through a change in internal relations. We may thus conclude that handedness is a function solely of internal relations."

If it does nothing else, I think this argument shows that internalism is more plausible than holism. We are about to see, however, that there are serious challenges to intrinsicism as such. To see what they are, we must take an excursion through the fourth dimension.

VI. THE FOURTH DIMENSION

Writers on incongruent counterparts often bring in the idea of a fourth dimension, typically believing that this idea has some adverse bearing on Kant's argument, but typically also not saying very much about just what this bearing is. I shall try to remedy this defect in what follows.

It must be emphasized that what we are about to discuss is a fourth dimension of *space*. It has nothing to do with time. If time is to be treated as a dimension, let it be the *fifth* dimension; it is not what we shall be concerned with here.[24]

What for our purposes is the salient fact about the fourth dimension was first pointed out by the mathematician Möbius in 1827. It is this: in a four-dimensional space, a right hand could be "flipped over" so as to become its own incongruent counterpart. The point may be put more generally and in mathematical language as follows: for any reflective mapping of an n-dimensional object, there are combinations of translational and rotational mappings in (n + 1)-dimensional space that will give the same result. Reflective mappings take an object into its incongruent counterpart; translational and rotational mappings give the same result as motions. We can therefore put the point in Bennett's dramatic way: in four-dimensional space a right hand could become a left by *sheer travel*.[25]

Now this is food for thought indeed. "Sheer travel" means *rigid motion*, that is, motion that leaves the shape of the moved object unchanged. How, then, can sheer travel change the shape of a hand? The answer, of course, is that it cannot. If sheer travel changes right into left, there must not have been any difference in shape between right and left to begin with.

What then becomes of the distinction I drew in Section I between the shape and the direction senses of 'right' and 'left'? Well, there is still a difference between right and left hands that amounts to something more than a difference in the sides of one's body to which they are attached, but it now amounts to something less than a difference in shape. For lack of a better term, I shall call it a difference in *orientation*. Do not be misled by the term: not just any rotation will produce a change in what I am calling orientation. For instance, rotations confined to three-dimensional space will not alter the orientation of a hand.

All of this can be more readily grasped by means of lower-dimensional analogies. Consider two back-to-back L-shaped figures in a plane:

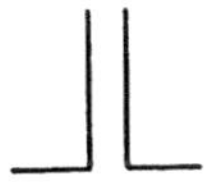

To a Flatlander — a being unable to perform or visualize any motions through the third dimension — the L's might appear to be of different shapes. He might call one a left L and the other a right L. If we were to pick up a right L and put it back down as a left L, he might think that a right L had undergone a sudden and mysterious change of shape. But he would be wrong; we who inhabit three dimensions can see that the L's differed merely in orientation, not in shape. And of course the status of hands in four dimensions would be just the same as that of L's in three.

Four-dimensional spaces are not the only spaces in which a right hand could become left by sheer travel; the same is also true in *non-orientable* spaces. The most familiar example of a nonorientable space is the two-dimensional Möbius strip, in which a right L becomes left after one circuit. In similar fashion, a right *hand* would become left after one circuit around a *three*-dimensional nonorientable space. But for our purposes it will not be necessary to give separate consideration to nonorientable spaces. The reason for this is that a nonorientable space of n dimensions is possible only in an ambient space of n + 1 dimensions. (If you haven't got three dimensions to work in, you can't make the twist necessary to create a Möbius strip.) The possibility of three-dimensional nonorientable spaces is thus dependent on the possibility of four-dimensional spaces, and therefore provides no new avenue of objection of Kant's argument.

Let us now see what the consequences of the four-dimensionality of space would be for the "isms" I have discussed so far.

VII. INTERNALISM AND THE FOURTH DIMENSION

The argument for internalism I presented at the end of Section V used the premise that the *only* way to change a right hand into a left is to rearrange its parts. We now see that in a four-dimensional space this would be false, since one could achieve the same result by a simple rotation. For the same reason internalism itself would be false, for one could change handedness without changing that upon which, according to internalism, handedness depends.

What, then, is wrong with the purported definitions I gave of right and left in terms of internal relations? I said that a hand is right if and only if: were its fingers to point up for an observer facing its palm, its thumb would point to the observer's right. We can now observe that this definition would meet the same fate in four dimensions as would the following in three:

An L is a right L if and only if: if its long leg were pointing up for an observer, its short leg would point to the observer's right.

For an observer situated as specified by the definiens, the short leg *might* point to the right, but it equally well might point to the left; it all depends on the observer's vantage point. So it is false that the leg *would* point right, whence it follows that *no* L has the property defined.

Similarly, in a four-dimensional space no *hand* would be right in the sense defined earlier, since for an observer situated as specified by the definiens, the thumb of the hand might point in either direction.[26] So no hand would be right in the sense defined, which shows that the definition fails to capture whatever it is that we do express by the terms 'right' and 'left'.

One could avoid this difficulty by adopting the following style of definition instead:

If the fingers of h point up for o and o is facing the palm of h, then h is *right for o* if and only if the thumb of h points to o's right.

The property defined by this definition *is* sometimes instantiated, but it is also clearly a property involving an external relation, so there is no help here for the internalist.

We may note that the considerations here adduced to refute internalism apply also to holism. The monadic properties of a hand are as unaffected by motion as its internal relations, so if mere motion can change right into left, holism is false.

VIII. EXTERNALISM AND THE FOURTH DIMENSION

In Section IV I accused Kant's externalist critics of confusing the shape and the directional senses of 'right' and 'left'. By taking it for granted that right and left are shapes, I also made Kant's case against externalism look compelling indeed, for the shape of a thing is surely a property it would have even if nothing else existed. We see now, however, that our dismissal of externalism was premature, since in a four-dimensional world right and left would be matters not of shape, but orientation. We must find out what happens to Kant's thought-experiment if handedness is merely a matter of orientation.

When two objects are the same or different in color, this is presumably because each of them *has* a color. Moreover, the color of each is a property it would have even if the other did not exist. Can we say similarly that if two hands are the same or different in orientation, this is because each of them *has* an orientation, and that the orientation of each is a property it would have even if the other did not exist?

A piece of terminology will help us to clarify what is at issue here.

A relation R is *grounded in its terms* (or *grounded*, for short) if and only if: it is necessarily the case that whenever anything x bears R to another thing y, there are nonrelational properties F and G such that (i) x has F, (ii) y has G, and (iii) x's bearing R to y is entailed by the conjunction of (i) and (ii).

A relation thus grounded is one that supervenes upon certain non-relational characters of its relata. It is an "internal" relation in one traditional sense of that term. *Being the same in color as* is a plausible example of a grounded relation, since it obtains only in virtue of its terms' both being red or both being blue, etc. On the other hand, *being six feet from* is a plausible example of an ungrounded relation; it could cease to hold without there being any intrinsic changes in its relata.[27]

The question I asked one paragraph back can now be put thus: is *being the same (or different) in orientation* a relation that is grounded in its terms? I think externalism is put in the best light if we see it as answering this question in the negative. Consider the following passage from Gardner:

It is meaningless to speak of [a hand] as left or right if there is no other asymmetric structure. . . . It is only when two asymmetric objects are present in the same space, and a choice of labels has been made with respect to one that labels applied to the other cease to be arbitrary.[28]

In one sense, of course, *all* labels are arbitrary; there is no inherent connection between a word and the feature it stands for. But Gardner must obviously have something deeper in mind. I think what he is getting at is this: it is arbitrary to label one hand in a pair 'left' not because word-feature correlations are arbitrary, but because there is no *feature* of a hand to fix a label to. The only fact of the matter is that the two hands are the same or different in orientation, and this sameness or difference is an ungrounded relation. It is, in Leibniz's phrase, "an accident with a leg in each of two substances" — *and a foothold in neither.*

How plausible is Gardner's view? To test our intuitions, let us imagine a universe that contains nothing but arrows, some pointing up and others pointing down. Up- and down-pointing arrows are related in a space of two or more dimensions as left and right hands are related in a space of four dimensions: they are the same in shape, but opposite in orientation. Now we can say of two arrows that they are the same or oppositely directed, but can we say of a *single* arrow that it has a direction? That its pointing a certain way is a purely nonrelational fact about it? Intrinsicists would have to answer these questions "Yes,"[29] but I think most people would sooner side with Gardner and say that a solitary arrow could not point any way at all.[30]

Of course, an intermediate position is possible. One may agree that the idea of a purely nonrelational orientation makes no sense, yet feel at the same time that two arrows cannot point the *same* way unless each points *some* way. "Pointing some way" cannot be purely nonrelational, yet must somehow be a fact about arrows taken singly. It must therefore be a fact about the relation of the arrow to a *special* entity – Absolute Space.

This line of reasoning is tailor-made to secure Kant's own absolutist position. Perhaps it is what anchors his claim, mentioned in Section V, that the concept of direction presupposes Absolute Space. But it is unraveled by the following dilemma. Is the relation of an arrow to Absolute Space that constitutes its pointing some way a grounded relation or not? If it *is*, then why can't the nonrelational features of two arrows that ground their respective relations to Absolute Space ground their relation to each other, without the help of Absolute Space?[31] This would give us a form of intrinsicism. If on the other hand the relation is *not* grounded, then why can't the relation of pointing the same way be an ungrounded relation that holds directly between the two arrows, again without the help of Absolute Space? This would land us in externalism. Either way there is no need to inroduce Absolute Space as a special term of relations. This, it seems to me, is the fundamental dilemma facing absolutism.

Returning to Gardner and summing up, his case for externalism is best seen as resting on two contentions: when we think of hands as being embedded in four-dimensional space, we see (1) that being left or right is a matter of orientation, and when we contemplate one-dimensional oriented objects such as arrows, we see (2) that orientation is an ungrounded relation. Putting (1) and (2) together, we see that there is nothing about a hand in isolation that can make it left or right. So Kant's famous thought experiment disintegrates.

IX. MUST KANT'S CRITICS PRESUPPOSE THAT THE FOURTH DIMENSION IS ACTUAL?

At this point there may be readers who are inclined to expostulate as follows: "It is no doubt true that *if* there were a fourth dimension, Kant's argument would break down in the way you say. But what follows from this? Nothing, unless you are prepared to assert that there really *is* a fourth dimension, that it is a genuine feature of the space in which we live and move. If the fourth dimension is only a mathematical abstraction, Kant's argument against externalism remains sound." The same could also be said in defense of my argument for internalism.

If partisans of the fourth dimension must indeed affirm its actuality in order to make their point, they are swinging on a thin vine. Carpenters cannot build houses with more than three beams meeting at right angles. Planets attract each other in inverse proportion to the *square* of the distance between them, not, as one would expect in four-dimensional space, in inverse proportion to the cube.[32] Three numbers, latitude, longitude, and altitude, are all it takes to fix one's position on earth. On empirical grounds if no other, the existence of a fourth dimension is dubious.

Moreover, even if there did turn out to be a fourth dimension, that would not be enough to block Kant's argument: there would also have to be a fifth dimension, a sixth, and so on *ad infinitum*. For suppose there were only n dimensions, for some natural number n greater than three: in that case, even though right and left hands would no longer be incongruent counterparts, suitably asymmetric objects of n dimensions *would* be. Concerning these objects Kant's argument could proceed as before, this time without any higher dimension for the critic to appeal to. So it appears that in order to prevent Kant's argument from being successful at any level, the critic would have to maintain that physical space has not just four, but *infinitely many* dimensions.

This, at any rate, is the extreme to which the critic would be driven if the *actuality* of the fourth dimension were essential to his case. But perhaps all that is really essential is the more modest assumption that the fourth dimension is *possible*.[33] If so, he is on much firmer ground. The established opinion among philosophers and physicists nowadays is that although there may be every reason to believe that our space is *in fact* three dimensional, there is no *necessity* that this be so. It is a logically contingent matter how many dimensions space has, just as it is logicallly contingent whether Euclidean or some other geometry holds of it. Let us therefore see what happens to the positions discussed so far if the fourth dimension is supposed to be, not indeed actual, but possible.

X. INTERNALISM RECONSIDERED

"Only by rearranging the parts of a hand can you convert it from right to left": this premise of the internalist's argument, though false as we saw in a world of four dimensions, remains true in one of three. Does this mean that if four-dimensional spaces are only possible, not actual, internalism is restored?

The answer is no. Internalism maintains handedness to be a function of internal relations in the following sense: any right (left) hand has internal relations of such a sort that any hand that had internal relations of the same sort would *necessarily* also be right (left). In other words, the internal relations of a hand *entail* its handedness. But we have now seen that there are possible spaces in which a solitary hand would be neither right nor left; hence a hand with the same internal relations as a given right hand might fail to be right, for it might be the sole occupant of such a space. So the rightness of a right hand is not entailed by its having the internal relations it does, contrary to internalism.

What, then, is wrong with my internalist argument? The answer is that it is simply invalid. In the actual (three-dimensional) world, a hand whose internal relations are held constant cannot indeed be right at one time and left at another; but from this it does not follow that a hand whose internal relations are held constant cannot be right in one *world* and *nonright* in another. The latter is what internalism would require.

There is evidently just one assumption on which internalism could still be defended, and that is the old-fashioned assumption that space is necessarily three-dimensional. The defense would go as follows: (1) a hand's having a certain set of internal relations *together with* its inhabiting a three-dimensional space *does* entail its being a right hand; (2) if P and Q together entail R, and Q is a necessary truth, then P by itself entails R; therefore, (3) if space is necessarily three-dimensional, having a certain set of internal relations is enough by itself to entail being a right hand — just as internalism maintains.[34]

XI. CONGRUENCE AND SUPERPOSABILITY

Before we reconsider externalism we must take up one other issue. Kant says that objects are congruent only if they "allow of superposition."[35] Now exactly what requirement is he laying down? We can distinguish two alternatives. According to the stronger, two objects are congruent only if they can be superposed or made to coincide *in the space they actually inhabit*; according to the weaker, objects are congruent if there is a *possible* space in which they could be made to coincide. For Kant, of course, who believed only one space to be possible, this distinction would not have mattered, but we moderns are obliged to consider it.

The stronger of the two conditions for congruence seems to me to be *too* strong. Is it not a sufficient condition for the congruence of two objects that either of them *could* (without changing shape, of course) exactly occupy the region now occupied by the other, regardless of whether any route through the surrounding space would permit such an interchange to occur? Recall the back-to-back L's we discussed earlier, and pretend that the plane in which they lie is the whole of space. In this case there would be no motion through space that would enable either L to take the place of the other, but wouldn't they be congruent nonetheless? Or imagine that there is a space shaped like an hourglass, each chamber of which contains a ball too large to pass through the neck. Mightn't the balls be congruent, even though they can never trade places? In short, it is the possibility of *fitting there*, not the possibility of *getting there*, that should count.[36]

Consider in this connection Wittgenstein's remark at *Tractatus* 6.36111:

> The right hand and the left hand are in fact completely congruent. It is quite irrelevant that they cannot be made to coincide.

If Wittgenstein's 'cannot' means 'cannot, given the nature of the surrounding space' (as opposed, say, to 'cannot, given the laws and forces that are operating'), then he and I are advocating the same point.

XII. EXTERNALISM RECONSIDERED

We can now take up the question: how does externalism fare under the assumption that the space we inhabit is actually three-dimensional, but possibly four-dimensional? We get one answer if superposability (that is, superposability in the actual space) is required for congruence and another if it is not.

If superposability is *not* required (as I recommended in the last section), the case for externalism can proceed more or less as it did on the assumption that a fourth dimension is actual. For if the mere possibility of a space in which two objects could be made to coincide suffices for their congruence, and if four-dimensional spaces are indeed possible, then right and left hands are congruent even in a three-dimensional world. This means that the difference between them is merely one of orientation, in which case the argument of Section VIII stands.

Suppose, however, that congruence *does* require superposability, as is more commonly held.[37] In that case right and left will be incongruent for want of a space in which to superpose them. If we assume that objects of the same size are incongruent only if they differ in shape, and that two objects differ in shape only if each of them *has* a shape, it will follow that right and left are shapes, just as one would think who had no notion of higher dimensions. This will enable Kant's thought-experiment against externalism to be reinstated, for a hand surely has whatever shape it does regardless of what other material things exist.[38]

The only way out I can see for the externalist at this juncture is to deny that two hands *differing* in shape must each *have* a shape. In other words, he must hold that sameness or difference in shape, just like sameness or difference in orientation, is an ungrounded relation. This move saves externalism, but at considerable cost. For the externalist is now declaring impossible not just a universe containing a solitary right hand, but even a universe containing a solitary sphere or cube.[39] A solitary object must be devoid of shape!

XIII. SUMMARY OF RESULTS TO DATE

Which of our "isms" is most reasonable depends on the answers to several questions, as indicated in the diagram below. I am assuming that orientation is an ungrounded relation; if a parting of the ways were permitted on this question, too, the diagram would become more complicated, internalism appearing as an alternative fork alongside externalism at the first two places where externalism now appears alone.

Is a fourth dimension possible? — no → Internalism (Section X)

yes ↓

Is a fourth dimension actual? — yes → Externalism (Section VIII)

no ↓

Is superposability required for congruence? — no → Externalism (Section XII)

yes ↓

Are congruence and incongruence grounded relations? — no → Externalism (Section XII)

yes ↓

Not externalism (Section XII), but what? For the answer, read through Section XVI.

The reader will note that absolutism appears nowhere in the chart. Are its prospects really that dim? It is time to examine a vigorous contemporary defense of Kant.

XIV. NERLICH'S ARGUMENT FOR ABSOLUTISM

Those who bring up the fourth dimension in connection with incongruent counterparts generally do so for the purpose of refuting Kant, but there is a notable exception. Graham Nerlich has argued that the fourth dimension is a help to Kant's cause, not a hindrance, and that his case for absolutism, though mistaken in some of its assumptions, is correct in both strategy and outcome.[40]

Nerlich's argument can be summed up briefly as follows. Kant's premise that a solitary hand would have to be left or right is false, since spaces are possible in which it would be neither. It is nonetheless true that a solitary hand would have to be either *enantiomorphic* or *homomorphic*. An object is enantiomorphic if it *could* have an incongruent counterpart, homomorphic otherwise; for a hand, being enantiomorphic is equivalent to having the disjunctive property of being either right or left. Now whether a hand is enantiomorphic depends on the nature of the surrounding space: in an orientable space of three dimensions a hand will be enantiomorphic, whereas in a four-dimensional or nonorientable space it will be homomorphic. Furthermore, if a hand *is* enantiomorphic, and thus either left or right, *which* of these it is will depend in part on how it is "entered" in the containing space.[41] So there are some features a hand *must* have (enantiomorphism or homomorphism) and others it *may* have (leftness or rightness) that are partly determined by its relation to the ambient space.

Nerlich's argument as he presents it depends on an assumption I have called into question, namely, that objects are congruent only if there is a way to bring them into coincidence. Without this assumption he could not maintain that hands are enantiomorphic in three dimensions and homomorphic in four, since left and right hands would be congruent with each other (and thus individually homomorphic) in spaces of any dimension. But Nerlich's argument can be restated without this assumption, for there is another property that hands (strictly, pairs of hands) possess in three-space, but not in four-space. Two hands are either alike or different in orientation. If the hands are embedded in three-space, their mutual orientation will be permanent; if they are embedded in four-space, it will be alterable by motion. So here is a feature of hands that is partly determined by the nature of the surrounding space: their being permanently fixed in orientation or not.

Does this mean that the absolutist is right, and that we must countenance Absolute Space as an entity in its own right? Such a conclusion would be premature. As Lawrence Sklar has pointed out, everything now hinges on whether the features of space on which enantiomorphism depends can be spelled out using only the relationist's resources — material objects and the actual and possible relations among them.[42] If they can be, the facts noted by Nerlich will be no argument for absolutism. (Compare: whether a man is of greater than average height could be said to depend on his relation to the Average Man, but this by itself is no argument that there *is* such a thing as the Average Man; it all depends on whether our purported references to that entity can be paraphrased away.)

So the case for absolutism now depends entirely on whether the relationist can give a satisfactory account of dimensionality — one that does not mention space itself, but only material objects and the relations among them (their actual and possible configurations).[43]

XV. CAN THE RELATIONIST DEFINE DIMENSION?

One of the oldest criteria of dimensionality is this: how many mutually perpendicular lines can meet in one point? One finds versions of this in Ptolemy, Galileo, Leibniz, and Kant.[44] Let us see if the relationist can make use of this criterion.

Note first that if the relationist talks of lines, he must mean *material* lines — lengths of matter rather than lengths of space. If the notion of a material line is problematic, he will have to talk about rods.

Note second that it would not do simply to say this: space is n-dimensional iff there are in fact somewhere n lines or rods meeting at right angles, but nowhere any more. If all the matter in the universe were compressed into a single plane, there would nowhere be three such lines; but space would still be three-dimensional provided one *could* erect a line perpendicular to the plane. As Leibniz remarked, "Space . . . is nothing at all without bodies, but the possibility of placing them."[45]

What this suggests is that the relationist should define dimensionality as follows:

Space has n dimensions if and only if: it is possible to bring n rods, but no more, into mutual perpendicularity.

Note that this is a contextual definition, the term 'space' not occurring on the right-hand side.

I should now like to present a dilemma that I think confronts the definition just given and any similar attempt to define dimensionality in terms of possibility. Is the possibility we are talking about *physical* possibility or *logical* possibility? If *physical* possibility is what is meant, the analysans does not give a sufficient condition for the analysandum. We just noted that there could be room (*Raum*) for a rod even where none is actually placed; could there not just as well be room for a rod even where it is physically impossible to put one? The laws of nature might be such that whenever three rods are brought into mutual perpendicularity, a force is set up that prevents the entry of a fourth at right angles to the original three. In this case the analysans would be true and the analysandum false.[46]

On the other hand, suppose what is meant is *logical* (or metaphysical) possibility. In that case the analysans, if true at all, would be *necessarily* true, since no statement ascribing a logical modality is contingent. But then the analysandum likewise would have to be necessarily true if true at all. In short, it would no longer be contingent how many dimensions space has, which flies in the face of the established view.[47]

I see only one way for the relationist who wants to uphold the contingency of dimension to avoid this dilemma, and that is to go between the horns by positing a new kind of possibility: *spatial* possibility. Spatial possibility would of course have to be *sui generis*, not reducible to either physical or logical possibility. It would be intermediate in strength between them, in the sense that the physical possibilities would be a proper subset of the spatial possibilities, which would in turn be a proper subset of the logical possibilities. The analysis could then be given as follows:

Space has n dimensions if and only if: it is spatially possible for n rods, but no more, to be mutually perpendicular.

Since statements of spatial possibility and impossibility would be logically contingent, the dilemma above would be avoided.

Of course, the absolutist is likely to retort that 'spatially possible' can only mean 'possible given the nature of the surrounding space', in which case the relationist has not really avoided commitment to Absolute Space after all. In reply, the relationist could perhaps point to the relationship between 'it is logically possible that P' and 'there is a possible world in which it is true that P': it can be argued that the former of these is ontologically and conceptually basic, and that the latter is merely a picturesque way of speaking. Well, then: if logical possibility can be taken as basic, not needing to be explained in terms of a framework of possible worlds, why cannot spatial possibility be taken as basic, not needing to be explained in terms of a framework of substantival space?

I shall not take sides on the question whether spatial possibility is a legitimate primitive. My purpose in introducing the idea is only to exhibit the move I think a relationist must make if he wishes to keep dimensionality contingent. If, as a referee has suggested, the move is "extravagant," then we should either embrace Absolute Space or affirm that the number of dimensions is no contingent matter.

To sum up this section: If the relationist *can* define dimension, whether by appeal to spatial possibility or in some other way, Nerlich's argument will give us no reason to believe in Absolute Space. If on the other hand the relationist *cannot* define dimension, there will be reason to believe in Absolute Space, though not necessarily in absolutism as here defined, that is, the view that right and left consist in relations to this Space. (One of our other "isms" may still be the correct account of right and left, depending on the questions canvassed above (Section XIII). Perhaps features other than right and left, for example, being permanently fixed in mutual orientation, will consist in relations to Absolute Space.)

XVI. THE INCONGRUITY OF CONGRUENCE

We left unsettled in Section XIII which position should be filled in at the bottom of the flow chart. It may have occurred to the reader at the time that there is something anomalous about the combination of assumptions leading to that slot. If congruence and its complement are grounded relations (which was one of the assumptions), there must be intrinsic properties of congruent objects that guarantee their congruence. If congruence requires superposability (which was another of the assumptions), then whether two objects are congruent depends on the nature of the space they inhabit. How can congruence be grounded in intrinsic features of objects and at the same time be dependent on the nature of the surrounding space? This is the incongruity of congruence.

A resolution of this paradox is possible if dimension is defined in terms of spatial possibility, as proposed in the last section.[48] To see how it works, we must first distinguish two senses in which a property may be intrinsic. Let us say that a property is *intrinsic*$_1$ if something's having it does not depend on the existence of any outside entity, extrinsic$_1$ otherwise. An intrinsic$_1$ property is one that is not existentially relational in the sense defined in Section V. And let us say that a property is *intrinsic*$_2$ if something's having it does not depend on the obtaining of any extraneous contingent fact, extrinsic$_2$ otherwise. Every intrinsic$_2$ property is also intrinsic$_1$, but not conversely: *being such that there are no unicorns* is intrinsic$_1$, but not intrinsic$_2$.

Now what sort of property is it to inhabit a space of (say) three dimensions? Call this property P. If the absolutists are right, P relates its bearer to an outside concrete entity, and is therefore extrinsic$_1$. But suppose for a moment that a relationist account of n-dimensional space can be given in terms of spatial possibility. In that case, something's having P would depend to be sure on an extraneous fact, but it would be a purely *modal* fact, not an existential one. P would therefore merely be $extrinsic_2$, not $extrinsic_1$.[49]

Here, then, is one resolution of the paradox: we can grant that congruence depends on the nature of the surrounding space and still cite intrinsic properties of objects that guarantee their congruence just so long as properties like P count as intrinsic. (P alone won't ground any relation of congruence, but it will be a necessary conjunctive part of any property that does.) And P *does* count as intrinsic in the relevant sense: on our current hypothesis it is $intrinsic_1$, and $intrinsic_1$ properties are all we need to ground relations.[50]

Now we can say something about what position to enter at the bottom of the chart. It is not any position to which I have so far given a name. It is not either variety of extrinsicism, since there is no dependence of handedness on any outside entity. Nor is it either of the varieties of intrinsicism, since handedness is not a function of intrinsic properties *alone*. Instead it is a function of $intrinsic_1$ properties together with extraneous, contingent, nonexistential fact.

XVII. CONCLUDING REACTIONARY POSTSCRIPT

I have assumed until now that spaces of more than three dimensions are at least possible. This is the prevailing view nowadays, but I would like to close with a protest against it.

It is sometimes thought that the possibility of higher dimensions is established by the mere fact that geometries of arbitrary dimension are formally consistent. This is not enough to settle the question, however, since formal consistency is no guarantee of metaphysical possibility. There are perfectly consistent systems of logic in which the law of excluded middle is denied or deleted; the mere existence of such systems does nothing to refute those who regard that law as necessary and exceptions to it as impossible.

But what reason is there to doubt the possibility of higher-dimensional spaces? None except this: such spaces defy visualization. The standard reply to such doubts is Edwin Abbott's little classic, *Flatland*, first published in 1884. Abbott's Flatlanders, whom we encountered briefly in Section VI, are two-dimensional beings "so restricted in sight and motion that they cannot look out of, or rise or fall out of, their thin, flat universe." They are as much incapable of visualizing the third dimension as we are the fourth; but of course if they were to deny the possibility of the third dimension (as nearly all of them do), they would be profoundly mistaken.

Writing on the philosophy of space in the century since *Flatland* has tended to discount altogether the significance of what can or cannot be visualized. In my opinion, much of this dismissal, though perhaps warranted in the end, is premature, since it overlooks an important distinction. The Flatlanders are said to be unable to visualize three-dimensional structures, or to see that a third dimension is possible, and such inability, I admit, is not a good reason for doubting the possibility of anything. But the Flatlanders' inability to see how certain things are possible must be distinguished from the ability to see positively that certain things are *im*possible. As Pierre Bayle noted, "There is a great deal of difference between not understanding the possibility of a thing and understanding the impossibility of it."[51]

To illustrate the distinction, consider first the question "Could there be a closed curve containing no four points that form the vertices of a square?" I cannot see how such curves are possible, but I should not be greatly surprised to learn that they exist. (It is an unsolved problem whether any do.) The reason, of course, is that I do not see that such curves are impossible. Here, then, is a case of lack of intuition into possibility unaccompanied by any positive intuition into impossibility.

Consider now the question "Could there be a cube with more than 12 edges?" I think nearly anyone who visualizes a cube and counts its edges will answer this question with a confident "No." Why the confidence? Because one does not merely fail to see how a higher number of edges is possible; one positively sees that a higher number is impossible. So here we have an example of positive intuition into impossibility.[52]

Now I would like to suggest that in connection with higher dimensions there may be intuitions of this positive kind. Visualize a trio of mutually perpendicular lines, and then try to add a fourth at right angles to each of the original three: I assume you will fail. Is this merely because you cannot see how to fit in the fourth line? Or do you not rather see that the thing cannot be done?

Another example may be more convincing. Imagine an intact sphere or a closed box, and then try to imagine an unbroken path that begins on the inside and winds up on the outside without ever penetrating the surface. In other words, try to imagine a continuous path by which a beetle could crawl out of a sealed box without burrowing through a wall. I assume again that you will fail. Now is it merely that you cannot discover such a path? Or do you not rather see that no such path is possible? Yet such paths would have to exist if space were four-dimensional! (This is shown in the Appendix.)

I do not know how the reader will react to these examples, but for my part, the intuition that three mutually perpendicular lines leave no room for a fourth, or that a sphere intercepts all paths from inside to outside, is as compelling as the intuition that red excludes blue from the same surface. Such intuitions may not be infallible, but I do not see why they should be held any less trustworthy than intuitions generally.

APPENDIX: POINCARÉ'S DEFINITION OF DIMENSION

In 1912 Poincaré proposed an illuminating inductive way to define the concept of dimension. He defined a *cut* as an aggregate of elements removed from a continuum and said that a cut *divides* a continuum iff the continuum minus the cut does not remain "all in one piece." He assigned to points and finite collection of points dimension zero, then proceeded to define the other dimensions as follows.

> If to divide a continuum C it suffices to consider as cuts a certain number of elements [of zero dimensions] we say that C is *of one dimension*. . . . If to divide a continuum C, cuts forming one or several continua of one dimension suffice, we shall say that C is a continuum *of two dimensions*; if cuts suffice which form one or several continua of two dimensions at most, we shall say that C is a continuum *of three dimensions*; and so on.[53]

Unfortunately, this definition was shown to be inadequate almost immediately by Brouwer's counterexample of the double-sheeted cone (two cones with a common vertex). This object is two-dimensional, but it can be divided by the removal of a single point (the common vertex), and thus counts as one-dimensional under Poincaré's definition. Mathematicians have since gone on to define dimension using other concepts, for example, the concept of neighborhoods.[54] It seems to me, however, that Poincaré's approach need not be abandoned; it needs only slight revision in order to be accepted.

In the definitions that follow I assume as understood the notion of a *continuous* object. A continuous object has more than one part and is, in Poincaré's phrase, "all in one piece." If we were willing to take the notion of *touching* for granted, we could define 'x is all in one piece' as follows: any two parts of x that compose it touch each other.[55] In any case, an object can fail to be continuous either because it is partless (like a point) or because it is scattered (like an archipelago).

I begin by defining a notion similar to Polincaré's cut:

x and y *are separated* in w by z if and only if: (i) x, y, and z are constituents of w; (ii) x and y have no constituents in common with z; and (iii) every continuous constituent of w that has x and y as constituents also has a constitutent in common with z.

The idea is that you can't get from x to y while remaining in w except by passing through z. Two points in a plane are separated in this sense by a line running between them and across the whole plane; they are also separated by a circle around either of them. Two points in a line are separated by a point between them, and two points in a circle, though not separated by any one point, are separated by an appropriately placed pair of points. (As the last example shows, the separator may be scattered.)

I now go on to define dimensionality. I presuppose a domain of discourse limited to concrete individuals, which may be either material objects or portions of space, as you please. Since an object (or portion of space) must either contain a continuous constituent or not, and in the former case must either itself be continuous or not, there are three cases to consider. One of them gives us the base clause for a recursive definition.

C1. If w has no continuous constituent, w is of dimension 0.[56]
C2. If w is itself continuous, w is of dimension n iff (i) for any two points x and y in w, there is a z of dimension (n − 1) such that x and y are separated in w by z; and (ii) w is not of any lower dimension, that is, it is *not* the case that for any two points x and y in w, there is a z of dimension (n − 2) such that x and y are separated in w by z.
C3. If w is a scattered object with a continuous constitutent, w is of dimension n iff (i) some continuous constituent of w has dimension n, and (ii) no continuous constituent of w has dimension greater than n.[57]

Let us now compare this definition with Poincaré's. His notion of a cut can be expressed in my terms as follows: z is a cut that divides w iff for some x and y, x and y are separated in w by z. Poincaré's inductive clause, which says that an object has dimension n if there is a cut of dimension (n − 1) that divides it, therefore amounts to this: an object is of dimension n if there is *some* pair of points or parts within it that are separated by a cut of dimension (n − 1). This is what makes Poincaré's definition open to Brouwer's counterexample. But if we require instead, as C2 does, that *any* pair of points or parts within an object of dimension n be separated by an (n − 1)-dimensional cut, the counterexample is avoided. (Points on the same side of the common vertex are separated only by cuts of dimension 1.)

Finally, let us note some of the consequences of the definition. Letting n = 4 in C2, we get the result that in a four-dimensional space any two points would be separated by something three-dimensional, but not in general by something two-dimensional. In other words, there would be pairs of points such that a sphere (or other two-dimensional surface) enclosing one of them would not be enough to separate it from the other; you could get from one to the other by a continuous path that nowhere penetrated the sphere. So in four-dimensional space (as claimed in the text), a beetle could crawl out of a closed box without ever encountering a wall, and you could remove the yolk from an egg without cracking the shell.[58]

NOTES

[1] Immanuel Kant, "On the First Ground of the Distinction of Regions in Space," in John Handyside, trans., *Kant's Inaugural Dissertation and Early Writings on Space* (Chicago: Open Court Publishing Company, 1929). This volume, pp. 27–34.

[2] In his Inaugural Dissertation of 1770 (Handyside, p. 60) and in Section 13 of the *Prolegomena to Any Future Metaphysics,* he also uses the example of spherical scalene triangles in opposite hemispheres with an arc of the equator as their common base and corresponding sides equal. These are *two-dimensional* figures – a fact worth pointing out, since in Jill Vance Buroker's *Space and Incongruence* (Dordrecht, Holland: D. Reidel Publishing Company, 1981), it is asserted that the problem of incongruent counterparts arises only because incongruent counterparts are three-dimensional objects (p. 55). Buroker's book is the most extensive study of incongruent counterparts to date; I have discussed some of its main points in "Incongruent Counterparts and Things in Themselves," a paper published in the Proceedings of the Sixth International Kant Conference. Reprinted in this volume, pp. 341–51.

[3] The ambiguity is pointed out by Jonathan Bennett in "The Difference Between Right and Left," *American Philosophical Quarterly* 7 (1970), pp. 175–191. This volume, pp. 97–130.

[4] A child may know which of her hands is the left one, yet not be able to tell which of her mittens is the left one.

[5] I have heard it said as an objection to this that a right glove *will* fit a left hand if you turn it inside out first. This is true, but irrelevant, since in turning a glove inside out you do violence to it.

[6] This sentence shows what I mean by 'internal' and 'external' relations. No connection is intended with the various other distinctions sometimes marked by the same pair of terms, such as the distinction between essential and accidental relations and the distinction between grounded and ungrounded relations. (The latter distinction will be important to us, however, from Section VIII on.)

[7] "The right hand is similar and equal to the left, and if we look at one of them alone by itself, at the proportions and positions of its parts relatively to one another and at the magnitude of the whole, a complete description of it must also hold for the other in every respect" (Handyside, p. 26; this volume, p. 31).

[8] "If we conceive the first created thing to be a human hand, it is necessarily either a right or a left" (Handyside, p. 27; this volume, p. 32).

[9] "In the constitution of bodies differences are to be found . . . which are grounded solely in their relation to absolute, primary space" (Handyside, p. 28; this volume, p. 32).

[10] Holism is not the only position open to one who denies Kant's first premise. One could hold that a hand is right or left in virtue of its relation to something besides either material things or Absolute Space — for example, *human consciousness*. Something like this was in fact Kant's own later view; I discuss it in the paper cited in note 2. Yet another position that denies Kant's first premise is identified in Section XVI below.

[11] It should be emphasized that what I am calling 'absolutism' implies a *substantival* version of the absolute theory of space, since it countenances Absolute Space as a concrete entity in its own right. Absolutism is thus to be distinguished from the *adjectival* version of the absolute theory, which regards the spatial properties of bodies (such as position and motion) as *purely qualitative*, not consisting in relations either to other bodies or to substantival space. In the matter of right and left, a proponent of the adjectival view would no doubt embrace one of the varieties of intrinsicism. (For more on the substantival versus the adjectival view, see C. D. Broad, "Leibniz's Last Controversy with the Newtonians," *Theoria* 12 (1946), pp. 143—168. What I have designated simply as the adjectival view Broad calls the "Adjectival Qualitative" theory.)

[12] This schema is a bit oversimplified. There need not be one "big" relation that relates all the parts of w; some parts could be related by one relation and others by another, etc. Nor is it required that the enumeration of parts be finite.

[13] Such a principle is proposed by Wittgenstein in the *Tractatus* (London: Routledge and Kegan Paul, 1961) at 2.0201 and also by Wilfrid Sellars in *Science, Perception, and Reality* (London: Routledge and Kegan Paul, 1963), pp. 26—27. I assume that the second occurrence of the word 'complexes' in Wittgenstein's formulation should be replaced by 'constituents'.

[14] Immanuel Kant, *Critique of Pure Reason*, translated by Norman Kemp Smith (New York: St. Martin's Press, 1965), A440 = B468.

[15] What if we revised MR so as to allow "reductions" in which the reduced property showed up again as a property of some of the parts of the original whole? It would then no longer conflict with the intuitions cited above about color and extension, but neither would it conflict with holism about right and left. Holism is not opposed to the trivial reduction that says a hand is right if and only if it has an inner right-shaped core and an outer sheath that fits it like a glove.

[16] Peter Remnant, "Incongruent Counterparts and Absolute Space," *Mind* 72 (1963), pp. 393–399 (this volume, pp. 51–59), and Martin Gardner, *The Ambidextrous*

Universe, revised edition (New York: Mentor Books, 1969). Chapter 17 (this volume, pp. 61—74). Remnant's position is actually more cautious than Gardner's. He says only that he would be prepared to assert the indeterminacy of a solitary hand *if* internalism proved false.

[17] Handyside, p. 28. This volume, p. 32.

[18] Handyside, p. 23; this volume, p. 29. And in the Inaugural Dissertation he refers to the difference between incongruent counterparts as the difference between "things which lie towards one quarter and things which are turned toward the opposite quarter" (Handyside, p. 60; this volume, p. 35).

[19] Handyside, p. 20. This volume, p. 27.

[20] Remnant and Gardner both dismiss the appeal to direction for this reason.

[21] Franz Brentano, *Psychology from an Empirical Standpoint*, edited by Linda McAlister (New York: Humanities Press, 1973), p. 272.

[22] John Earman, "Kant, Incongruous Counterparts, and the Nature of Space and Time," *Ratio* 13 (1971), pp. 1—18. This volume, pp. 131—149.

[23] It will differ in some ways from a normal right hand; for example, the shortest finger will no longer be farthest from the thumb. If you think such things matter, rearrange the fingers, too.

[24] The reader must also be warned that much talk by physicists about alleged extra dimensions of physical reality is not talk about dimensions in the sense that concerns us here. For example, in "The Hidden Dimensions of Spacetime" (*Scientific American*, March, 1985, pp. 74—81) D. Z. Freedman and P. van Nieuwenhuizen report that spacetime may have as many as *eleven* dimensions, of which seven are not readily observable because they are "curled up" inside very small spheres. These can hardly be *spatial* dimensions, since objects of higher spatial dimension can never be contained within objects of lower spatial dimension. (A line could not be "curled up" inside a point.) For more on the relevant notion of dimensions, see the Appendix.

[25] Bennett, p. 178. This volume, p. 104.

[26] Similar remarks apply to the clock definition. A hand placed against the clock as the definiens requires might have its thumb on 3 as easily as on 9; it all depends on the orientation of the hand in four-space.

[27] It is hard to give an example of an ungrounded relation that is altogether uncontroversial. The example I have cited would be challenged by proponents of the adjectival theory of space mentioned in note 11; according to this theory, the relative distance of two bodies would be grounded in pure positional qualities. And Leibniz would have denied that there are *any* ungrounded relations.

[28] Gardner, pp. 147—148. This volume, pp. 64—65.

[29] There are two sub-possibilities: either the arrow as a whole has an intrinsic irreducible upness (holism), or the upness of the arrow is a matter of the head's being absolutely above the shaft (internalism). 'Absolutely above': that is to say, the aboveness is genuinely dyadic, requiring no relation to any outside entity. Internalism of this variety seems to be implicit in Huyghens's response to Newton's thought-experiment of the spinning bucket. See Hans Reichenbach, "The Theory of Motion According to Newton, Leibniz, and Huyghens," in *Modern Philosophy of Science* (London: Routledge and Kegan Paul, 1959), pp. 46—66, especially p. 64.

[30] Leibniz would have rejected the intrinsicist answers for the same reason he rejected Absolute Space: a universe consisting of a single arrow pointing up would be distinct

from, yet indiscernible from, a universe consisting of a single arrow pointing down, which violates the Identity of Indiscernibles. Intrinsicists might reply by pointing out that if the two universes Leibniz objects to are really one and the same, so must be the various phases in the history of a solitary spinning arrow, in which case the arrow couldn't be spinning after all. Indeed it couldn't, would be Leibniz's rejoinder. We are now getting into the issues surrounding Newton's "bucket experiment," which I cannot discuss here.

[31] A referee has suggested the following way of sharpening the first horn of the dilemma: The absolutist maintains that the relation between the arrows, xRy, derives from two further relations xR'S and yR*S. If these relations are themselves grounded, we have it that the original relation is entailed by the conjunction Fx & GS & Hy & KS. Now the facts GS and KS will arguably be *necessary* facts, so in a sense they drop out: anything entailed by the fourfold conjunction will be entailed by Fx & Hy alone. But this means that the original relation is adequately grounded in features just of x and y.

[32] For an explanation of the connection between gravitation and dimension, see Max Jammer, *Concepts of Space*, second edition (Harvard University Press, Cambridge, Mass.: 1969), p. 179.

[33] And more generally, that for any n, it is possible that space has n dimensions. This is not to say that it is possible for space to have infinitely many dimensions, but only that there are infinitely many numbers of dimensions that it would be possible for space to have.

[34] Kant himself *did* believe that space is necessarily three-dimensional. So why wasn't he an internalist? The answer, as we saw in Section V, is that he thought relations of direction are possible only given the existence of Absolute Space. (For Kant's commitment to the necessary three-dimensionality of space, see Handyside, p. 60 (this volume, p. 35), and the *Critique of Pure Reason*, B41. In one early paper (1747), he evidently regarded three-dimensionality as contingent; see Handyside, pp. 11—12.)

[35] Handyside, p. 26; this volume, p. 31. As Buroker notes (p. 53), this makes Kant's notion of congruence potentially divergent from Euclid's, since figures with equal sides and angles may not be superposable.

[36] Some people may want to say that what matters is not *motions* in the actual space, but *mappings* in the actual space. In the case of the balls in the hourglass space, but not in the case of the L's in the two-dimensional space, there is a translational mapping of one object onto the other that preserves distances between corresponding points. To these people I put my point somewhat differently: why should only translational (and rotational) mappings be relevant to establishing congruence? I note that *reflective* mappings are allowed as establishing congruence in Walter Prenowitz and Henry Swain, *Congruence and Motion in Geometry* (Boston: D.C. Heath and Company, 1966), pp. 26—29.

[37] For example, in *Kant: An Introduction* (Cambridge: Cambridge University Press, 1978), C. D. Broad says of figures like my back-to-back L's that they are "incongruent if we suppose them to be confined to a plane" (p. 42). Graham Nerlich and Lawrence Sklar, whose work I discuss in the next section, operate on the same assumption.

[38] It is not strictly necessary for me to assume here that right and left are shapes, but only that congruence and incongruence are grounded relations. The grounding properties, whether shapes or something else, will be nonrelational, which suffices to refute externalism.

[39] Incredible though it be that a solitary object must lack shape, this is evidently a consequence of Grünbaum's doctrine of the "intrinsic amorphousness of space." Grünbaum holds that whether two pieces of a manifold are equal in length depends on conventions regarding the behavior of some outside entity used as a measuring rod. Since an object is a cube only if its edges are equal in length, it follows that a solitary object could not be a cube. See Adolf Grünbaum, *Philosophical Problems of Space and Time*, second edition (Dordrecht, Holland: D. Reidel, 1973), Chapter 16.
[40] Graham Nerlich, "Hands, Knees, and Absolute Space," *The Journal of Philosophy* 70 (1973), pp. 337—351.
[41] Nerlich adds that it will also depend on how the conventions for 'right' and 'left' have been fixed (p. 345). Revised version printed in this volume, pp. 151—172.
[42] Lawrence Sklar, "Incongruous Counterparts, Intrinsic Features, and the Substantiviality of Space," *The Journal of Philosophy* 71 (1974), pp. 277—290; this volume, pp. 173—186. Sklar also includes among the relationist's resources "possible objects."
[43] Nerlich would say it also depends on whether the relationist can characterize that relation to space that makes an enantiomorphic hand right or left.
[44] For references, see G. J. Whitrow, "Why Physical Space Has Three Dimensions," *British Journal for the Philosophy of Science* 6 (1955), pp. 13—31.
[45] G. H. Alexander, ed., *The Leibniz-Clarke Correspondence* (Manchester: Manchester University Press, 1956), Leibniz's third letter, Section 5.
[46] Here is another argument to the same end: Suppose the law excluding a fourth rod were to lapse. In that case would space suddenly become four-dimensional? No, you cannot add regions to space as you can rooms to a house. The lapsing of the law would rather be the opening up of pre-existing territory, which shows that space cannot be equated with the physical possibility of placing bodies.
[47] Using 'L' for 'it is necessary that', 'M' for 'it is possible that', 'P' for 'Space has n dimensions', 'Q' for 'There are n rods meeting at right angles', and 'R' for 'There are more than n rods meeting at right angles', the argument just given can be symbolized as follows. (1) P = Df MQ & ~MR (the proposed analysis). (2) MQ & ~MR iff L(MQ & ~MR) (valid in the modal system S5). (3) P iff LP (by substituting in both sides of 2 in accordance with 1). Q.E.D.
[48] I do not say that this is the only way of dealing with the paradox. It will not arise to begin with if one denies any of the assumptions leading to the bottom slot, and I have already urged (in Section XI) that there is at least one of them we should deny — that congruence requires superposability.
[49] What if dimensionality were analyzed in terms of *logical* possibility? In that case (since dimensionality would no longer be contingent), something's having P would not depend on extraneous *contingent* fact, so P would be *in*trinsic in both senses.
[50] How will Nerlich, who holds that P is $extrinsic_1$, avoid the incongruity? Not being able to reconcile the assumptions in the manner I have suggested, he will have to deny one of them. We have seen that he holds that superposability is required for congruence, so he will have to deny that congruence is grounded. This is in effect what he does. Congruence and incongruence for him are not grounded in nonrelational properties; nor are they ungrounded *simpliciter* as they are for Gardner; instead, they are grounded in relations of the terms to a special entity, viz., Absolute Space. This is analogous to the view I discussed briefly in Section VIII, and I believe it is vulnerable to the same dilemma I raised there.

[51] I have discussed this distinction at greater length in "Conceivability and the Cartesian Argument for Dualism," *Pacific Philosophical Quarterly* 64 (1983), pp. 35—45.

[52] "But if something had 13 edges, we wouldn't *call* it a cube." Yes — because we see that it would't *be* a cube. If the objector means to suggest that it is somehow a matter of definition that cubes have 12 edges, I remind him that the definition of a cube is 'regular solid with six square faces'; there is no mention of edges.

[53] Henri Poincaré, *The Value of Science*, in *The Foundations of Science*, translated by George Bruce Halsted (New York: The Science Press, 1913), pp. 242—243. See also the passage from *Science and Hypothesis* on pp. 52—53 of the same volume.

[54] For a brief review, see Richard Courant and Herbert Robbins, *What is Mathematics*? (Oxford: Oxford University Press, 1941), pp. 248—251, or Whitrow.

[55] I owe this suggestion to Richard Potter. I find a similar definition in Eli Hirsch, *The Concept of Identity* (Oxford: Oxford University Press, 1982), p. 97. The relevant sense of 'compose' is this: y and z compose x iff (i) y and z are parts of x, (ii) y and z have no parts in common, and (iii) every part of z has a part in common with x or y.

[56] Thus points have dimension 0; so do finite collections of points; so even do certain nondenumerable collections of points, such as the Cantor point set described on pp. 248—249 of Courant and Robbins.

[57] If it seems wrong to assign a dimension to a scattering of things of various dimensions, we could proceed instead as follows. First, we define the concept "set of parts" for a scattered object. If w is a scattered object, S is a set of parts for w iff (i) no two members of S have any constituents in common; (ii) every member of S is a constituent of w; (iii) every constituent of w overlaps or is identical with some member of S; (iv) every member of S is continuous; and (v) no member of S is a constituent of any continuous constituent of w. We can then replace C3 by the following: if w is a scattered object with a continuous constituent, w is of dimension n iff every member of w's set of parts is of dimension n. (Otherwise, w is of "mixed" dimension and has no dimension as a whole.)

[58] For more in this vein, see various of the essays in *The Fourth Dimension Simply Explained*, edited by H. P. Manning (Gloucester, Mass.: Peter Smith, 1977). The essays in this volume were chosen from entries to a *Scientific American* contest in 1909.

Acknowledgments: For illuminating discussions of the material in this paper I would like to thank Felicia Ackerman, Thomas Banchoff, Philip Quinn, Ernest Sosa, Stephanie Troyer, and members of colloquium audiences at Brown University, Franklin and Marshall College, Miami University, and the University of Massachusetts at Amherst.

JOHN EARMAN

ON THE OTHER HAND . . . : A RECONSIDERATION OF KANT, INCONGRUENT COUNTERPARTS, AND ABSOLUTE SPACE

In his 1768 essay 'Concerning the Ultimate Foundation of the Differentiation of the Regions in Space', Kant used incongruent counterparts in an attempt to refute a Leibnizian-relationist account of space. It is hard to imagine that scholars could be more divided on how to understand Kant's argument and on how to assess its effectiveness (compare Alexander 1984/85; Broad 1978; Buroker 1981; Earman 1971; Gardner 1969; Lucas 1984; Nerlich 1973, 1976; Remnant 1963; Sklar 1974; Van Cleve 1987; Walker 1978; and Wolff 1969). Two years later in 1770 incongruent counterparts resurface in Kant's Inaugural Dissertation, this time as part of a proof that our knowledge of space is intuitive. They appear yet again in the *Prolegomena* (1783) and the *Metaphysical Foundations of Natural Science* (1786) as part of the argument for transcendental idealism. Not surprisingly, scholars are also at odds on how to explain the shifts in the roles Kant wanted incongruent counterparts to play and how to assess the importance of these matters for the development of his critical philosophy (compare Alexander 1984/85; Allison 1986; Bennett 1970; Broad 1978; Buchdahl 1969; Buroker 1981; Walker 1978; Winterbourne 1982; and Wolff 1969).

The present work is concerned primarily with Kant's 1768 argument. It aims both at a better understanding of Kant's argument and a sharper formulation of the ways incongruent counterparts are and are not relevant to the controversy over absolute vs. relational conceptions of space.[1] A few remarks are offered on Kant's post-1768 use of incongruent counterparts.

1. KANT'S ARGUMENT AGAINST RELATIONISM

Although Kant is his usual cryptic self in 'Concerning the Ultimate Foundation of the Differentiation of Regions in Space,' an argument against Leibniz's relational conception of space is readily extractable.

(K1) "Let it be imagined that the first created thing were a human hand, then it must necessarily be either a right hand or a left hand." (1768, p. 42)

It follows, supposedly, that the relational theory is not adequate since

(K2) On the relational theory the first created thing would be neither a right hand nor a left hand since the relation and situation of the parts of the hand with respect to one another are exactly similar in the cases of right and left hands that are exact mirror images of one another.

Before turning to an evaluation of Kant's argument, it is worth remarking on how novel the argument was. The standard absolutist response to the relational conception of space was that motion is absolute rather than relational and that absolute motion must be understood as motion with respect to a space that is ontologically prior to bodies.[2] Kant thought that he could arrive at the same result — that, as he puts it, "*absolute space has its own reality independent of the existence of all matter*" (1768, p. 37) — by showing that the differences between right and left "are connected purely with *absolute and original space* . . ." (ibid., p. 43). Kant was, of course, aware of the more standard objections to relationism; indeed, he specifically mentions Leonhard Euler's[3] attempt to show that the relational theory cannot properly ground Newton's first law of motion, but brushes it aside as having a merely *a posteriori* character. Kant's aim was to "place in the hands not of engineers, as was the intention of Herr Euler, but in the hands of geometers themselves a convincing proof" of the reality of absolute space. (ibid., pp. 37—38).

Turning now to Kant's argument, the relationist may wish to attack either (K1) or (K2). Consider first the attack on (K1). The relationist may urge that whether or not a hand is right or left depends upon the relation of the hand to an appropriate reference body. Thus, contrary to Kant, for a hand standing alone there aren't two different actions of creative cause for God to choose between.[4] If a reference body is introduced then, as the relationist will readily agree, different acts of creative cause are required for the hand in question to have different relations to the reference body.

It is instructive to compare the situation here to that for continuous symmetry transformations. So as to be free of the problem of incongruous counterparts, imagine that the bodies in question are point masses. According to the absolutist, the operation of shifting all the bodies one mile to the East in the container space produces a different state of affairs, and different acts of creative cause are required for God depending upon which of these states He chooses to actualize. In the correspondence with Clarke, Leibniz charged that this result violates the Principle of Sufficient Reason since God would have no good reason to actualize one rather than another of these absolutist states (Alexander 1984, p. 26), and he proposed to rescue God from the Buridan's ass situation by maintaining that what the absolutist is providing is different descriptions of the same intrinsic (relational) state. Leibniz buttressed his resolution with the further claim that the states the absolutist counts as different are "indistinguishable" — at least they are not separated by any features which are in-principle observable. Of course, if we introduce a reference body which itself is not subjected to the shift, then observable differences will arise; but then the relationist will also want to count the states as distinct since in these states there will be differences in the relations of the original system of bodies with respect to the reference body.

Many commentators assume that Kant was acquainted with the Leibniz–Clarke correspondence, and some, such as Robert Paul Wolff (1969), even assume that Kant read Leibniz's operation of "changing East to West"[5] not as a continuous transformation — translation or rotation — but as mirror image reflection. I know of no definitive evidence that Kant did read the Leibniz–Clarke correspondence, but it seems more likely than not, given Kant's intense interest in Leibniz during this period and given the availability of two German editions of the correspondence (Köhler 1720, 1740). However, the German translation does not naturally suggest mirror image reflection; for instance, the relevant passage in the 1740 edition reads "durch eine Verwechselung des Aufgangs der Sonnen mit ihrem Niedergangs" (literally, through an interchange of the rising and the setting of the sun).

Moreover, if Leibniz's construction did in fact suggest the problem of incongruent counterparts to Kant, then it would be somewhat surprising if he did not consider that the relationist can transfer Leibniz's argument as applied to continuous symmetries to the case of reflections. One can speculate that Kant would have thought that the parallelism breaks down since he would have thought that in the case of hands the states which the absolutist wants to count as different are perceptually distinguishable — in one case the hand presents itself as a right hand, in the other as a left hand, and the difference, he may have thought, is not due to some difference in the relations of the hands to the observer. I will have more to say on the point below in Section 2.

Another response Kant may have intended can be discerned from his characterization of incongruent counterparts.

As the surface limiting the bodily space of the one cannot serve as a limit of the other, twist and turn it how one will, this difference must, therefore, be such as rests on an *inner principle*. (1768, p. 42; italics added)

C. D. Broad interprets Kant as arguing that since the unlikeness of incongruent counterparts depends upon an unlikeness of their bounding surfaces and since the surface of a body is something intrinsic to it, so the difference between incongruent counterparts must rest upon "a difference in their *intrinsic spatial properties* and not on a difference in *their spatial relations to some third body*." (Broad 1978, p. 38) But if this is Kant's line, then it is not at all apparent how introducing an absolute container space will help in distinguishing right from left. Suppose, in accordance with the absolute theory, that the spatial relations among the material parts of the body are parasitic upon the spatial relations of the points of absolute space occupied by the body, and consider two bodies, B and B′, which are shaped like human hands and are exact mirror images of one another. Broad imagines that the set of space points occupied by B and the set of points occupied by B′ might differ in some geometrical properties that are not manifested in the relations among the occupying particles of matter. But these would have to be mysterious properties indeed. If the spatial points occupied by B have the property of 'standing in a left-handed configuration' while those occupied by B′ have the property of 'standing in a right-handed configuration,' why can't the material points occupying the spatial points have corresponding properties?

This brings us back to (K2). If the relations among the material parts of a hand are construed narrowly to involve only, say, distance, line and angle, then it is certainly true that the relation and situation of the parts of the hand with respect to one another are exactly similar in the cases of mirror-image left and right hands. But this is equally true if space points are substituted for material points. The absolutist may hope to distinguish between right and left by construing 'relation and situation' of the space points occupied by the hand more broadly. But it is not at all clear why the relationist cannot entertain a similar hope by appealing to a broadened notion of relation and situation of the material points of the hand.

It can be shown that, in general, being right (left) handed cannot be purely a matter of the internal relations and situation of the material parts of a hand, no matter how broadly relation and situation are construed. Suppose, on the contrary, that it were. Then it follows that for any choice of closed path in space it is always possible to arrange a consistent sequence of hands around the loop, where the consistency condition is that immediately adjacent hands have the same handedness. For at each location on the loop it is sufficient to construct a material hand whose parts instantiate the list of properties and relations that constitute being (say) right handed. But non-orientable spaces show that such a consistent arrangement is not possible for every closed loop.[6]

Of course, the same argument suffices to show that being right (left) handed cannot be purely a matter of the internal relations and situation of the points of absolute space occupied by the body.[7] Nor does it help the absolutist to bring in relations the points of the hand have to particular points of the external space surrounding the hand since whatever relations a left hand has to those points are exactly mirrored by the relations its right handed counterpart has to similarly situated external points.

Perhaps, however, as Nerlich (1973, 1976) suggests, the absolutist can show that the difference between right and left must make reference to the relation the hand has not just to particular points of the external space but, to use Kant's phrase, its relations to "space in general as a unity."[8] To see the point, suppose that space is equipped with a metric which conforms to the 'axiom of free mobility' so that it is meaningful to speak of the rigid transport of body.[9]

Def. An object O is an *enantiomorph* just in case there is a neighborhood N of O such that (i) N is large enough to admit reflections of O, and (ii) the result of every reflection of O in N differs from the result of every rigid motion of O in N.[10]

The restriction to local reflections and local rigid motion is necessary if we want to be able to distinguish between handed and non-handed objects regardless of whether space is globally orientable.[11] On the other hand, no such restriction seems justified when it comes to defining incongruent counterparts.

Def. Objects O and O' are *incongruent counterparts* just in case their surfaces cannot be made to coincide by any rigid motions but coincidence can be achieved by means of a combination of rigid motions and a reflection.

This second definition does appeal to space as a unity in that it quantifies over all motions of a certain type in the space. It has as a consequence that in non-orientable spaces — where space as a unity has a kinked structure — there are no incongruent counterparts. But that is no surprise; indeed, this consequence is just a way of restating the essential element of the definition of a non-orientable space.

What is the relevance of all of this to Kant's attempted refutation of the relational theory of space? Very little. Nerlich is incorrect in holding that Kant is right if we interpret him as saying that the enantiomorphism of a hand depends upon the relation between it and the absolute container space considered as a unity. For, in the first place, my definition of enantiomorphism does not quantify over all mapping in the space. And, in the second place, all that has been shown is that the absolutist has an account of enantiomorphism — something never in doubt — and not that the relationist cannot produce such an account. As for the notion of incongruent counterparts, the definition does appeal to space as a unity in Nerlich's sense. But again Kant has not been shown to have or to have anticipated some insight into the right—left distinction that is beyond the ken of the relationist. Perhaps the point is that relationism is inadequate to deal with the global structure of space (orientability in particular and perhaps other global properties as well). If true, that would be a damning indictment of relationism, but it remains to be substantiated. And even if it could be substantiated, relationism would be condemned for reasons quite remote from those put forward in Kant's 1768 essay.

In closing this section it is worth emphasizing the difficult position into which Kant maneuvered himself in the 1768 essay. The difference between right and left must, he asserts, rest on an inner principle. The relationist is incapable of supplying this principle since the "difference cannot . . . be connected with the different way the parts of the body are connected with each other." (Kant 1768, p. 42) But the absolutist can do no better in this regard since the difference cannot be connected with the different way the points of absolute space occupied by the body are connected with each other. Rather than remarking on this obvious tension, Kant masks it by going on to assert that the "complete principle of determining physical form" rests on the relation of the object to "general absolute space." (Kant 1768, p. 41) But if Kant's second assertion is not to be incompatible with the first, it can only mean that the inner difference — for which neither the absolutist nor the relationist can account — reveals itself through the external relation of the hand to space in general as a unity. The case of the hand standing alone shows, Kant thinks, that the relationist is blocked from claiming that the difference is revealed through the relation of the hand to other objects. But this only shows that of two inadequate theories, the relational theory is more inadequate. I will return to Kant's conundrum in Section 4. But first I want to turn to a more systematic examination of the possibility of a relational account of the difference between right and left.

2. A RELATIONIST ACCOUNT OF THE DISTINCTION BETWEEN RIGHT AND LEFT

Leibniz's criticism of Newton in the Leibniz—Clarke correspondence suggests that, from the relationist perspective, the proper way to construe absolutist talk about space is to read it as providing representations of relational models of reality. These representations involve the fictional entity of space points and, consequently, the representation relation is one—many, with many absolutist representations corresponding to the same relationist model.[12] In investigating the implications of the difference between right and left for the absolutist—relationist controversy, I propose to explore the possibility of extending this version of relationism to cover the cases at issue.

In asking whether the relationist can account for the difference between right and left, it is useful to distinguish between three items: (1) being an enantiomorph; (2) being incongruent counterparts; and (3) being right (left) handed. In considering the possibility of Leibnizian accounts of (1)—(3) it is convenient to suppose that space is two-dimensional.

(1) *Being an enantiomorph.* Consider the following relational descrip-

tion of an object. The object has a long shaft and two shorter shafts which are attached perpendicular to the long shaft. One of the cross shafts is attached to an end of the long shaft, while the second cross shaft, which is shorter than the first, is attached to the long shaft at a point between the midpoint and the end to which the first cross shaft is attached. Many different absolutist representations of this relationist description are possible, principally (a) and (b) of Figure 1 (where the points of the page represent points of absolute space) and rigid rotations and translations of (a) and (b). In response to Kant's example of a hand standing alone, the relationist can say that Kant was partially correct; namely, a hand standing alone is a hand — it has a handedness. More specifically, the relational hand-in-itself possesses various absolute representations — viz. (a) and (b) — but on any such representation the hand is an enantiomorph, as defined in Section 1 above. The relationist strategy then is not to directly define concepts like 'enantiomorph' and 'incongruent counterparts' but rather to establish that an acceptable relationist description guarantees that any allowed absolutist representation will fulfill the absolutist definition of the concept.

(2) *Being incongruent counterparts.* Now consider the following relational description. There are two objects, one red, the other green. Both have the characteristics as described in (1). The lengths of the corresponding long and cross shafts in the two objects are the same. The objects are so situated with respect to one another that the free ends of the longer of the cross shafts are almost touching, their long shafts are parallel, and the corresponding cross shafts each lie on a line. Absolute representations consist in (c) an (d) of Figure 2 and rigid rotations and translations of (c) and (d). Again the relational description assures that on any absolute representation the red and green objects are incon-

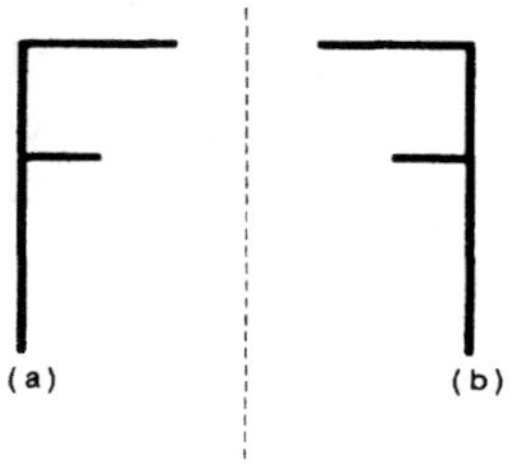

Fig. 1.

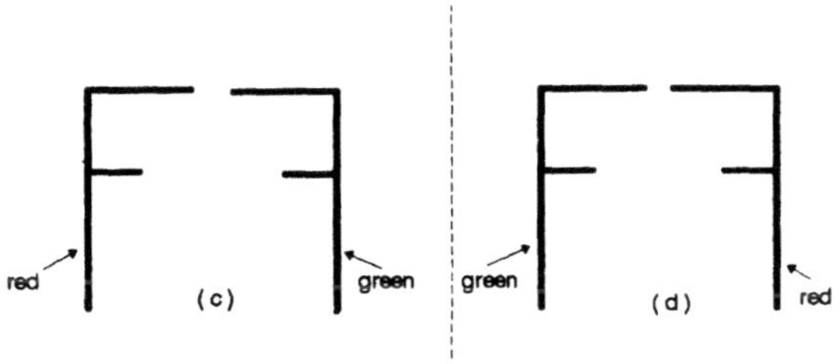

Fig. 2.

gruent counterparts as defined in Section 1. At least this will be so if the objects are embedded in an orientable space. Thus, the question arises as to whether the orientability of the embedding space can be indicated in relational terms. It might seem that the answer is obviously in the affirmative in the present case. For the only way a space can fail to be orientable is for it to be non-simply connected (a theorem of topology), and such a multiple connectedness would necessarily involve a multiple relatedness of objects. Such a multiplicity is absent in the above relational description, which we may take to be a complete catalogue of the relations between the red and green object. The hitch is that multiply connected spaces need not be non-orientable so that the absence of multiple relatedness in a relational model is a sufficient but not a necessary signal that the embedding space should be orientable. I will not speculate here on how the relationist might respond to this difficulty.[13]

(3) *Being right* (*or left*) *handed.* Here the relationist can repeat the suggestion made in Section 1. The difference between right and left in the which is which sense requires the choice of a reference body (say the red one in Figure 2) and the stipulation that it is (say) right handed. Then being right or left handed is just a matter of bearing the appropriate relations to the reference standard.

A seeming difficulty (already hinted in Section 1) with the above version of relationism is that a minimal condition for counting (a)—(b) and (c)—(d) as equivalent descriptions of the same reality is that they be observationally indistinguishable. But in one sense they are patently distinguishable — they look different. The relationist will respond that talk of appearances presupposes an observer and that the introduction of an observer involves several distinct cases. If the observer is non-enantiomorphic, (a) and (b) of Figure 1 are replaced by (â) and ($\hat{b}$) of Figure 3. If the observer is enantiomorphic the replacements can be either ($a^{\dagger}$) and ($b^{\dagger}$) or (a^*) and (b^*). Similar replacements apply to (c) and (d) of Figure 2. In the ($a^{\dagger}$)—($b^{\dagger}$) case the introduction of the observer breaks the relational equivalence of (a) and (b), and so the relationist can entertain the hypothesis that the phenomenal contents of the visual experiences of the ($a^{\dagger}$)—($b^{\dagger}$) observers are different. Perhaps Kant would want to claim that the phenomenal contents of the experiences of the (â)—($\hat{b}$) and the (a^*)—(b^*) observers are or can be different, a claim the relationist must deny since (â)—($\hat{b}$) and (a^*)—(b^*) are relationally equivalent.

Walker (1978) might seem to be endorsing the Kantian claim when he writes that "A left-handed and a right-handed glove look different regardless of their relations to other things, and a left-handed and a right-handed universe would look different too" (this volume, p. 189). Nor, according to Walker, is the difference in appearances to be accounted for in terms of the relations of the gloves to an asymmetrical body of the observer; for "the difference between the gloves is immediately obvious without reference to my body or to anything else; I should perceive it in just the same way if my body were itself symmetrical about the plane that forms the axis of symmetry between the gloves" (this volume, pp. 189—190). But what the opponent of relationism must claim in the (â)—($\hat{b}$) case is that the mirror-image non-enantiomorphic observers, who have the receptor sites of their sense organs left—right reversed, nevertheless have different visual experiences. Even on the absolutist's own terms the case for this claim is weak. To support the Kantian claim the absolutist must maintain that spatial perceptions are a function not only of the spatial relations of

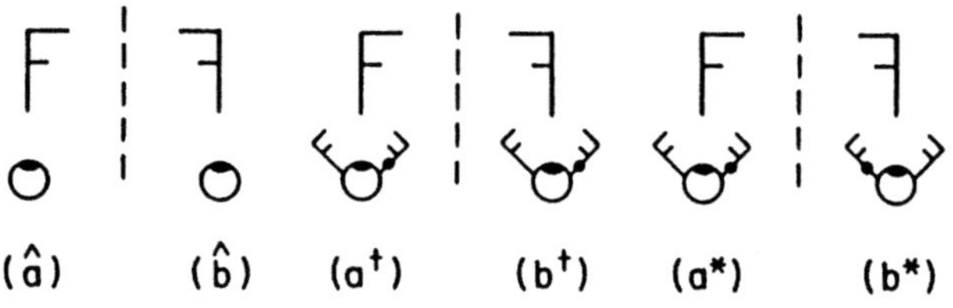

Fig. 3.

observer to object but also of the relation of the observer to a preferred orientation of the space containing the object–observer system. As a hypothesis about perception, this seems farfetched. And the postulation of the preferred orientation is unmotivated, at least at this juncture. In rejecting the relationist notion that all motion is the relative motion of bodies, the absolutist can point to the need to postulate a preferred family of reference frames – the inertial frames – in order to achieve simple but predictively accurate laws of motion. By contrast, nothing in the fundamental phenomena of physics seems to call for a preferred orientation of space. At least this was so until quite recently, as will be discussed below in Section 3. But the laws in question concern exotic weak interactions of elementary particles, interactions that are presumably irrelevant to the human perceptual process.

I conclude that on the version of relationism I have proposed, the introduction of enantiomorphs and incongruent counterparts does not by itself alter the dialectics of the absolute–relational controversy. Nevertheless, it might be thought that although what I called the Kantian claim does not help to establish that objects are literally embedded in absolute space, it suggests that objects as they present themselves to us in perceptions are spatial in a sense that outstrips the relational account and, thus, the claim may be useful in helping to explain how the mature Kant came to understand the implications of incongruent counterparts. Unfortunately, as will be discussed in Section 4, it is no help whatever in this regard, for after 1768 Kant used incongruous counterparts not as an objection to the Leibnizian view *per se* but rather as a reason to reject both the Leibnizian and the Newtonian views.

3. PARITY NON-CONSERVATION

In the 1768 essay, Kant mentions various contingent left–right asymmetries, viz. that most hops wind round their poles from left to right while most beans twist in the opposite direction; but nothing in his argument against relationism relies on any non-essential properties of incongruent counterparts. While this feature of Kant's argument promises to make it powerful, it may in fact account for its ineffectualness. I now propose to ask whether adding contingent but lawlike features will help.

The discussion that follows is conducted under the assumption that parity non-conservation experiments in elementary particle physics indicate that mirror image reflection fails to be a symmetry of some of the fundamental laws of physics.[14] The experiment of Crawford et al. (1957) tracks first the decay of a negative pi meson (π^-) and a proton into a neutral hyperon (Λ^0) and neutral K meson (K^0), and then the subsequent decay of the hyperon into another pi meson and a proton. The momentum vectors for the initial decay process lie in a plane while the momentum vector for pi meson in the subsequent decay is oblique to this plane. The possible results, (e) and (f), are pictured in Figure 4 (protons not shown).[15] If parity is conserved, the mirror image processes (e) and (f) should have the same probability, but in fact (e) dominates. Nature's preference for (e) can then be used to define almost surely[16] a right handedness: perform a large number of repetitions of the decay experiment to identify the dominant decay mode, and then $\vec{p}_{\pi_{inc}}$, $\vec{p}_{\Lambda}$, and $\vec{p}_{\pi_{decay}}$, taken in that order, define a right handed triad.

The failure of mirror image reflection to be a symmetry of laws of nature is an embarrassment for the relationist account sketched in Section 2, for as it stands that account does not have the analytical resources for expressing the lawlike left–right asymmetry for the analogue of Kant's hand standing alone. Putting some 20th century words into Kant's mouth, let it be imagined that the first created process is a $\pi^- + p \rightarrow \Lambda^0 + K^0$, $\Lambda^0 \rightarrow \pi^- + p$ decay. The absolutist

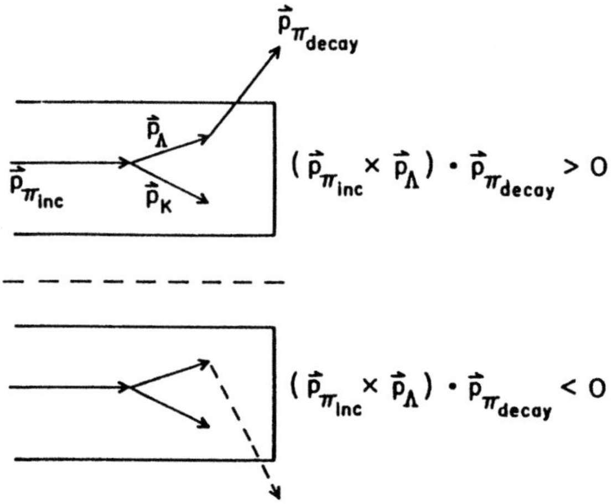

Fig. 4.

has no problem in writing laws in which (e) is more probable than (f), but the relationist of Section 2 certainly does since for him (e) and (f) are supposed to be merely different modes of presentation of the same relational model. Evidently, to accommodate the new physics, relational models must be more variegated than initially thought.

The variegation of relationist models of particle decay processes can proceed by the addition of intrinsic properties R^* and L^*. The relationist may not be able to describe R^* and L^* in traditional relationist terms, but he can give a functional specification of these properties in terms of their roles in the lawlike dispositions that ground the non-conservation of parity. The relationist can say that in the relevant experimental setup, R^* outcomes are more probable in the propensity sense than L^* outcomes and that, almost surely, these propensities will be reflected in the long run relative frequencies of multiple repetitions of the decay experiment.

But saying only this much about R^* and L^* is not enough to remove the suspicion that we have been treated to a sleight of hand exercise. The relationist has to tell us enough about these properties to convince us that his account deserves to be called relationist in the minimal sense that R^* and L^* are not merely devices for naming absolutist states. At the same time, however, there has to be a close enough relation between R^*—L^* and absolutist states so as to establish an explanatory connection between the differential propensities for R^* and L^* and the observed asymmetry between (e) and (f). The problem here has two aspects, the first and most fundamental being how and why R^* and L^* connect with (e) and (f) at all, the second being why R^* is manifested as (e) and L^* is manifested as (f) and not the other way round. The relationist may refuse to accept the second challenge and claim that there is no truth to the matter as to whether (e) or (f) is the correct representation of the dominant R^* decay mode. In this connection it is interesting to note that Lee and Yang (1956) speculated that there are actually two species of elementary particles π_R^-, Λ_R^0, K_R^0 and π_L^-, Λ_L^0, K_L^0 with the corresponding particles having the same masses, charges, and spins. Parity conservation would be maintained if the two species transform into one another under mirror image reflection and if they exhibit the opposite asymmetries in their decay modes. The apparent violation of parity conservation would be due to the fact that we inhabit a region where the R species is predominant. Although there is no experimental evidence to support this speculation, the two species do exist for the relationist as different representations or pictures of the relationist models. That what we actually see conforms to one rather than to another of these pictures might be the basis for an objection to relationism, but it is the same objection already considered in Section 2.

Now suppose for sake of argument that some suitable connection between R^*—L^* and (e)—(f) has been established. It might seem then that parity non-conservation implies that being right or left handed can be explicated in terms of the internal properties of a system, contradicting the construction of Section 1 involving non-orientable spaces. The conflict is only apparent. If, as I have assumed, parity non-conservation entails a violation of left—right symmetry in the basic laws of nature and if laws are universal in the sense that they hold good in all regions of space and time,[17] then it follows that actual space is in fact orientable.[18] In a hypothetical non-orientable space, either the laws exhibit no left—right asymmetry or else there are no universal laws. The absolutist may still wish to object that it remains mysterious why it is that in non-orientable spaces the properties R^* and L^* are either not possessed or alternatively are possessed but are not manifested as right—left asymmetries. Whether or not the relationist can produce a satisfying reply depends upon two matters left hanging above — the relationist's explanation of the connection between R^*—L^* and spatial representations, and the ability of the relationist to account for orientability properties of space.

4. INCONGRUENT COUNTERPARTS AND THE INTUITIVE NATURE OF SPACE

The most conservative interpretation of the shifting roles Kant assigned to incongruent counterparts would locate the cause of the shift in the expansion of Kant's possibility set.[19] In 1768 the possibilities Kant considered included only Newtonian substantival space and Leibnizian relational space; after 1768 the possibility set is enlarged to include the view that space and time belong to "a form of intuition and therefore to the subjective constitution of our minds, apart from which they could not be ascribed to anything whatever" (Kant 1781, A23/B39). On the conservative interpretation, Kant used incongruent counterparts both before and after 1768 to show that space is non-relational. Thus, to be consistent with the 1768 essay, Kant must have developed after 1768 an independent reason for rejecting the Newtonian view. Such a reason is to be found, for example, in the Inaugural Dissertation, where the Newtonian view is denigrated as follows:

> The former [Newtonian] empty figment of reason, since it imagines an infinity of real relations without any things which are so related, pertains to the world of fable. (Kant 1770, p. 62)

One problem with this conservative reading is that after 1768 Kant never used incongruent counterparts as an argument against Leibnizian relationism *per se*. In the Inaugural Dissertation, Kant's complaint against Leibniz and his followers is that

> they dash down geometry from the supreme height of certainty, reducing it to the rank of those sciences whose principles are empirical. For if all properties of space are borrowed only from external relations through experience, geometrical axioms do not possess universality, but only that comparative universality which is acquired through induction. . . . (Kant 1770, p. 62)[20]

It is true that in both the Inaugural Dissertation and the *Prolegomena*, incongruent counterparts are used as an argument against relational space — but not as an argument against relational space vs. absolute space (as in 1768) but as an argument against any conception of space, relational or absolute, that would make space something objective. Another difficulty is that post-1768 Kant used incongruent counterparts to argue directly that space is something intuited and not merely conceived (Inaugural Dissertation) and that space is not a quality inherent in things in themselves (*Prolegomena*). The argument does not have the indirect pattern required by the conservative reading.

More importantly, the conservative reading neglects the real possibility that incongruent counterparts were not merely recycled to buttress Kant's mature view of space but rather played a direct role in generating this view. Several commentators have remarked that the Inaugural Dissertation is foreshadowed in the penultimate paragraph of the 1768 essay, where Kant says that "absolute space is not an object of external sensation, but rather a fundamental concept, which makes all these sensations possible." (1768, p. 43). What needs to be emphasized more strongly is the impetus provided by the tension in the 1768 argument (see Section 1 above). Kant could hardly have failed to have been aware, if only unconsciously, that Newtonian absolute space squared no better than Leibnizian relational space with his claim that the difference between right and left rests on an inner principle. The resolution called either for an abandonment of this claim or else the exhibition of a *tertium quid* to Newton and Leibniz. But Kant was unwilling to abandon the claim, and if space is regarded as the objective structure of things in themselves, there is seemingly no third alternative. Thus, the 1768 essay contained both the problem and the germ of a solution. The notion that space is something that "makes all of these sensations possible" became Kant's doctrine that space is a form of outer intuition, allowing an escape from the corner into which he had painted himself in the 1768 essay and allowing incongruent counterparts to be seen, as they were from 1770 onward, as an objection to both the absolute and relational views in so far as they deny that space belongs to the subjective constitution of the mind.

This reading of Kant makes more natural the otherwise startling and abrupt shift that occurs between 1768 and 1770. Unfortunately, confirmation is to be found not so much in the Inaugural Dissertation as the *Prolegomena*. In the latter, Kant repeated his original claim that the differences between right and left are "internal differences."[21] But instead of using this claim as a refutation of relationism, Kant combined it with the further claim that the differences are ones which "our understanding cannot describe as internal," and presumably this is so whether the understanding is using the concepts of Leibnizian or Newtonian theory. The upshot is that

> These objects are not representations of things as they are in themselves and as some mere understanding would know them, but sensuous intuitions, that is, appearances whose possibility rests upon the relation of certain things unknown in themselves to something else, namely to our sensibility. (Kant 1783, p. 30)

Thus, incongruent counterparts are not being used to adjudicate between relational and absolute space but to reject both in so far as they are supposed to apply to things in themselves. Substituting space as the form of external intuition for Newtonian absolute space allowed Kant to retain the 1768 idea that "region is related to space in general as a unity, of which each extension must be regarded as a part." Thus, in the *Prolegomena* we find:

> Space is the form of the external intuition of this sensibility, and the internal determination of any space is possible only by the determination of its external relation to the whole of space, of which it is a part (in other words, by its relation to external sense). (Kant 1783, p. 30)[22]

If this less conservative reading of Kant's philosophical development is accurate and if the analysis of the difference between right and left given in the preceding sections is correct, we have another example of a major philosophical system that is rooted in a mare's nest of confusions.

5. CONCLUSION

Whatever the correct account of the role of incongruent counterparts in Kant's philosophical development, it is hard to find in his writings the suggestion for even half-way plausible argument to the effect that the very existence of incongruent counterparts establishes the reality of absolute space. The addition of the consideration of parity non-conservation makes an interesting but not decisive difference. If he is willing to add enough epicycles to his theory, the relationist can deflect any objection launched by the absolutist. This judgment is inductively well confirmed, and the case of incongruent counterparts provides but another instance. The need to add epicycles is not necessarily an indication of falsity, but the accumulation of enough epicycles may cause one to lose interest in the theory.

NOTES

[1] There are a number of different senses in which space can be or fail to be 'absolute.' For present purposes, to say that space is absolute is to say that space as a collection of points or regions is ontologically prior to the material bodies it contains and that spatial relations among bodies are derivative of the spatial relations holding among the space points occupied by the bodies. See Sklar (1976) and Friedman (1983) for further discussion.

[2] The Newtonian version of this argument was mistaken in holding that adequate laws of motion must appeal to an absolute concept of velocity or change of position. However, adequate laws do need to appeal to absolute acceleration, and the absolutist can proceed from there; but see Sklar (1976, pp. 229—232).

[3] Euler's paper appeared in the *History of the Royal Berlin Academy*, 1748. An English translation is to be found in Koslow (1967).

[4] The full quotation reads: "Let it be imagined that the first created thing were a human hand, then it must necessarily be either a right hand or a left hand. In order to produce the one a different action of the creative cause is necessary from that, by means of which its counterpart could be produced." (1768, p. 42)

[5] Leibniz's argument to "confute the fancy of those who take space to be a substance" is couched not in terms of the shift transformation I used above but in terms of "changing East to West," (Alexander 1984, p. 26) an operation that can be interpreted either as rotation by 180° or as mirror image reflection.

[6] By 'space' I mean (in absolutist terms) something having, at a minimum, a manifold structure. The manifold may carry additional structures, e.g. affine or metric, and while such additional structures are needed for the discussion below, they are not needed for the definition of orientability. If such a space is n-dimensional, it is said to be *orientable* just in case there exists a continuous, non-vanishing field of n-ads of linearly independent tangent vectors. Equivalently, choose any closed loop in the space and erect an n-ad of linearly independent vectors at some point on the loop. Then carry the n-ad around the loop by any means of transport that is continuous and keeps the vectors linearly independent. Then upon return to the starting point the transported n-ad should not differ from the original by the reflection of any axis.

[7] Van Cleve (1987) argues that orientable $n + 1$ dimensional spaces serve the same function here as do non-orientable n dimensional spaces. I think that this contention is incorrect. First, he maintains that "a nonorientable space of n dimensions is possible only in an ambient space of $n + 1$ dimensions." (p. 45; this volume, p. 212) This contention is false. One way to conceive of a non-orientable n space is to embed it in an orientable $n + 1$ space, just as one way to picture a curved n space is to embed it in a flat $n + 1$ space. But non-orientable and curved spaces exist in their own right — they possess intrinsic characterizations and there is no need to define them by means of higher dimensional embedding spaces. Second, he takes higher dimensional orientable spaces to show that "there are possible spaces in which a solitary hand would be neither right nor left; hence a hand with the same internal relations as a given right hand might fail to be right, for it might be the sole occupant of such a space. So the rightness of a hand is not entailed by its having the internal relations it does, contrary to internalism." (p. 52; this volume, p. 217) But an n-dimensional 'hand' in an $n + 1$ space is not a *hand* in even the most minimal sense; namely, it is not an enantiomorph (see definition below). How then can considerations about higher dimensions, which convert *hands* into 'hands' show anything about *hands*? Perhaps the answer is that higher dimensional spaces show that the enantiomorphism of an object depends upon the dimensionality of the space and, in that sense, is not an internal property of the object. But that sense is irrelevant to the absolute-relational controversy. If the relationist cannot give a satisfactory account of the dimensionality of space, then relationism fails even before the question of incongruent counterparts arises. If the relationist can produce a satisfactory account, then the question becomes whether n-enantiomorphism in an n-dimensional space is a matter of the internal relations of the body. I show in Section 2 that the relationist can give a positive answer to this question. I also show why being right or left handed is not the sort of property a relationist would want to count as being internal.

[8] The full quotation reads: "The region, however, to which this order of the parts is directed, is related to space outside, but not with reference to its localities, for this would be nothing else than the position of just those parts in an external relation; region is rather related to space in general as a unity, of which each extension must be regarded as a part." (1768, p. 37)

[9] Such a space must be of constant curvature.

[10] This definition presupposes that space is locally orientable, a presupposition assured by the restriction to spaces that are manifolds. 'Reflection' means a single reflection. The product of an even number of reflections is equivalent to a rotation and/or translation. See note 14.

[11] Here I depart company with Nerlich (1976, p. 35; this volume, p. 157) whose definition of enantiomorphism is non-local.

[12] Various technical versions of the representation relation have been proposed; see Friedman (1983), Mundy (1983, 1986), and Earman (1979, 1987). The technicalia need not be considered here.
[13] Perhaps this is the place where, as Sklar (1974) suggests, the relationist must make use of possibilia. However, I do not think that the core of the absolute-relational controversy turns on the use or status of possibilia.
[14] In three spatial dimension the parity operation is $x \rightarrow -x$, $y \rightarrow -y$, $z \rightarrow -z$. In spaces of an even number of dimensions the parity operation does not correspond to mirror image reflection since the product of an even number of reflections is equivalent to a rotation and/or translation.
[15] This picture is taken from Sakurai (1964).
[16] There is a possible, but measure zero, set of cases where the relative frequencies do not conform to the propensity probabilities.
[17] The appropriate sense of universality here is not captured by conditions of the syntactic form of the law statement (e.g., that laws are written in universally quantified form or that their mathematical expression is generally covariant). Rather the requirement is that the law be invariant under space—time translations. Such invariance conditions are best expressed in semantic model-theoretic terms. See Rynasiewicz (1986).
[18] When space—time and CPT invariance are taken into account, a more complicated conclusion emerges. See Earman (1971).
[19] Although something of a straw man, the conservative interpretation is nevertheless a useful straw man. It is not too far removed from the reading Broad (1978) gives.
[20] See also Kant (179?) where it is claimed that Leibniz cannot account for the fact that 'Space is three dimensional' is "an apodictic a priori proposition."
[21] In both the 1768 essay and the *Prolegomena* Kant uses the same word, 'inner,' which is variously translated as 'inner' and 'internal.'
[22] Broad attempts to preserve a conservative reading of the *Prolegomena* by taking Kant to be arguing, first, that incongruent counterparts show that space is absolute and, second, that absolute space has a property that is incompatible with it being a thing in itself, i.e., "In absolute space the existence and nature of every part would be dependent upon the existence and nature of the whole." (Broad 1978, p. 41) But Broad's reconstruction is a very strained reading of Kant's announced claim that the "paradox" of incongruent counterparts itself leads one to suspect that "the reduction of space and time to mere forms of sensuous intuition may perhaps be well founded." (1770, p. 29) In addition, the *Prolegomena* does not contain the quotation Broad attributes to Kant but rather the assertion "That is to say, the part is possible only through the whole, which is never the case with things in themselves . . ." (1783, p. 30) The phrase 'That is to say' indicates that Kant was not giving a separate argument but only emphasizing the previous line that "Space is the form of external intuition of this sensibility . . ."

REFERENCES

Alexander, H. G.: 1984, *The Leibniz-Clarke Correspondence* (New York: Barnes and Noble).

Alexander, P.: 1984/85, 'Incongruent Counterparts and Absolute Space', *Proceedings of the Aristotelian Society* **85**, 1—21.

Allison, H.: 1986, *Kant's Transcendental Idealism* (New Haven, CN: Yale University Press).

Bennett, J.:1970, 'The Difference Between Right and Left', *American Philosophical Quarterly* **7**, 175—191. This volume, pp. 97—130.

Broad, C. D.: 1978, *Kant: An Introduction* (Cambridge: Cambridge University Press).

Buchdahl, G.: 1969, *Metaphysics and Philosophy of Science* (Cambridge, MA: MIT Press).

Buroker, J.: 1981, *Space and Incongruence* (Dordrecht: D. Reidel).

Crawford, F. S. et al.: 1957, 'Detection of Parity Non-Conservation in Λ Decay', *Physical Review* **108**, 1102—1103.

Earman, J.: 1971, 'Kant, Incongruous Counterparts and the Nature of Space and Space-Time', *Ratio* **13**, 1—18. This volume, pp. 131—149.

Earman, J.: 1979, 'Was Leibniz a Relationist?' in P. French and H. Wettstein (eds.), *Midwest Studies in Philosophy*, Vol. 3 (Minneapolis, MN: University of Minnesota Press).

Earman, J.: 1986, 'Why Space Is Not A Substance', to appear in *Pacific Philosophical Quarterly*.

Earman, J.: 1989, *World Enough and Space-Time: Absolute vs. Relational Theories of Space and Time* (Cambridge, MA: MIT Press).

Friedman, M.: 1983, *Foundations of Space-Time Theories* (Princeton, NJ: Princeton University Press).

Gardner, M.: 1969, *The Ambidextrous Universe* (New York: Mentor Books).

Kant, I.: 1768, 'Concerning the Ultimate Foundation of the Differentiation of Regions in Space', in Kerferd and Walford (1968). Editor's note: The Handyside translation is reprinted in this volume, pp. 27—33.

Kant, I.: 1770, 'On the Form and Principles of the Sensible and Intelligible World', in Kerferd and Walford (1968).

Kant, I.: 1781, *Critique of Pure Reason*, trans. by N. K. Smith (London: Macmillan, 1933).

Kant, I.: 1783, *Prolegomena to Any Future Metaphysics*, trans. J. W. Ellington (Indianapolis, IN: Hackett, 1977).

Kant, I.: 179?, *What Real Progress Has Metaphysics Made in Germany since the Time of Leibniz and Wolff?* trans. by T. Humphrey (New York: Arabis Books, 1983).

Kerferd, G. B. and Walford, D. E. (eds.): 1968, *Kant: Selected Pre-Critical Writing* (New York: Barnes and Noble).

Köhler, H.: 1720, *Merkwürdige Schriften welche . . . zwischen dem Herrn Baron von Leibniz und dem Herrn D. Clarke über besondere Materien der natürlichen Religion in Franzos, und Englisher Sprache gewechselt und . . . in teutscher Sprache* (Jena).

Köhler, H.: 1740, *Des freyherrn von Leibnitz kleinere philosophische schriften . . . nebst . . . herrn Thümmings antwort auf herrn Clarkens leztes schreiben und . . . herrn prof. Köhlers discurs über das licht der natur . . .* (Jena).

Koslow, A.: 1967, *The Changeless Order* (New York: G. Braziller).

Lee, T. D. and Yang, C. N.: 1956, 'Question of Parity Conservation in Weak Interactions', *Physical Review* **104**, 254—258.

Lucas, J. R.: 1984, *Space, Time and Causality* (Oxford: Oxford University Press).
Mundy, B.: 1983, 'Relational Theories of Euclidean Space and Minkowski Spacetime', *Philosophy of Science* **50**, 205—226.
Mundy, B.: 1986, 'Embedding and Uniqueness in Relational Theories of Space', *Synthese* **67**, 383—390.
Nerlich, G.: 1973, 'Hands, Knees, and Absolute Space', *Journal of Philosophy* **70**, 337—351. Revised version reprinted in this volume, pp. 151—172.
Nerlich, G.: 1976, *The Shape of Space* (Cambridge: Cambridge University Press).
Remnant, P.: 1963, 'Incongruent Counterparts and Absolute Space', *Mind* **72**, 393—399. This volume, pp. 51—59.
Rynasiewicz, R. A.: 1986, 'The Universality of Laws in Space and Time', *PSA 1986*, Vol. 1 (East Lansing, MI: Philosophy of Science Assoc.).
Sakurai, J. J.: 1964, *Invariance Principles and Elementary Particles* (Princeton, NJ: Princeton University Press).
Sklar, L.: 1974, 'Incongruous Counterparts, Intrinsic Features and the Substantivality of Space', *Journal of Philosophy* **71**, 227—290. This volume, pp. 173—186.
Sklar, L.: 1976, *Space, Time, and Space-Time* (Berkeley, CA: University of California Press).
Van Cleve, J.: 1987, 'Right, Left, and the Fourth Dimension', *Philosophical Review* **XCVI,** 33—68. This volume, pp. 203—234.
Walker, R. C. S.: 1978, *Kant* (London: Routledge and Kegan Paul).
Winterbourne, A. T.: 1982, 'Incongruent Counterparts and the Intuitive Nature of Space', *Auslegung* **9**, 85—98.
Wolff, R. P.: 1969, *Kant's Theory of Mental Activity* (Cambridge, MA: Harvard University Press).

GRAHAM NERLICH

REPLIES TO SKLAR AND EARMAN

I am glad to have this chance to thank Sklar and Earman for their acute and searching criticisms of my thoughts about hands and for a deeper and wider debt to all I have learned from their distinguished writings on space and time. I thank the editors, too, for this opportunity to make some sort of reply to them.

REPLY TO SKLAR

Sklar shows beyond question that I overstated Kant's (and my own) case. But he surely overstates his own in saying, as I think he does, that we have *nothing* to learn about the ontology of space from the case of hands. The solitary hand grabs us. The very existence of this volume suggest that some thing about it rightly captures our attention.

Kant throws down a gauntlet (left?) to relationists. It's a challenge. It is not that it leaves the relationist with nothing to say (though I imprudently claimed so). But he needs to say something elegant, explanatory, illuminating – plausible independently of the plausibility of his wider position. He needs to match the elegance, intuitiveness and generative power that realism offers us. If that is the point about hands, then they must have a unique feature which relationism can handle only maladroitly.

And so they do. The one hand has some character – enantiomorphy or homomorphy. Handedness is unique, for left and right hands can be similar up to reflection and plausibly reflection preserves all properties which are properly intrinsic. Hands, that is to say, may be incongruent *counterparts*. So what makes them differ is something extrinsic, a relation which they bear differently to space globally in terms of its topology. So hands engage with properties of space that are both deep and global in a simple and intuitive way unmatched by any other example known to me.

Sklar dissents but I think he is wrong. Consider the challenge (in his Section 2) that same-sided and opposite-sided figures pose all the problems posed by hands. They don't, since the difference in the figures as wholes lies in differing relations among the parts: we can describe this — simply but not trivially — in terms of purely local relations. They are not alike up to a reflection as hands are: the difference of 'left' and 'right' resists any local non-trivial description. Further, as Chris Mortensen pointed out to me, if we move either of these figures *as a whole* (i.e. as deformable surfaces) by any continuous motion we will never map one onto the other. Sklar does not speak of each figure as one object but as objects, though I do not see how that can help him. It is only by moving the parts through separate motions that we can get the outside bar inside, whatever the dimensions or topology of the space. Clearly this just illustrates the point that, unlike these figures, handedness does not depend on the relations of parts to one another. Same-sideness is not an intrinsic feature of object arrangements; it does not depend on the orientability or any other global property of the space save for its dimensions; we can transform same-sided into opposite-sided figures only by moving the objects through distinct motions in a space of higher dimensions.

Sklar's relationist resorts to possible relations among possible things; I have already grumbled a little about that. But there is something more, darkly mentioned in my sections 1 and 8, and worth expanding. We do not generally ascribe a character to things because of their *possible* relations to *possible* things unless we ascribe it as a disposition. Handedness is not a disposition. It is not causal, not a power. It springs from the way a hand is (has been?) embedded in space. The one hand's being handed, analyzed as Sklar suggests, would license our each being a "twon" (on my facetious usage), for isn't it possible for each of us to have been born in the same birth as a possible other? But there's no such character; twons are factitious. Why should we think of spatial attributes, *construed as Sklar wants*, as any less fanciful?

Hands are neither left not right, enantiomorphic or homomorphic, intrinsically. Sklar writes as if there were some intrinsic orientation of a hand but I find that obscure. What is intrinsic is a matter of the relations of the parts of the object to one another. It is no matter of relations between things and what is outside them. I argued, and he does not directly dispute, that these intrinsic relations (of distance and angle) are the same for left and right hands and that all intrinsic properties are preserved in reflection. The other relations he describes as intrinsic all involve some relation of a hand to the wider space outside it and are not, for my money, intrinsic.

The difference between a left hand and a right is a difference in the way the hands are embedded in some wider space containing both. I called it a difference in the way in which they are entered in the space and drew attention to Kant's metaphor for this — that the difference involves different actions of the creative cause. However seriously Kant took that metaphor, I gave it no role but that of a metaphor. It is not clear to me whether Sklar recognizes that a difference does lie in the differing entries or embeddings. But that difference is a crucial one.

So what is the peculiar virtue of hands? They are the most familiar and domestic of things. We easily tell the difference between left and right but can't easily describe or analyze it. It is surprising and revealing that this domestic difference takes us at one bound to a deep, global property of space without intervening relations of a hand to any other object. It relates handedness to the nature of space but not by simply giving that property to the bit of space which the hand fills. A realist does *not* offer to solve the problem by ascribing handedness to a piece of space in such a way that we can then simply transfer the property in question to the object itself, thus cancelling all reference to space. Whether it be the hand or the handspace, we must relate it to space considered as a unity, shaped in some definite way. It shows how very simply handedness can direct us to the global shape of space and how a feature which looks as if it lies in mere relations of one object to others, does not lie there.

If we are realists about space we have an immediate, intuitive, and, I find, delightfully elegant account of what is at issue. For we can find a host of rich analogies between familiar shapes of material things and the shape of space itself to make it plain at a stroke what is going on in the case. I tried to illustrate that sense of insight and discovery in Section 7 of the paper by talking of paper strips and the like. That is the great intuitive advantage of realism. It has the power to generate an understanding, to lead the imagination readily on to further insights.

Of course the relationist has his version of handedness and it was my mistake to dispute that. But notice that Sklar, throughout, puts what he has to say in terms of space itself and then goes on to tell us, correctly enough, that the relationist has his tale to tell of it. I guess that is because he recognizes how readily we will grasp that way of presenting things. By contrast the relationist version is laborious, opaque to intuition, inelegant and long — so I think the comparison I suggest here is the right way to form one's judgement. There is little question in my mind to which conclusion it will lead us.

REPLY TO EARMAN

Earman's splendid paper stresses how little we have been able to make of Kant's remarks about an inner principle: it also states tangibly and more plausibly than before (for my money) how a relationist might reject the premise that any hand must be determinate as to which, left or right, it is. I have nothing to add about the idea of an inner principle nor to the deeper exegesis of Kant, therefore.

Unlike Earman, I think the case I presented has everything to do with how to argue for a realist position from the case of hands. Kant offers a challenge. It works if its intuition about the character of a lone hand is more winning than that of the relationist. It isn't itself an argument as to *why* the relationist can't meet the challenge, though perhaps it is widely thought that it should be. Lack of that argument doesn't mean that it is *not* a challenge. Surely challenge arguments are pertinent to the claims of reductionists that they can say, and say well, all they need to say. Challenges needn't be proofs: they are vulnerable to any plausible, elegant, non-circular relationist analysis. But, even at this late date, the challenge has not plainly been met, and Earman's writing on the issue as he's done suggests that he may agree with that.

I see little point in the notion of local enantiomorphy that Earman (and Sklar) define (and Earman calls enantiomorphy). I have two objections to it: it is too realist to help the debate, yet is not a real property of anything.

(i) To define (local) enantiomorphy you still have to quantify over all the rigid-motion paths in some space or other. Of course you can take just a neighbourhood as the space in question and ignore its wider containing space. That does nothing to diminish the absolutist, realist or substantivalist character of the definiens. A neighbourhood is just the same sort of metaphysical entity as space globally, even if it is smaller. And you still have to talk about all of it.

(ii) In any event, the local 'property' isn't really a property: local enantiomorphy (or local anything else of the kind) is a *recessive* 'property' relative to the corresponding global one. My actual concrete left hand is 'locally enantiomorphic' relative to a neighbourhood, a part of space, which contains nothing but it. And, relative to that neighbourhood, it's 'neither right nor left', since that can be determined only when we say in which of two ways the neighbourhood is actually embedded in our space as a whole (in which, further, there are other hands and a convention in use for 'left,' etc.). But this does not mean that there is any sense at all in which the hand is not left. Its character relative to a wider space always dominates over any local definition. Further, this same concrete hand is not really a left hand unless space is globally orientable. Again, the global character always dominates, the local one recedes. Kant's challenge is to explain the dominant character, the actual property which, intuitively, one hand would have. That's why what counts is space in general as a unity.

This does not deny that, in the generous region of spacetime we can observe, we find that we cannot move a left hand into congruence with a right.

Rather than focussing on enantiomorphy (local or other), the relationist should start with asymmetry (see the end of my section 1). At least that *looks* like an intrinsic property. If, after all, it, too, is a property which one has to define by motions, these can be confined, perhaps, to what falls within the space filled by the hand. Asymmetry carries no blatant realism about space. Furthermore, it is not recessive. Nor does it entail (by itself) enantiomorphy, thus inviting a charge of begging the question. Yet, given an orientable space or neighbourhood, enantiomorphy follows. My gambit is just to give the relationist asymmetry free of charge. I don't see how it offers him less than enantiomorphy does. Then his problem is how to account for the plausible intuition about the *one* hand's *dominant* character, whatever it is.

If I'm right about which concepts count, then the challenge is to define handedness for *one* hand. It was not to incongruent counterparts in Earman's p. 240 sense that Kant appealed to refute the Leibniz-relationist account of space. Kant's (or, if not his, then my) challenge is to account for a propery of one hand alone. Of course, if there are two hands, the property Earman calls incongruent counterparthood can be ascribed to each of them in terms of relations it bears to the other and of some deeming-it left ceremony (just as he describes). But, as he says of the substantivalist's account of things, that was never in doubt and is beside the point. I'm puzzled why Earman thinks he has met the challenge of the one hand. It may well be that I miss his drill here. He neither defines nor even talks about the properties which Kant, and I, see as the problematic ones — (dominant) leftness, enantiomorphy, homomorphy — of the one hand.

A last point against (local) enantiomorphy. Earman claims without argument that his F's, as he describes them on pp. 241—242, are enantiomorphs. But I think this is mistaken. Given his definition of enantiomorphy, his description of an F would have to entail an existential conclusion that *there is* an orientable neighbourhood in which the F is embedded, one large enough to permit its reflection. How can the given description entail that? As I said before, in postulating possible worlds we standardly trail space tacitly in the wake of objects; but that is no legitimate help to relationists. Earman's description would seem to entail that F's are asymmetrical. I think this allows him quite as much of what he needs as he can get from local enantiomorphy.

I now think that relationism can always complete its analysis. For I think an account can be given of the shape of space in terms of the structure of its parts — neighbourhoods, paths, points and so on. A relationist will have his story about these and the relations among them. He can just mimic what the realist says. So he is invulnerable to the charge of incompleteness even in so testing and revealing a case as the one hand. His real problem is circularity and to this charge I think he really is vulnerable — specifying spatial relations is not logically independent of specifying a space. But to pursue that theme further would abuse the editors' hospitality.

WILLIAM HARPER

KANT ON INCONGRUENT COUNTERPARTS

Consider your right hand and a mirror image duplicate of it. Kant calls such pairs incongruent counterparts. According to him they have the following puzzling features. The relation and situation of the parts of your hand with respect to one another are not sufficient to distinguish it from its mirror duplicate. Nevertheless, there is a spatial difference between the two. Turn and twist them how you will, you cannot make one of them occupy the exact boundaries now occupied by the other. In his 1768 paper, 'Concerning the Ultimate Foundations of the Differentiation of Regions in Space', Kant uses these claims to argue against relational accounts of space and goes on to argue that the difference between incongruent counterparts depends on a relation to absolute space as a whole. In his 1770 *Inaugural Dissertation* he argued that this difference could not be captured by concepts alone but required appeal to intuition. In the *Prolegomena* (1783) and again in the *Metaphysical Foundations of Natural Science* (1786) Kant appealed to these puzzling features of incongruent counterparts to support his transcendental idealism about space.

I. KANT'S 1768 ARGUMENT: BASICS

(a) *The Argument Against Relationalism*

There is so much controversy over what Kant's argument against relationalism is supposed to be that it may be worth looking at the relevant texts in some detail.[1]

It is already clear from the everyday example of the two hands that the figure of a body can be completely similar to that of another, and that the size of the extension can be in both, exactly the same; and that yet, however, an internal difference remains: namely, that the surface that includes the one could not possibly include the other. As the surface limiting the bodily space of the one cannot serve as a limit for the other, twist and turn it how one will, this difference must, therefore, be such as rests on an inner principle. This inner principle of difference cannot, however, be connected with the different way in which the parts of the body are connected with each other. For, as one sees from the given example everything can be perfectly identical in this respect (K & W, p. 42; AK II, p. 382, lines 24—36).

As I see it Kant is making two interesting claims about pairs such as a hand and a mirror duplicate of it.

Premise (1) The spatial relations among themselves of the parts of one are not sufficient to distinguish it from the other.

Premise (2) Nevertheless, there is a spatial difference between them — the surface limiting the bodily space of the one cannot serve as a limit of the other, twist and turn it how one will.

These two claims are premises in Kant's argument. They formulate what he takes to be the central puzzle about incongruent counterparts. They certainly raise questions. What does Kant mean when, as in the present passage, he claims that the differences between incongruent counterparts are internal and must rest on an inner principle? How is it that these (inner) spatial differences are not to be found in the spatial relations of the parts among themselves (these are surely the spatial relations internal to the object)? Clearly, more needs to be said about these first two premises. For the present, however, I shall go on to the next passage in Kant's text in order to extract another premise.

Kant introduces this premise through a thought experiment wherein the reader is asked to imagine a universe consisting of a lone hand.

Premise (3) Let it be imagined that the first created thing were a human hand, then it must necessarily be either a right hand or a left hand (K & W, p. 42; AK II, p. 382, lines 36—37, p. 383, line 1).

The passage continues as follows:

> In order to produce the one a different action of the creative cause is necessary from that, by means of which its counterpart could be produced (K & W, p. 42; AK II, p. 383, lines 1—3).

This continuation seems to be a colorful way of emphasizing the fundamental claim of premise 2 — that there really is a difference between incongruent counterparts.

The point of the thought experiment premise is to rule out the suggestion that the difference between incongruent counterparts is due to a difference in their spatial relations to other objects. This leads, in the next paragraph, to a reductio argument against relational accounts of space such as that of Leibniz and his followers.

> If one accepts the concept of modern, in particular, German philosophers, that space only consists of the external relations of parts of matter, which exist along side one another, then all real space would be, in the example used, simply that *which this hand takes up*. However, since there is no difference in the relations of the parts to each other, whether right hand or left, the hand would be completely indeterminate with respect to such a quality, that is, it would fit on either side of the human body. But that is impossible (K & W, p. 43; AK II, p. 383, lines 5—12).

The premises we have extracted are sufficient to refute a relationalist account of the sort Kant considers here. By Premise 3 the lone hand must be either right or left, and the only spatial relations available are those of the parts of the hand among themselves. By Premise 2 the difference between right and left must be accounted for, but by Premise 1 the relations of the parts among themselves are not sufficient to do it. Thus a defender of relational accounts of space must either reject one or more of Kant's premises or offer a different sort of account.

When we examine Kant's premises in more detail we shall want to see whether we can find an interpretation which makes this argument sound as well as valid. Our attempt to interpret the various premises can, perhaps, be guided somewhat by the role they play in the argument. This suggests that the real point of the difference between incongruent counterparts asserted in Premise 2 is just that the difference is among the facts that an adequate theory of space must be able to account for. I believe that Kant's remarks about inner differences resting on an inner principle and his remarks about different actions of the creative cause can all be regarded as ways of making this same point.

With this in mind we can, also, appreciate better what Kant is doing in the very last sentence of the present passage, where he tells us that an indeterminate hand is impossible because it would have to fit on either side of a human body. This is not, as Peter Remnant (1963) would have it, a proposal for an operationalist definition of left or right in terms of possible relations of a hand to a human body.[2] Rather, it is more reasonably construed as a vivid illustration of the fact that there is a physical difference in shape between right and left hands. It is designed, as are the other illustrations of this point, to support both

Premises 2 and 3. In later sections we shall look more closely into the question of how well these premises can be supported.

(b) *Absolute Space?*

In each of the three conjuncts of the one-sentence paragraph following the reductio argument in the text, Kant goes on to draw additional conclusions from his discussion of incongruent counterparts.[3] Here is the first of these conjuncts:

> From this it is clear that the determinations of space are not consequences of the situations of the parts of matter relative to each other; rather are the latter consequences of the former (K & W, p. 43; AK II, p. 383, lines 13—15).

The first part repeats Kant's rejection of accounts which attempt to reduce determinations of space to spatial relations among parts of matter. It certainly follows from Kant's foregoing reductio argument. The second part, however, goes on to turn the priority around. This claim of the priority of space over spatial relations among parts of matter is an additional conclusion.

C1 The spatial relations among parts of matter depend on prior determinations of space.

It doesn't follow from our three premises as straightforwardly as the reductio argument did. More needs to be said if we want to interpret Kant as having a valid argument to this conclusion.

The second conjunct in the one sentence paragraph under consideration expands upon this idea of the priority of space.

> It is also clear that in the constitution of bodies differences, and real differences at that, can be found; and these differences are connected purely with *absolute* and *original* space, for it is only through it that the relation of physical things is possible (K & W, p. 43; AK II, p. 383, lines 15—18).

This passage tells us that the differences between incongruent counterparts are

> . . . real differences at that,

which suggests that they are differences a theory of space ought to be able to account for. This fits well with the role of Premise 2 in our explication of Kant's reductio argument. The passage goes beyond the general priority claim of C1 in two ways. It suggests specifically that it is these real differences between incongruent counterparts that depend on space and also that this space they depend on is *absolute* and *original*. I think it is useful to read this passage with some passages from Kant's introductory remarks in mind.

> . . . region is related rather to space in general as a unity, of which each extension must be regarded as a part . . . (K & W, p. 37; AK II, p. 378, lines 2—4).

It is clear that Kant sees the difference between incongruent counterparts to be a difference of region (see Sections VI and VII below), so we can suggest that Kant intends the difference between incongruent counterparts to be related to absolute space in general as a unity. Accordingly, we can render the additional conclusion under consideration as,

C2 The difference between incongruent counterparts is related to *absolute, original* space in general as a *unity*.

This conclusion has been the focus of many commentators who see Kant as providing an argument for absolute space.

Just as with C1, an interpretation that gives Kant a valid argument to C2 will have to either add additional premises or unpack more from Kant's explicit premises. Van Cleve (1987) achieves a valid argument by introducing an additional premise,

> A hand is left or right (as the case may be) either (a) solely in virtue of the *internal* relations among the parts of the hand, or (b) at least partly in virtue of the *external* relations of the hand to something outside it — if not other material objects, then absolute space (Van Cleve 1987, p. 37),

that is not to be found in Kant's text and that is incompatible with Kant's later views on space. Graham Nerlich (1973) followed the other strategy. As we shall see (Section II), he proposes to interpret the last part of Kant's Premise 2 about the difference between incongruent counterparts,

> . . . the surface limiting the bodily space of the one cannot serve as a limit of the other, twist and turn it how one will . . .

in such a rich way as to give conclusion C2 directly, without much need at all for the details of Kant's own argumentation.

Nerlich and Van Cleve have in mind a substantival version of

Newton's absolute space. Kant's introductory remarks certainly suggest that this is what he had in mind as well,

> a clear proof can be found that *absolute space has its own reality independently of the existence of all matter and that it is itself the ultimate foundation of the possibility of its composition* (K & W, p. 37, Kant's emphasis; AK II, p. 378, lines 9—11)

as does the following passage from his concluding paragraph:

> A reflective reader will therefore regard the concept of space in the way geometers regard it, and also as perceptive philosophers have taken it up into the theory of natural science, as other than a mere entity of reason [Gedankending] (K & W, p. 43; AK II, p. 383, lines 24—27).

There is no doubt that Newton is intended as the perceptive natural philosopher; nevertheless, there is some room for doubt about Kant's commitment here to absolute space in any fully substantival sense. Such doubts are suggested by the clause Kant adds to the end of the sentence.[4]

> Nonetheless, there is no lack of difficulties surrounding the concept when one tries to grasp with the ideas of reason its reality, evident enough to inner sense (K & W, p. 43; AK II, p. 383, lines 27—29).

This passage also suggests something of Kant's later account of space. It especially suggests his discussion in the Inaugural Dissertation (1770) of the need to appeal to intuition to distinguish one incongruent counterpart from another.

Consider the last of the three conjuncts in the one sentence paragraph we have been exploring:

C3.1 It is also clear that since absolute space is not an object of external sensation, but rather a fundamental concept, which makes all these sensations possible in the first place, . . . (K & W, p. 43; AK II, p. 383, lines 19—20).

Here we have a conclusion about a role as the foundation for our external sensations, for the absolute space Kant has under consideration. This suggests that Kant is already on his way toward his later doctrine of space as a form of outer sensibility. Kant goes on to complete the sentence with the following conclusion about the role of pure space as a foundation for perception of forms of bodies.

C3.2 . . . , we can only perceive through the relation to other

> bodies that which, in the form of a body, purely concerns its relation to pure space (K & W, p. 43; AK II, p. 383, lines 21–23).

I think it's clear that it is the absolute space Kant has been considering that is supposed to play the role he attributes to pure space here. Here again we have Kant's absolute space playing a fundamentally epistemological role. This role is akin to his later role for space as the pure form of outer intuition.

Kant's discussion of incongruent counterparts cannot avoid raising epistemological issues about a foundational role for space for our perception of bodies. If his argument for C2 is correct then such perceptible features as the shape of a right hand depend upon the relation of that hand to space as a unity. I shall attempt to show how consideration of such issues lead rather naturally to Kant's later discussions in the *Inaugural Dissertation, Prolegomena* and the *Metaphysical Foundations of Natural Science.*

II. INCONGRUENCE, ORIENTABILITY AND DIMENSIONALITY

In Premise 2 Kant claims that the difference between a hand and an incongruent counterpart of it is such that

> — the surface limiting the bodily space of the one cannot serve as a limit of the other, twist and turn it as one will.

This suggests the idea that no *rigid motions* can make one come to occupy the space now occupied by the other.[5] Rigid motions can be represented by mappings. A rigid motion is some combination of translations and rotations. Another sort of mapping, a reflection about a plane parallel to the surface of the palm, represents a space that could be filled by an incongruent counterpart of a hand. According to Graham Nerlich (1973), the point Kant is getting at is captured by limitations on the sorts of mappings our space allows for.

> We can assert the following: Every reflective mapping of a hand differs in its outcome from *every* rigid motion of it. That *is* a matter of space in general and as a unity.
>
> Though this terminology is much more recent than Kant's, the ideas are old enough. I see no reason to doubt that they are just what he intended. Space *in general as a unity* is exactly what is at issue (Nerlich 1973, p. 343; this volume, p. 157).

Nerlich uses two dimensional 'toy model' spaces to give a fascinating illustration of how it is that topological features of a space — its dimensionality and orientability — are what provide the relevant limitations on the mappings it allows for.

Since Nerlich's discussion, the contrast between such two-dimensional orientable spaces as the surface of a flat strip of paper and such non-orientable two-dimensional spaces as a Möbius strip has become a standard topic for scholars investigating Kant's discussions of incongruent counterparts. An orientable two-dimensional space allows for two-dimensional incongruent counterparts. No rigid motions within such a space can make a left-handed

d

into a right-handed

b.

In a non-orientable two-dimensional space, however, rigid motions within the space can 'flip' a left-handed figure over so that it becomes congruent to its right-handed counterpart.

Nerlich points out that the orientability of the surface of a flat strip and the non-orientability of the surface that results from twisting it into a Möbius strip are matters of how the surface is pathwise connected globally. This suggests that for such a 'toy model' orientable two-space as a flat strip we can construct a rather direct argument for a version of Kant's conclusion C2,

C2(b) The difference between incongruent counterparts is related to space in general as a unity.

Consider our left- and right-handed figures

d b

on a clear vinyl plastic strip. Imagine the figures can be moved about and rotated on the surface of this strip. These translations and rotations represent in the model the 'rigid motion' maps allowed by the two-space.[6] As long as the strip is flat (or bent into a cylinder) no combination of these motions within the space can make one occupy the boundaries now occupied by the other. On the flat strip these left and right hand figures have the difference Kant attributes to incongruent counterparts. Now, join the far ends of our strip together with a half-twist to form a Möbius strip. Our two figures no longer have the differences — they are no longer incongruent. One can be made to occupy the exact boundaries now occupied by the other by moving it around the strip and then rotating it.

Another way we can make the two figures congruent is to pick one off the strip and turn it over in our three-space. It is a mathematical commonplace that figures which are incongruent counterparts in an n-dimensional Euclidean space can be made to coincide in an $(n + 1)$-dimensional Euclidean space. The higher dimension space affords new rigid motions which can turn the figures over to map them onto their counterparts. So, the incongruence of incongruent counterparts depends on the dimensionality of the space that affords rigid motions as well as upon the orientability of this space.

Nerlich thinks it is defensible to claim that our space has the relevant topological features to support three-dimensional incongruent counterparts.

> It seems to be pretty clearly the case that, as a matter of fact, there is no fourth spatial dimension that could be used to turn hands "over" so that they become homomorphs. No evidence known to me suggests that actual space is a non-orientable manifold (Nerlich 1973, p. 344; this volume, p. 159).

Perhaps our space is a globally orientable three-dimensional manifold, as it would be if it were a Euclidean three-space as Kant thought. Nerlich has not offered anything that would positively settle that his conjecture about the global topology of our space is correct.[7] What does seem clearly to be the case is that our space is locally three-dimensional and orientable.

James Van Cleve (1987) tentatively suggests thought experiments which he takes to show positively that a fourth dimension is impossible in our space.

> Now I would like to suggest that in connection with higher dimensions there may be intuitions of this positive kind. Visualize a trio of mutually perpendicular lines, and then try to add a fourth at right angles to the original three; I assume you will fail. Is this merely because you cannot see how to fit in the fourth line? or do you not see that the thing cannot be done?
>
> Another example may be more convincing. Imagine an intact sphere or a closed box, and then try to imagine an unbroken path that begins on the inside and winds up on the outside without ever penetrating the surface.

I assume again you will fail. Now is it merely that you cannot discover such a path? or do you not rather see that no such path is possible? Yet such paths would have to exist if space were four-dimensional (this is shown in the Appendix) (Van Cleve 1987, p. 64; this volume, pp. 226—227).[8]

Kant was committed to the idea that ostensive geometrical constructions provide intuitive warrant that can show impossibility. I think it not unlikely that something like Van Cleve's thought experiments could be made to satisfy the sort of constraints necessary to make them into something Kant would have counted as an *a priori* constructive demonstration that space has only three dimensions.

I have elsewhere (Harper 1984, 1986) attempted to argue that Kant's account of the *a priori* warranting power of ostensive constructions in geometry should be taken far more seriously than those myriads who have been unduly influenced by Reichenbach (1927) and his followers are wont to take it.[9] My investigation has revealed little that would support any application of Van Cleve's thought experiment to settle the global topology of our space,[10] but it does suggest that such considerations may well settle the dimensionality of the local space in which we can observe nearby objects.

Nerlich is using a global conception of incongruent counterparts. We can define this notion in terms of mappings.

Def. (GIC) Objects are *global incongruent counterparts*

iff

(1) No combination of translations and rotations can map one onto the other (they are incongruent).

(2) Some combination of translations and rotations with a reflection does map one onto the other (they are perfect mirror duplicates).

As we have seen this global conception makes the difference between incongruent counterparts depend on the topology of the space as a whole. We can, however, define a perfectly reasonable local conception of incongruent counterpart that does not have this implication.[11]

Def. (LIC) Objects are *local incongruent counterparts*

iff

There is a neighborhood N containing both objects such that

(1) *N* does not admit any combinations of translations and rotations that map one onto the other.

(2) *N* does admit some combination of translation, rotations and reflection that does map one onto the other.

According to this local conception our two figures next to each other on the strip

d b

can count as local incongruent counterparts whether or not the far ends of the strip are left flat, attached to form a cylinder, or attached with a half-twist to form a Möbius strip.[12]

Consideration of local incongruent counterparts suggests to me that we ought to be very careful about drawing a conclusion as strong as C2(b) about the actual space we live in from Kant's observations about incongruent counterparts. It seems to be pretty clearly the case that hands (and mirror duplicates) are local incongruent counterparts. This is one sense in which Kant's claim in Premise 2 that there is a spatial difference between left and right hands

> – the surface limiting the bodily space of one cannot serve as a limit of the other, twist and turn it as one will

can be interpreted so as to be true of our actual space. Whether left and right hands are also global incongruent counterparts is, as far as I know, not settled by what anyone knows about the global topology of our universe.

III. NERLICH'S LONE HAND ARGUMENT

Nerlich discusses Kant's thought experiment about imagining a lone hand. He finds Kant's claim (Premise 3) that such a lone hand would have to be determinately left or right problematic.

> Clearly enough Kant's claim that a hand must be either left or right springs from his assumption that space must have Euclidean topology, being infinite and three dimensional. The assumption is false and so is the claim (Nerlich 1973, p. 345; this volume, p. 160).

According to Nerlich, whether a hand is left or right depends on more than its shape and the relevant topological features of the space:

whether the thing is *left* depends also on *how it is entered* in the space and on how the convention for what is left has been fixed (p. 345; this volume, pp. 159—160).

He suggests, however, that a retreat to the more general property of being an *enantiomorph* is feasible.

Rather than insist that the hand be determinately either left or right, we insist rather that it be determinately either enantiomorphic or homomorphic (p. 345; this volume, p. 160).

Roughly, an object is an enantiomorph just in case is could have an incongruent counterpart.

We can use maps to define the property of being an enantiomorph, much as we did the relation of being incongruent counterparts. Nerlich has in mind global enantiomorphs. He also wants to be able to consider spaces no larger than the volume of a lone hand. To this end he allows contraction as well as rigid motion maps so that a reflection of a contraction of a hand could count as a smaller incongruent counterpart which is wholly inside the volume of space taken up by the hand.[13]

Def. (GE) an object is a *global enantiomorph*

iff

no combination of (contractions) translations or rotations can map it onto any reflection of (any contraction of) itself.

In our space a hand is an enantiomorph while a sphere is a homomorph. In an orientable two-space our figure b is an enantiomorph, but a figure c is a homomorph; in a non-orientable space all figures are homomorphs.

Consider a version of Premise 3 wherein the more general property of being an enantiomorph replaces the more determinate property of being, say, right.

(3′) Imagine the first created thing were a human hand; then it must be determinately either an enantiomorph or a homomorph.

The advantage Nerlich sees for the more general property of being an enantiomorph is that he finds this version of Premise 3 quite plausible, even if Kant's version fails. Being an enantiomorph is fixed by the shape of the hand, together with the topology of the space.

Which of these new determinate characters the hand bears depends, still, on the nature of the space it inhabits, not on other objects. The nature of this space, whether it is orientable, how many dimensions it has, it absolute and primitive (Nerlich 1973, p. 345; this volume, p. 160).

We do regard the topology of our space as objectively fixed. So being an enantiomorph will be objectively fixed, even if whether we take the hand as right rather than left depends on the way we take it to be embedded in the space, and even if this latter were merely a matter of convention.

Suppose we try to use this version of Premise 3 to construct a corresponding version of Kant's reductio argument. Nerlich (1973) doesn't fill in these details. If he has in mind to restrict the space so that to conceive the lone hand universe is to conceive the hand as the lone object in a space which has the same topology as our actual space, then he begs the question against the relationalist, unless he has some independent argument to show that the topological structure of our actual space could not be recovered from spatial relations among the objects that are in it.[14] Premise 3 would just be an idle wheel. Nerlich (1973) suggests that he has something better in mind.

But we can never dream up a *hand* without the space in which it is extended and in which its parts are related. To describe a thing as a hand is to describe it as a spatial object. We saw that the range of spaces a knee might inhabit to be wide; the same goes for spaces in which a hand might find itself. So dreaming up a hand does not determine *which* space accompanies it, though Kant thought it did. But it does not follow that there could be a hand in a space that is indeterminate (with respect to its global connectivity, for example). We can *describe* a hand, leaving it indeterminate whether it is white or black. But there could not *be* a hand indeterminate in respect of visual properties.

No considerations mentioned yet admit of a hand that could be neither enantio- nor homo-morphic. There seems to be no glimmer of sense to that expression (p. 346; this volume, pp. 160—161).

So, when we dream up a hand we need to dream up a space to contain it, but there may be quite a variety of alternatives. Each of these alternatives, however, will determinately settle whether or not its lone hand counts as an enantiomorph.

You have dreamed up a lone hand and a space to contain it. You must conceive this space to be rich enough to accommodate the shape of the hand and to have some fixed global topology. But, there is latitude about what this global topology is fixed to be. It may be fixed so that it admits of some combination of (contractions) rotations and translations that map the hand into a reflection of itself. For example, it may be a non-orientable three-space. On this way of fixing the space the hand is a homomorph. Alternatively, the topology can be fixed so that no such maps are allowed, as in the case of an orientable three-space. Whatever way the topology is fixed it will be determinate whether or not the hand is an enantiomorph.

Now suppose the global topology is fixed so that the hand is a homomorph. Clearly, Premise 1 is satisfied. The other alternative ways of fixing the global topology allowed by the need to accommodate the shape of the hand show that the relations among themselves of the parts of that hand do not determine this specific way of fixing the topology. We can also use Premise 3, as Kant does, to limit the candidate spatial relations among parts of matter to the relations among themselves of the parts of this hand. Therefore, on this alternative we have a spatial fact (the hand is a homomorph) that is not determined by any spatial relations among parts of matter. A similar argument can be constructed for any alternative where the global topology is fixed so that the hand is an enantiomorph. Perhaps this is the argument Nerlich has in mind.

John Earman (1987) has taught us the standard empiricist reply to this sort of argument. What is settled about the topology of the space when you carry out your thought experiment to dream up a lone hand is only whatever constraints on the topology of the space are required to admit the empirically determinate features of the hand. The empiricist will happily admit that you must regard it as true of the space you dream up that its global topology must be fixed in some determinate way, but he will limit what is empirically true about what you dream up to what is true on every allowable way of filling in these details. In effect he will use these alternative absolutist representations as alternative allowable valuations in a supervaluation semantics.[15] On this account the disjunction the hand is enantiomorphic or homomorphic is empirically true. It comes out true on every allowable absolutist representation. But each of the disjuncts is empirically indeterminate, since some allowable representations make one true, while other allowable representations make the other true.

What this does to the reductio argument is to undercut the use of Premise 2 to secure the requirement that a relationalist account is inadequate unless it can specify that the hand is, say, a homomorph rather than an enantiomorph. A relationalist account will be adequate if it can specify how the family of allowable topologies can be determined from the relations among themselves of the parts of the hand. But, this is just the same old problem of whether the relationalist can account for topological features of space. Once again the special argument about the lone hand is doing no work.

Earman (1987) also considers local enantiomorphs. One of the interesting advantages of his local definition is that, on it, we can plausibly argue that the lone hand must be an enantiomorph.

Def. (LE) an object *O* is a *local enantiomorph*

> iff
>
> there is a neighborhood *N* of *O* large enough to admit reflections such that no combination of rotations and translations within *N* can map *O* into any reflection of itself.

All of the allowable spaces have to accommodate the shape of the hand. There may be considerable latitude about the global topology of such a space, but surely, any space that accommodates that of the hand must be locally orientable and three dimensional. If this is correct then the hand will be a local enantiomorph in all the allowable ways of setting the topology of the space.

What about Premise 1? The shape of the hand determines that all the allowable topologies for the space be three dimensional and orientable for some neighborhood of the hand and also that it counts as a local enantiomorph. What about the relations among themselves of the parts of the hand? Do they determine that the hand count as a local enantiomorph? I think the answer would have to be yes, even if these relations of the parts among themselves do not determine the leftness or rightness of the hand. So here again we have a tension between Premise 1 and Premise 2. Being a local enantiomorph rather than a local homomorph is a real difference, one of the facts that relationalists must be able to account for as Premise 2 requires; but, in this case it would seem that they can account for it. Premise 1 fails, for the property of being a local enantiomorph.

IV. NERLICH ON THE DIFFERENCE BETWEEN RIGHT AND LEFT

Now that we have seen that Nerlich's attempt to fall back on the more general property of being an enantiomorph fails we have motivation to look more closely at his rejection of Kant's claim that a lone hand must have one of the more determinate properties, rightness or leftness. Nerlich appeals to his toy model two-spaces to motivate the difficulty he sees with Kant's claim. He calls the two dimensional mirror duplicate figures in his model *knees.*

But, what differentiates a thing which is an enantiomorph from one of its incongruous counterparts is a matter of how it is *entered into the space* — how we cut the hole for it in the white strip. Whether we call a knee in an orientable strip a left knee or a right is wholly conventional; it does not differentiate the knees themselves at all, but simply marks a difference in how they are entered (p. 351; this volume, p. 166).

Suppose we establish a convention to call one figure, say a (b), 'right-handed,' and to call the other one 'left-handed.' We can change this figure from being right-handed into being left-handed by changing the way it is *embedded* in the orientable two-space corresponding to the surface of this page. We can do this by taking it out of the surface and then reentering it after turning it over. This change in its orientation with respect to the page in no way changes its *shape.* In this example being right-handed rather than left-handed is not fixed by the shape of the figure alone, but requires in addition specification of its orientation (the way it is embedded) in the two-space.

Nerlich's view is that the same thing holds for hands in our three-space. According to him, the shape of a hand does not determine whether it is left or right, rather this further determination of handedness depends not only on the shape, but also on the way the hand is embedded in our three-space. This flies in the face of our everyday understanding of shape. What could be more obvious than the empirical differences in shape between right and left hands? Before we dismiss Nerlich's view as incredible, however, we should understand that there is a well-developed account of intrinsic shape that has just the features Nerlich has in mind.

This is an account based on point by point determinations of intrinsic curvature using the methods of differential geometry. Nerlich (1976) appealed to such a notion of shape when he argued for the claim that space has an intrinsic shape in his interesting book *The Shape of Space.* One can give an intrinsic characterization of the point by point Gaussian curvature of a given differentiable manifold, independently of any considerations of whether and how that manifold may be embedded in any larger manifold structure. Consider a differentiable manifold corresponding to the surface of a given hand. We can represent the intrinsic shape of the hand by the point by point intrinsic curvature of this structure. This would count as an example (perhaps simplified) of what Nerlich has in mind as a hand's own handspace. One feature of such an intrinsic representation of shape is that a right hand cannot be distinguished from a left hand unless we specify something of how their separate manifolds are embedded in a common manifold.

An oddity here is that, though a hand filling its handspace can be a dilated incongruous counterpart of a small hand in its space, we could never compare or contrast one hand in its own handspace with another in its own handspace. That requires mutual embedding in a common space. I conclude that leftness is *not* a primitive relation with respect to which hand parts are configured (Nerlich 1973, p. 347; this volume, p. 162).

On this account to say that there is a real difference between a left and a right hand we have to say that there is a fact of the matter about how they are mutually embedded in a common space.

When we calculate the intrinsic shape of the manifold corresponding to the surface of our given hand, say a right hand, we calculate its Gaussian curvature at each point. The Gaussian curvature at a point is equivalent to a product of a maximum and minimum curvature at that point. It is positive if the curvature is uniformly convex or uniformly concave and it is negative at any saddle point. We can use the normal vector at the point to assign signs to distinguish concave and convex curvatures, but the Gaussian curvature is invariant with respect to any such assignment of signs (Spivak 1979, Vol. II, pp. 86–114; Sklar 1979, pp. 36–38; Nerlich 1976, pp. 55–61, 203–212).[16] Thus, we can take inside toward outside as + or indifferently take outside to inside as +. The intrinsic Gaussian curvature structure we arrive at will be the same either way. If we have fixed our assignment so that inside to outside is +, then the differentiable manifold that corresponds to changing the signs uniformly is the manifold corresponding to the result of turning our original surface inside out like a glove. This, of course, corresponds to the surface of a left hand. The intrinsic point by point Gaussian curvature is indifferent to any of these interpretations. If we want to get a difference between a right and left hand we have to specify their embedding into a common manifold sufficiently to be able to coordinate the convention for sign assignments to normal vectors.

There are easily recognizable empirical differences between left and right hands. They are incongruent. One cannot be made to occupy the boundaries now occupied by the other, twist and turn them how you will. These spatial differences between them are invariant with respect to rigid motions in our local space. In the case of hands the differences are so profound that they can be apprehended visually from almost any perspective from which a hand can be observed. Such differences are also perceptible to touch and there is the normal agreement between the information afforded to touch and that afforded to sight. I think there is a fairly robust concept of shape according to which the observable differences between right and left hands are recognizable as differences in shape. Let us call this 'empirical shape' to distinguish it from Nerlich's technical account of 'intrinsic shape'.

The differences in empirical shape between a right and a left hand are observable differences. A person who has learned to use 'left' and 'right' ostensively on actual examples doesn't need to compare one hand directly with another in order to see which hand is right. Such a person can recognize the observable features of empirical shape that make a lone hand right rather than left.

V. EARMAN ON THE OBSERVABILITY OF THE DIFFERENCE BETWEEN RIGHT AND LEFT

John Earman (1987) has recently offered an interesting argument against the claim that the differences between left- and right-handed figures are observable differences. He considers three two-space cases, with figures of the following sort:

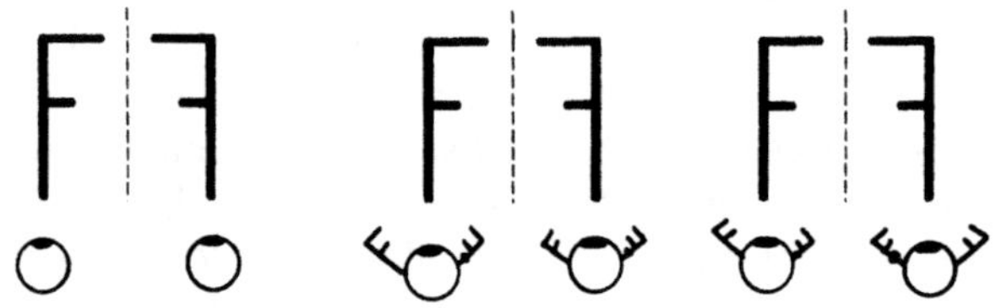

(â) (b̂) (a^+) (b^+) (a^*) (b^*)

Here is what he says about them:

In the (a^+)—(b^+) case the introduction of the observer breaks the relational equivalence of (a) and (b), and so the relationist can entertain the hypothesis that the phenomenal contents of the visual experiences of the (a$^+$)—(b$^+$) observers are different. Perhaps Kant would want to claim that the phenomenal contents of the experiences of the (â)—(b̂) and the (a*)—(b*) observer are or can be different, a claim the relationist must deny since (â)—(b̂) and (a*)—(b*) are relationally equivalent (1987, p. 13; this volume, p. 244).

Consider the Kantian claim that in the (â)—(b̂) case these will be a perceptible difference between the left and right hand figures. You can be the observer looking at an orientable two-space like this page. You observe figure (â). It looks like this:

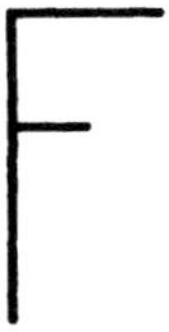

If the figure you observed had been (b̂) instead, you would have noticed a difference. It would have looked like this:

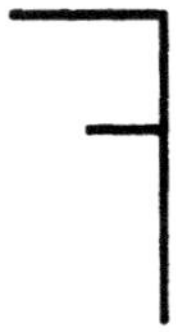

Walker (1979, p. 47; this volume, pp. 189—190) claims that such an observable difference does not depend on any enantiomorphism in the observer's body. This is, surely, correct. Your hands, feet, ears and presumably any other enantiomorphisms of your body had nothing to do with your capacity to observe the difference. This difference would have shown up equally well to a system consisting of a single eye, or on a printed photograph of the page.

After noting Walker's claim, Earman responds as follows:

But what the opponent of relationism must claim in the (â)—(b̂) case is that the mirror-image non-enantiomorphic observers who have the receptor sites of their sense organs left—right reversed, nevertheless have different visual experiences. Even on the absolutist's own terms the case for this claim is weak (this volume, p. 244).

To see what he is getting at, let the two-space be a clear vinyl strip (as we did in the discussion of Nerlich). Going around to observe the strip from the other side left—right reverses your receptor sites (your retinal cells or, in the case of a camera, the sensitive points on the surface of the film) with respect to the figures on the strip. To an observer on the other side of the strip the mirror image figure (b̂) has exactly the same shape as figure (â) does from our side of the strip. The same right—left reversal of receptor sites with respect to a given figure can be achieved by turning the figure over on the strip, while the observer remains on our side of it. Could an observer find any difference in shape between figure (â) with its present orientation and figure (b̂) flipped over? I think that, as Earman suggests, the case for supposing that our observer could make any such discrimination is very weak indeed.

Walker (1979) would agree with this, as his discussion of non-orientable space-travelling Alice makes clear.

> If Alice follows a glove round a non-orientable, three-dimensional space and comes back to where she started — or if a two-dimensional Alice follows a knee round a Möbius band — it will look the same to her all the way, and when she gets back again it will still look to her as it did before she started out. Only everything else in that region will now appear to her right-left reversed; the exactly similar glove she left behind will have apparently changed its handedness; she will seem to have come back to a mirror image of the place she started from (this volume, p. 191).

Here, as in the vinyl strip two-space, we have perception of handedness depending on orientation of object and observer. Earman and Walker agree with Nerlich that what we have been calling empirical shape depends on how objects and observers are embedded in space.

Earman goes on to argue that to maintain that an observer could make such discriminations between right- and left-handed shapes would require maintaining that the observer has a capacity to discriminate a preferred orientation of the space containing the object-observer system.

> To support the Kantian claim the absolutist must maintain that spatial perceptions are a function not only of the spatial relations of observer to object but also of the relation of the observer to a preferred orientation of the space containing the object-observer system. As an hypothesis about perception, this seems far-fetched, and the postulation of the preferred orientation seems *ad hoc* (this volume, pp. 244—245).

Earman's argument shows that handedness is observable only if a preferred orientation (embedding) in space is specified. I think we must agree that it would be far-fetched indeed to suppose that embedding is, itself, observable. Does this mean that we have to agree that the differences in shape between right and left hands are not observable after all?

If the orientation that counts for settling handedness is specified demonstratively then there may be no need for the observer to actually perceive what it is in order to be able to perceive the handedness of an object. Can we take up this suggestion to circumvent Earman's argument? The problem that looms is that of being able to live with whatever observer-relativity this will introduce into our conception of empirical shape.

We are committed to ruling out certain sorts of observer-relativity of empirical shape. Three-dimensional empirical shape is an invariant structure that affords appropriate information to each of the indefinitely large array of alternative perspectives from which the object could be observed. This rules out relativity of shape to any of the changes in orientation that we could bring about by observing the object from any of these alternative perspectives provided for by our space. We also regard the empirical shape of a rigid object as invariant with respect to the rigid motions afforded by our space. This rules out relativity of empirical shape to those changes of orientation that could be brought about by any of the rigid motions of the object available in our space. Fortunately, none of these changes in orientation that could be brought about by the motions available to observer and object in our space would count as a change in *embedding*. That is, none of them would count as the profound sort of change in orientation that could transform something recognizable as a right hand into something recognizable as a left hand. So, we have, as yet, no reason to doubt that we could live with whatever observer-relativity would be introduced into our conception of empirical shape if we take up the suggestion that these profounder aspects of orientation that support the observability of three-dimensional handedness are to be fixed demonstratively.

Such a profound change of orientation as would count as a change in embedding could be produced by motions only if our space were either non-orientable or had a fourth spatial dimension. Kant surely thought that the space in which we meet objects of experience has neither of these features. He my have been right. As we saw Nerlich point out, there is no evidence that either feature is available, even if we allow the global topology connecting us with remote regions to count. If we limit our concern to local structure, then it is pretty clearly the case that our space is orientable and has no fourth spatial dimension. There just don't seem to be any local motions of observer or object that could transform something recognizable as a right hand into something recognizable as a left hand.

Suppose Walker's Alice story really could take place in our space. This is just the sort of example where Nerlich and Walker, at least, would agree that there is no fact of the matter for Alice and her stay at home friends to disagree about when she says the glove she carried with her is still left-handed while they say that it is now right-handed. Nerlich (p. 351; this volume, p. 166) suggests that what counts as the handedness of the glove is wholly conventional. Certainly, how the words 'left-handed' and 'right-handed' are used is conventional, but, once the convention for their use has been settled ostensively, is there no fact of the matter about which objects count as left- or right-handed? Walker (this volume, p. 192) suggests that the only support for the stay-at-homes is their numbers. Must we admit that there is no more to it than this? If so, then there is no reason to care enough about the disagreement to require that a theory of space ought to be able to settle it. Let us see how our conception of empirical shape handles this sort of case when we take up the suggestion that local embedding is to be fixed demonstratively. I think this will support a conception of local handedness that is both observable and also objective enough to be worth bothering about.

The two-space analogue to the Alice story is a transparent Möbius strip where observers and objects are able to change their local embedding by, and only by, transport around the strip. Let us have three observers: observer (A) is about to leave on her trip, observer (B) is the stay-at-home, and observer (C) is embedded on what counts locally as the other side of the strip (perhaps (C) has recently returned from a trip around it). They now set the convention for their use of the words 'right-handed' and 'left-handed' ostensively. All agree that Earman's figure (a) counts as 'right-handed' while figure (b) counts as 'left-handed'. To observes (A) and (B) these figures

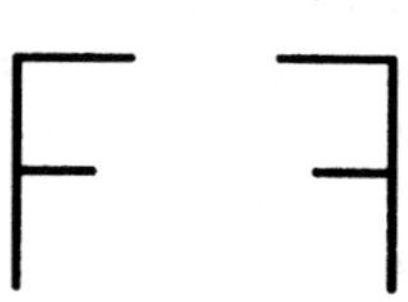

(a) (b)

look just as they do to you here on the page, but to observer (C) they look like this:

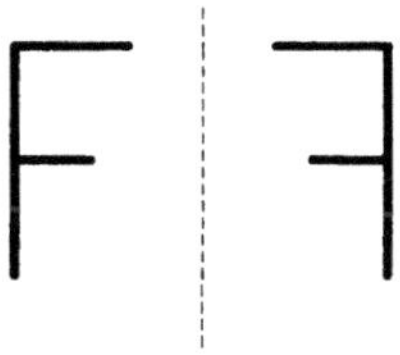

(d) (ɒ)

Observer (C) is looking at them from the back of the strip; he reads letters mirror-wise so that to him an (a) looks like (ɒ) and a (b) looks like (d).[17]

On our suggestion for using demonstrative reference to embedding to support observability of handedness, the preferred orientation settling local handedness was fixed demonstratively to be whatever embeddings the observers actually had on the occasion when they set their conventions for 'left-handed' and 'right-handed' by ostension.

Once the convention is settled, so are the facts about what count as right-handed and left-handed objects in this vicinity. Our observers all agree, for example, that the figure (x) which (A) plans to take on her trip, and which looks like this

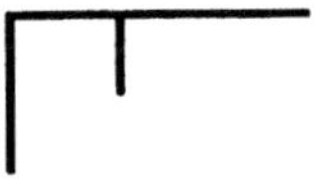

(x)

to (A) and (B) and like this

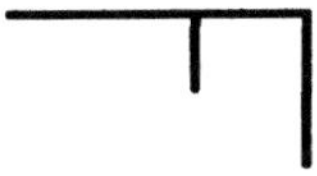

(x)

to (C), is left-handed. Moreover, they are correct about this. The preferred orientation fixed demonstratively when the convention for using 'left-handed' was set by ostension allows the perceptible information afforded by the figure to these various observers to objectively settle its handedness.

Now, observer (A) has returned from her trip. Figure (x) still looks left-handed to her, but observers (B) and (C) correctly identify it as right-handed. They are correct because they are still observing from the embedding that was fixed demonstratively as their preferred orientations for observing handedness in this vicinity, when the conventions for using 'right-handed' and 'left-handed' were set by ostension. Observer (A) no longer has the embedding that correctly coordinates perceptible information afforded to her with objective judgments about handedness in this vicinity. If she is an experienced strip traveller, observer (A) will be able to make appropriate adjustments to bring her practice into line with her new embedding as soon as she returns to the vicinity. She will have plenty of information available to tell her that her embedding has changed since she was last here. Everything that stayed home while she travelled around the strip will appear right-left reversed.

As long as there is something to count as the facts of the matter about which observers and things travelled around the strip and which stayed at home, a space can support our suggestion for using demonstrative reference to embedding to make local handedness objective and observable, even if that space is globally non-orientable.[18]

VI. KANT'S LONE HAND ARGUMENT

Let us try the lone hand argument with Earman's two-space figures. Here is Earman's complete relational description of the figure:

> The object has a long shaft and two shorter shafts which are attached perpendicular to the long shaft. One of the cross shafts is attached to an end of the long shaft, while the second cross shaft, which is shorter than the first, is attached to the long shaft at a point between the midpoint and the end to which the first shaft is attached (1987, p. 10; this volume, p. 242).

This relational description may exhaust the content of the relations among themselves of the parts of such a figure: yet, it does not distinguish between left- and right-handed versions. Kant's Premise 1 is granted at the outset. Suppose now that a two-space observer is asked to imagine a lone figure of this sort. Such a figure cannot be imagined without imagining it to have a determinately left-handed or right-handed empirical shape. Kant's Premise 3 seems unobjectionable here. Moreover, it would seem that imagining the figure with left-handed shape is not the same thing as imagining it with right-handed shape, so

Premise 2 is apparently satisfied as well. The relational description leaves out these important observable features of empirical shape; therefore, an account of shape limited to the resources of a relationist's theory of space is inadequate.

Let's run this through again, with more detail, to see if the argument holds up when we explicitly take into account Earman's objections to the observability of handedness. Consider observer (B) in our two-space example. Suppose this

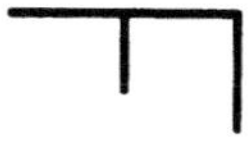

is what he imagines. He imagines it from a virtual point of view. What he imagines here is a right-handed figure viewed from a certain perspective by an observer with the sort of embedding he has. What he imagines carries information about what it would look like from other points of view. He imagines it as something that would look like this

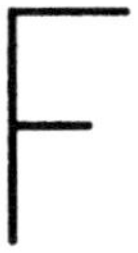

or this

if it were observed from perspectives rotated respectively 90° clockwise or 90° counterclockwise relative to the virtual point of view from which he actually imagines it. This may be the only sort of alternative perspective included in what he imagines if observer (B) has no experience with embedding changes. In this case it is quite clear that to have imagined a left-handed figure would have been to imagine something different.

Suppose now that an experienced strip traveller like (A) is asked to imagine a lone figure. She images a right-handed figure from her current embedding, after returning from her recent trip. This is what it looks like:

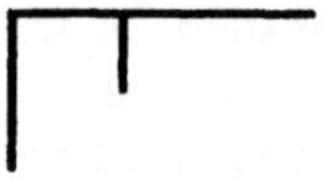

Remember, he receptors are now right-left reversed and she has adjusted her convention for judging right- and left-handed objects accordingly. Just as with observer (B), what she imagined would be different if she had imagined a left-handed figure instead.

She also shares with (B) the characteristic that what she imagines carries information about what it would look like from other perspectives. In her case, however, the range of alternative perspectives may be larger. It may include alternative embeddings, as well as more mundane changes of perspective. Specifically, it may be part of the information included about the right-handed figure she imagines from her virtual point of view that it would look like this

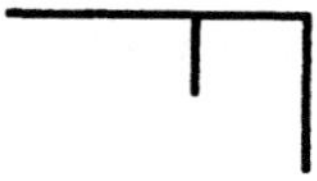

if observed from a point of view with opposite embedding. When sophisticated observer (A) imagines a lone figure she does so from a virtual point of view that includes specification of embedding as well as more mundane aspects of perspective. Once again, however, what she imagines when she imagines a right-handed figure will be different from what she imagines when she imagines a left-handed figure. Her current convention for judging right- and left-handed objects is the one she applies at the virtual point of view she actually imagines the object from.

Perhaps this is unfair to the relationists. Why can't observer (A) just imagine a lone hand from a perspective that is left indeterminate with respect to embedding? Embedding is not like rotational orientation; it is not fixed by the image. Observer (A) could have been asked to imagine a right-handed figure from an embedding opposite to her own. It might have looked like this:

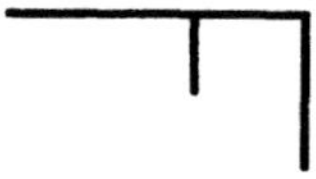

Which is exactly the way the figure would have looked had it been imagined as a left-handed figure from an appropriately rotated version of a perspective having her actual embedding. We can grant that observer (A) can imagine observing, say, a right-handed object from a point of view having opposite embedding from her own actual embedding, without granting that she can imagine a figure from a point of view left indeterminate with respect to embedding. The embedding of the imaginary point of view is coded by the convention for judging right- and left-handed figures in it. For (A) to imagine a right-handed figure from an embedding opposite to her own is more than to have an image of a certain sort; it is to have an image and apply to it the appropriate convention for judging what sort of object it is an image of. To simply entertain an image is not the same thing as to imagine a two-space object having the shape of one of Earman's figures.

As far as I can tell there is no need to grant that an observer like (A) could do something that would count as imagining a lone figure from a virtual point of view left indeterminate with respect to embedding. If, however, this were allowed as a possibility, then we would have to admit that, in the two-space version of the Alice story, there could be examples where imagining a lone figure did not require imagining a figure that is determinately left or right. This possibility would not let the relationists off the hook, however, because their theory of space still has to cover the more natural cases where to imagine a lone figure does require imagining it as determinately left or right.

The argument works even better for Kant's three-dimensional case of the lone hand. When you, Kant's reader, are asked to imagine a universe consisting of a lone hand, you are like observer (B) in the two-space example. You imagine the lone hand from a virtual point of view, and what you imagine carries information about how it would look from alternative points of view, of the sort that would be afforded to any shaped object in our three-dimensional space; but, you do not, and probably cannot, imagine any alternative point of view that would count as a change in embedding. Certainly, you cannot imagine a hand with a shape that is indeterminate between that of a left hand and that of a right hand.[19] Objects are presented, whether in perception or imagination, with shapes which are more determinate than any relationist description can specify.

Kant asks his reader to imagine a lone hand. I think his use of imagine here is quite deliberate. According to him a shape which is unimaginable could not be apprehended in perception as the shape of a possible object of experience. He would regard specification of an object which satisfied only Nerlich's 'intrinsic shape'-description of a hand, and which also failed to satisfy the requirements of empirical shape by being determinately either left or right, as a specification which could not possibly be realized by any object of experience. He may have been right about this. Could there *be* an actual object of about the size of a human hand which had the "intrinsic shape" of a hand and yet failed to have the more determinate handedness of empirical hand shape? We are not considering here a possible shape which may be unimaginably complicated such as a chilagon, or a possible shape for something such as a galaxy or a universe which is so large that no human observer could take it in, nor are we considering a shape for something such as an electron which is so small (and perhaps otherwise elusive) as to be imperceptible to observers like us. What we are considering with Kant's hand example is a possible shape for an object of a sort that we are familiar with and which is well inside the optimal range of our capacity to discriminate shapes by observation.

The differences between right and left are observable differences. They are also differences that make a difference. A left hand cannot be made to occupy the exact boundaries now occupied by a right hand twist and turn them how you will. A left glove does not fit on a right hand. A left hand would not fit properly on a right arm. Kant mentions other examples. Screws with left- and right-handed threads cannot be interchanged (p. 41). Hops wind round their poles from left to right; beans, however, twist in the opposite direction (p. 39). In the nineteenth century, biologists and chemists began discovering organic compounds and molecules with enantiomorphic structures where the specific 'handedness' of the structures made a difference in important biological processes (see e.g. H. Weyl 1951, pp. 30—38, or M. Gardner 1979, pp. 95—136). More recently, we have the celebrated parity nonconservation experiments which suggest that fundamental laws in elementary particle physics are not indifferent to the differences between right- and left-configured enantiomorphic processes (see M. Gardner 1979, pp. 176—208).

VII. REGIONS, REFERENCE FRAMES AND THE RELATIONS OF THE PARTS AMONG THEMSELVES

(a) *Two-Dimensional Relations in Three-Dimensional Space?*

As we have seen, Earman (1987) granted Kant's Premise 1 at the outset of his discussion. In an earlier paper (Earman 1971), however, he was unable to find an interpretation of this premise that would make Kant's argument coherent. He considered interpretations which make Premise 1 false. For example:

It seems true that, as Körner [1966, pp. 33—4] puts it, the relations of the parts of left and right hands are 'entirely similar' — at least if this is taken to mean that the relations are different but resemble each other in a systematic way; the relations for one hand mirror, in an obvious and literal way, the relations for the other hand. But on this interpretation it is false to say that the difference between right and left cannot lie in the differences between the internal spatial relations of the parts of left- and right-handed objects (Earman 1971, p. 7; this volume, pp. 136—137).

He also considered an interpretation that would make the premise true:

If we limit the internal spatial relations of the parts of the hand to those which can be defined purely in terms of the distances between points of the hand and angles between lines lying in the hand, then obviously Kant's claim is correct since mirror image reflection preserves these distances and angles (Earman 1971, p. 7; this volume, p. 137).

But he suggests that this interpretation does not make the premise support a coherent argument for Kant's conclusion:

But if one continues to limit oneself to these kinds of relations, then introducing points and lines of the space external to the hand will not help to distinguish between right and left since whatever the relation in this sense a right hand has to an external point or line, these relations can be mirrored by a left hand. Parroting Kant, one can then argue that the difference between right and left cannot lie in the relation of the hands to the encompassing space since these relations are the same; therefore, the difference must lie in something external to absolute space (Earman 1971, p. 7; this volume, p. 137).

Earman admitted that this parody is unfair to Kant (p. 7), since Kant did not intend the difference to consist in a relation to *points* in the wider space but to space as a unity. Nevertheless, he claims (p. 8) to be unable to see how this helps to make Kant's argument coherent.

Jill Buroker (1981) endorses a version of the interpretation of Earman (1971) which makes the premise true.

I think John Earman correctly identifies which relations Kant has in mind in referring to similarities of incongruent counterparts such as left and right hands (Buroker 1981, p. 54).

Now although Earman does not explicitly mention it, in taking these individual measurements one is limited to a two-dimensional plane, although that plane varies depending on the distances and angles being measured. Thus we can restate Kant's point about the relations among the internal parts of left and right hands as follows: *The physical magnitudes of two-dimensional distance and angles among parts of left and right hands can be identical and yet the hands are not substitutable in space* (p. 55).

My initial reaction is to sympathize with Earman's doubts about the sufficiency of this interpretation of Kant's premise. It would seem to render it too weak to support the sort of conclusion Kant was arguing for. Buroker's own remark about Kant's lone hand argument exactly expresses my reservation.

Would this argument convince a Leibnizian? No, because Kant is wrong in thinking that the hand would be neither right nor left. As we saw above, the relations among the parts of the hand are not the same for right and left hands if one takes into account relations in all three dimensions (Buroker, p. 59).

The conclusion that some relation to absolute space as a unity is necessary to distinguish right and left hands does not seem to follow from the premise that they cannot be distinguished by *two-dimensional* relations. Why can't the Leibnizian simply appeal to the three-dimensional relations of the parts among themselves?

(b) *Regions and Positions*

The inadequacy of this interpretation of Premise 1 is shown up further by its implications for Buroker's interpretation of Kant's account of the difference between regions and positions. According to Kant,

The position of the parts of any extended object, with respect to each other, can be sufficiently recognized from the object itself. The region, however, to which this order of the parts is directed, is related to space outside, but not with reference to its localities, for this would be nothing else than the position of just those parts in an external relation; region is related rather to space in general as a unity, of which each extension must be regarded as a part. It is no wonder if the reader finds these concepts still very incomprehensible; but they should become clear in due course (K & W, p. 37; AK II, p. 377, lines 26—28; p. 378, lines 1—7).

According to Buroker,

Kant is thinking of position as a relation among objects (or parts of objects) in two dimensions.

The *positions* of parts relative to one another are two-dimensional relations; the *region* containing all these positions is the three-dimensional volume of space containing the object (1981, p. 57).

I think this restriction of position to two-dimensional relations does not go far enough toward removing the reader's incomprehension. In particular, it makes it difficult to comprehend why Kant thought the idea of position was interesting enough to bother talking about. One thing we ought to want of any interpretation of Premise 1 is that it should make comprehensible Kant's distinction between position and region.

Kant prepares his reader by relating the concept of region to a coordinate system oriented on one's body.

Because of its three dimensions, three surfaces can be conceived in physical space. They all intersect each other at right angles. Since we know nothing external to us through the senses, except in so far as it stands in relation to ourselves, it is no wonder that we derive from the relation of these intersecting surfaces to our body the ultimate foundation of generating the concept of regions in space. The surface on which the length of our body stands vertically is called with respect to ourselves, horizontal; and this horizontal surface gives occasion for the differentiation of objects which we indicate by *above* and *below*. Two other surfaces can stand vertically on this surface and they can, at the same time, intersect each other at right angles, so that the length of the human body is conceived along the line of the intersection. One of these vertical surfaces divides the body into two externally similar halves and gives the foundation of the distinction between the *right* and the *left* half; the other vertical surface which stands perpendicularly to it, enables us to conceive the *front* and *back* side (K & W, p. 38; AK II, p. 378, lines 32—37; p. 379, lines 1—10).

This passage suggests that regions are differentiated by reference to coordinate systems oriented on our bodies. Such a suggestion is further supported by Kant's discussion of geographical knowledge.

The same holds true of geographical, indeed of our most ordinary knowledge of the position of places; such knowledge is of no help to us, so long as we are unable to place the so ordered things and the whole system of reciprocally related positions, according to regions through the relation to sides of our bodies (K & W, p. 39; AK II, p. 379, lines 35—37; p. 380, lines 1—2).

According to this suggestion, an example of the distinction between position and region is the difference between the knowledge one can get from a map (even a three-dimensional topographical model) before and after one has oriented the map with respect to one's own body. One can learn the relative positions of places (even in three dimensions) by reading them off from the map, but this information will not tell you how to get from here to there until you have located here and there on the map and oriented it to your body. Only after such an orientation can you transform your knowledge of relative positions of the places among themselves into knowledge of the regions in which those places are located.

Consider some of the claims Kant made in the puzzling passage where he introduced the difference between position and region. One claim is that region is related to space in general as a unity.

Region is related rather to space in general as a unity (K & W, p. 37; AK II, p. 378, lines 2—3).

In the sentence immediately preceding the first passage we quoted Kant tells us,

In the most abstract sense region does not consist of the relation of one thing in space to the next. That would really be the concept of position. Region really consists rather in the relation of the system of these positions to absolute world-space [*absoluten Weltraume*] (K & W, pp. 36—37; AK II, p. 377, lines 23—26).

Here we have region consisting in the relation of a system of positions to absolute space. When we put this together with Kant's claim that a system of positions is placed according to regions by specifying its relation to our bodies we get an interesting suggestion. We specify the relation of a system of bodies to space in general as a unity — to absolute world space — when we locate and orient the system of positions relative to our point of view. The idea here seems to be that we *demonstratively* specify the orientation and location of our point of view in absolute space so that specifying the location and orientation of a system of positions relative to our point of view demonstratively fixes its location and orientation in absolute space.[20]

The sentence where Kant tells us that regions are related to space in general as a unity goes on to tell us that each extension must be regarded as a part of that space.

Region is related rather to space in general as a unity, *of which each extension must be regarded as a part* (K & W, p. 37; AK II, p. 378, lines 2—4).

Today we might think of the various extensions which are all parts of the one general space as submanifolds of a single differentiable manifold. This idea of space (space-time) as a differentiable manifold is a salient part of Nerlich's conception of substantival absolute space. According to Nerlich (1976) this manifold has an intrinsic metrical structure sufficient to support an account of intrinsic shape based on Gaussian curvature.

We can use the idea that global space is a manifold with metrical structure to clarify Kant's remarks about positions. In the passages contrasting positions and regions Kant tells us,

> region does not consist of the relation of one thing in space to the next. That would really be the concept of position (K & W, pp. 36—37; AK II, p. 377, lines 23—25).

> The position of the parts of any external object with respect to each other, can be sufficiently recognized from the object itself (K & W, p. 37; AK II, p. 377, lines 26—28).

Position, here, is relative position of things with respect to one another. The relative positions among a group of things should not depend on how that system of relative positions is related to anything outside it. They should depend only upon features that are located inside of every convex neighborhood containing those positions. The key idea separating Kant's concept of position from his concept of region is that a system of relative positions should be independent of how it is embedded in global space.

(c) *The Relations of the Parts Among Themselves*

We can use these ideas about relative positions to motivate a more interesting interpretation of Kant's Premise 1. The relations among themselves of the parts of a hand are to be given by the system of their relative positions. A system of relative positions is to be independent of how it is embedded in global space. Gauss's account of intrinsic curvature of a surface has such independence. His surprising theorem, the famous *Theorema Egregium* (Spivak 1979, Vol. II, p. 112), tells us that the complete intrinsic point by point Gaussian curvature of a surface is invariant with respect to alternative embeddings of that surface into R^3. This invariance with respect to alternative embeddings is exactly what we want of any intrinsic descriptions of the relations among themselves of the parts of a hand.

Suppose we attempt to specify the spatial relations of the parts of a hand among themselves, making no reference to anything outside. Let us set up a coordinate system, say one such that the origin is at the base of the thumb, the thumb points in the $-x$ direction, the outstretched fingers point in the $+y$ direction, and their order from little finger to index finger is in the $+z$ direction. We can give the positions of all the parts relative to one another once we establish a scale, say using the length of the first thumb joint as one unit. This ought to count as a representation of the three-dimensional situation and relations of the parts among themselves.

We could have used reference frames centred on other parts of the hand with various conventions about scales and we could have used alternative orientations of axes (rotations). Relative to any one of these conventions about our reference frame we could represent the other reference frames and represent the information they carry about the relative positions of the parts among themselves. There is no reason to count any of these representations as any more correct than any other; they all represent the same information about the relative positions of the parts among themselves. This information is invariant with respect to the transformations that take us from one of these conventional representations to any of the others. We can set the appropriate parameters for taking us from one frame to another by appealing to the relations between parts of the hand. Thus, if an alternative frame uses the length of the nail of the thumb as its unit, the appropriate scale transformation of the unit for length is the ratio of the length of the first thumb joint to that alternative length. If we have a frame centred on the end of the index finger the appropriate transformation is a translation of origin to that location and if we have a rotated frame the relevant transformation is an appropriate rotation of axes.

We are allowed to appeal to relative positions of the parts of the hand among themselves, but we are not allowed to appeal to anything outside the hand. It would seem then that we ought to regard the sign assignment to, say, the $+x$ axis as a mere convention that could have been made in the opposite way, just as our scale assignment, location of origin, and orientation of axes are mere conventions. Given that we can appeal to the relations among the parts of the hand, we can settle the appropriate parameters to transform the information carried by one frame into the corresponding information carried by a reflected frame. Given that we are not to appeal to anything outside the hand, there would seem to be no more reason to count one way of settling the sign convention as any more correct than any other. The information about the relative position of the parts among themselves ought to be regarded as invariant with respect to any way of settling these conventions.

This gives a relatively clear and plausible interpretation of Premise 1 that makes it come out true. If the information about the relative positions of the parts among themselves is invariant with respect to sign assignments to axes then it is invariant with respect to reflections and this information will not distinguish between a hand and an incongruent counterpart of it.

VIII. EMBEDDING AS A RELATION TO SPACE IN GENERAL WHICH CAN BE SPECIFIED ONLY BY APPEAL TO INTUITION

Our defense of Premise 3 involved specification of a virtual point of view from which the figure is imagined. Can the relationist appeal to some aspect of this to get off the hook? I don't think so. For one thing the virtual point of view is not an imagined observer with a body located in the space with the imagined figure, so it does not count as an additional imagined part of matter that the parts of the figure could have some appropriate spatial relations with. To imagine such a thing would not be to imagine a lone figure. For another thing, even if we let the body of the observer into the imagined space, this need not provide for any enantiomorphic parts of matter the relationist could appeal to. As we have seen (Section VI) the capacity to discriminate right- from left-handed figures in no way depends on the observer having enantiomorphic bodily parts. Finally, even if the observer did have enantiomorphic parts, appealing to them would not help the relationist.

This last point can be made evident by reflection upon the way that observer (A) is able to reset her convention for identifying right-handed figures. The point is argued ably by Peter Remnant (1963), by Martin Gardner (1964, 2nd ed. 1969, reprinted 1979), and by Jonathan Bennett (1970). The deep problem Kant sees with the relational account of space is one it shares with any intrinsically specified version of a substantival absolutist account, this being its failure to connect properly with our practice of using space as a framework for making demonstrative reference to regions.

After introducing his readers to the idea of regions discriminated by reference frames oriented on our bodies, and introducing us to incongruent counterparts, Kant tells us what he wants to argue for.

We wish, therefore, to show that the complete principle of determining a physical form does not rest merely in the relation and the situation of the parts, with respect to each other, but also on its relation to general absolute space, as conceived by geometers; indeed, in such a way that this relation cannot be immediately perceived, though, perhaps the physical differences that rest uniquely and alone on this ground can be (K & W, p. 71; AK II, p. 381, lines 14—20).

I think that what Kant speaks of here as a relation to general absolute space required for determining physical form is the profound sort of orientation in space we have been calling embedding. Earman argued that a preferred embedding must be appealed to if the differences in shape between right- and left-handed objects are to be observable, and that such a preferred embedding is not itself observable. Here we see Kant agree to Earman's point. We have argued that the required appeal to a preferred embedding can be made demonstratively. I think the importance of the sort of relation to our bodies Kant insists upon for reference frames that can specify regions in space is that this relation to our bodies affords the possibility of appealing demonstratively to our actual embedding in space.

You are observing a right hand, and your usage of the words 'right-handed' and 'left-handed' has been fixed ostensively in such a way that it agrees with that of the rest of us. You know how to recognize a right hand when you observe one, you know how to recognize a left hand when you observe one, and you don't need to observe two hands together at once in order to see that they differ in handedness. Suppose you set up a co-ordinate system to give the relations among themselves of the parts of the hand you are observing. You set the convention for sign assignments to axes so that, let us say, the thumb and palm point in the $-x$ direction while the back of the hand points in the $+x$ direction, the fingers point in the $+y$ direction and their order from little finger to index finger is in the $+z$ direction. You are setting these assignments ostensively as you actually observe the hand. This allows your assignment of signs to axes to pick out demonstratively a preferred embedding that aligns your interpretation of it with the embedding that actually governs your recognition of left- and right-handed objects.

Once you have done this you can interpret the co-ordinate system so as to read off the handedness of the figure from its specification of the relations among its parts. The interpretation specified by fixing the embedding transforms the information given by the co-ordinate representation from one that specifies merely the relative positions of the parts among themselves to one which also settles the regions to which those parts are directed. As I set up the example, your assignment of signs to axes agreed with a standard convention used by most of us when we set up cartesian co-ordinates. This was not in any way essential to your capacity to interpret it by demonstratively specifying the relevant embedding. Had you assigned the opposite signs to the x axes, you would have had an equally good specification of the right-handedness of the figure you are observing. What matters is that the embedding fixed demonstratively when you assign your signs to axes aligns your interpretation of them with your practice for judging left- and right-handed objects.

If we are given a co-ordinate description of the relations among themselves of the parts of a hand, we cannot tell whether it counts as a description of a right hand or of a left hand unless we can settle the relationship between the assignments of signs to axes and the embedding in space that fixes our practice for judging whether a hand is right- or left-handed. If we are given a separate co-ordinate description for each of two hands we cannot tell whether they are incongruent counterparts without specifying whether their assignments of signs to axes agree or disagree in their relation to our embedding in space. If we are given a co-ordinate description where the parts of both hands are located with respect to a common reference frame, then we can tell whether or not they are incongruent even if we cannot tell which is right or left.

This last point is the basis of Remnant's (1963) argument that we cannot settle whether a lone hand counts as right or left even if we introduce an enantiomorphic human body into the space with it.

> . . . a 'standard' body has a green right arm and a red left arm. It will then have an incongruent counterpart with colors reversed.

> But when all there is in the universe is a hand and a handless body, then even though it is quite determinate which arm the hand belongs in, it remains completely indeterminate whether this is a right arm or a left arm and consequently indeterminate whether the hand is right or left (p. 298; this volume, p. 57).

Until we settle whether the enantiomorphic body in the space is right-handed or left-handed we cannot use it to settle the leftness or rightness of a hand which fits on its green arm. We are in just the situation that we are with respect to the two hands in a common frame when we don't have the embedding interpretation of the frame fixed.

This point is also a problem for Grünbaum's (1970, pp. 331—332) suggestion that we can use the sign of the Jacobian of the transformations which can take a co-ordinate description of a right hand into a co-ordinate description of a left hand as a purely conceptual characterization of the specific structural differences between left and right hands. It is a well known mathematical fact that the Jacobian of Euclidean rigid motion transformations is +1, while that of reflections is −1. What this gives us is a specification of the agreement or disagreement in handedness between two hands that are already represented in a common reference frame. It doesn't give a conceptual characterization of what makes one of them say left rather than right. This latter specification requires settling the interpretation of the assignments of signs to co-ordinates with respect to the embedding we appeal to in order to fix our conventions for 'right' and 'left'. I think Kant is quite correct that settling this requires appeal to intuition so as to demonstratively fix the required embedding.

Once we have the two hands represented relative to a common co-ordinate system we can certainly specify whether they have the same or opposite handedness without any further appeal to intuition. The Jacobian of the transformations taking one description into the other is just one of any number of equivalent ways of specifying whether they agree or disagree in handedness. Kant would never have doubted that such specifications are possible. He would have pointed out, however, that they do not settle which hand is to count as right or as left, which is what must be specified to say what the difference is. He may have also pointed out that the method cannot even be applied until the hands are represented relative to a common frame. Getting them to such a representation, he might argue, requires settling the relation between the ways they are each embedded even if it leaves open what their embeddings are. This, Kant might argue, is itself something that requires appeal to intuition to settle.

IX. KNOWLEDGE OF THE DIFFERENCE BETWEEN RIGHT AND LEFT IS NON DE DICTO KNOWLEDGE DE SE

In the *Inaugural Dissertation* (1770) Kant claims that incongruent counterparts

could be substituted for one another as far as concerns all the things which may be expressed in marks intelligible to the mind in verbal descriptions (K & W, p. 69; AK II, p. 403, lines 7—8).

The *Metaphysical Foundations of Natural Science* (1786) included the claim that the concept of the difference between incongruent counterparts

does not at all admit of being clarified by universal marks in the discursive way of knowing [Erkenntnissart] (Ellington 1970, p. 23; AK IV, p. 484, lines 3—5).

Kant has similar things to say in his relatively long discussion of incongruent counterparts in Section 13 of the Prolegomena (1983). Here are some examples. Speaking of similar and equal but incongruent spherical triangles he tells us[21]

nothing is found in either, when it is described alone and completely, which does not appear in the description of the other (AK IV, p. 285, line 35, p. 286, lines 1—2).

Here then is an *inner* difference, between the two triangles, which no understanding can give an account of as inner (AK IV, p. 286, lines 4—5).

In the last sentence of Section 13 he tells us[22]

We cannot make the difference between similar and equal things which are not congruent (e.g. oppositely wound [Helical] spirals) intelligible by any concept (AK IV, p. 286, lines 31—35).

I have taken these quotations out of context in order to single out a common theme running through all of Kant's later discussions of incongruent counterparts. For the present, I shall put aside consideration of other important issues raised by these discussions to concentrate on explicating this theme.

A few years ago David Lewis (1979, reprinted in Lewis 1983, pp. 133—156) argued that the objects of beliefs which self-ascribe a location in space and time do not correspond to propositions construed as sets of possible worlds.

We can and do have beliefs whereby we locate ourselves in ordinary time and space; whereby we self-ascribe properties that don't correspond to propositions (1983, p. 138).

He calls attitudes which can have propositions as objects '*de dicto*' and introduces '*de se*' as a suitable term for attitudes with self-ascriptions of properties as objects.

When there is a propositional object, we are accustomed to speak of an attitude *de dicto*. Self-ascription of properties might suitably be called belief or knowledge *de se*. My thesis is that the *de se* subsumes the *de dicto*, but not vice versa (1983, p. 139).

Lewis uses John Perry's (1977) case of Lingens lost in the library to illustrate *non de dicto* knowledge *de se*.

An amnesiac, Rudolf Lingens, is lost in the Stanford library. He reads a number of things in the library, including a biography of himself, and a detailed account of the library in which he is lost. . . . He still won't know who he is, and where he is, no matter how much knowledge he piles up, until that moment when he is ready to say, "*This* place is aisle five, floor six, of Main Library, Stanford. *I* am Rudolf Lingens" (1983, p. 138).

Lingens' book knowledge is knowledge *de dicto*. We can think of it as locating our world in logical space, so we can think of its object as a proposition construed as a set of worlds. Lewis' point is that such propositional knowledge, however detailed and complete it may be, does not include the knowledge *de se* which Lingens lacks. Only when Lingens locates his own point of view (e.g. as that of the person standing in the middle of aisle five, on floor six, etc.) does he acquire the knowledge *de se* expressed in the demonstrative sentences at the end of the passage.

Kant's discussion of regions makes it clear that distinguishing regions in space fits Lewis' characterization of *non de dicto* knowledge *de se*. Let us consider again the passage where Kant tells us that geographical knowledge of relative positions is of no help to us unless we can place it according to regions through relation to our bodies.

The same holds true of geographical, indeed of our most ordinary knowledge of the position of places; such knowledge is of no help to us, so long as we are unable to place the so ordered things and the whole system of reciprocally related positions, according to regions through the relation to sides of our bodies (K & W, p. 39; AK II, p. 379, lines 35—37, p. 380, lines 1—2).

In order to place the relative positions according to regions I must specify their relation to my body. When I specify the relation of a system of reciprocally related positions to *my* body I reciprocally specify where *this* body is in relation to the system of positions and I specify that this body is *mine*. This locates *my* point of view with respect to the system of positions. Such knowledge is a paradigm example of *non de dicto* knowledge *de se*.

We have seen that the essential requirement for discriminating incongruent counterparts is to place the relative positions of those parts according to regions. Using Lewis' terminology we can express what I take to be Kant's main point in the quotation under consideration. Incongruent counterparts cannot be distinguished by any amount of knowledge *de dicto* however detailed and complete that knowledge may be. Knowledge of the difference between incongruent counterparts is *non de dicto* knowledge *de se.*

Jonathan Bennett (1970, pp. 116—178; this volume, pp. 97—130) sees these passages as pointers to a similar, but not quite identical, theme.[23]

Behind Kant's words in the inaugural lecture I have detected the claim that an explanation of the meaning of 'right' and 'left' requires *showing*, i.e. demands an appeal to sensorily presented examples. I shall call this claim the Kantian Hypothesis (this volume, p. 100).

Showing requires that the party doing the showing make ostensive (or at least some relatively direct) reference to the same particulars that are the examples sensorily presented to the party being shown. To require showing certainly does require some appeal to *non de dicto* knowledge *de se* but to require appeal to *non de dicto* knowledge *de se* need not require showing. Any non-eliminable appeal to perception is an appeal to *non de dicto* knowledge *de se* even if it does not require that what is perceived be some particular object that another party is also referring to.

Bennett's ingenious exploration of his theme can help illuminate our somewhat different theme. His tactic is as follows:

So I shall invent someone — call him an Alphan — whose grasp of English is perfect except that he gives to 'right' the meaning of 'left' and vice versa. We have to see how he could learn of his mistake (this volume, p. 110).

He takes as a contrast case someone whose only mistake is a confusion about 'between'

The Betan's mistake concerns the word 'between': he gives to the form "x is between y and z" the meaning we give to "y is between x and z". (He thinks that the thing asserted to be between the other two is the thing whose name occurs between the names of the other two: any English sentence containing the form "x is between y and z" is a kind of *picture* of what the Betan thinks it means.) (this volume, pp. 110—111)

Bennett shows (this volume, pp. 114—120) that the Betan's mistake about 'between' is rather easily revealed to him, while the Alphan's reversal of 'right' and 'left' is almost completely resistant to correction without appeal to showing.

The problem for Bennett's Alphan is a version of what Martin Gardner (1979) calls the Ozma problem.[24] This is the problem of communicating our use of 'left' and 'right' to extraterrestrials with whom our only contract is by exchanges of signals.

To state it precisely: Is there any way to communicate the meaning of 'left' by a language transmitted in the form of pulsing signals? By the terms of the problem we may say anything we please to our listeners, ask them to perform any experiment whatever, with one proviso: *There is to be no asymmetric object or structure that we and they can observe in common* (Gardner 1979, p. 155; this volume, p. 77).

Gardner points out that the celebrated parity non-conservation experiments (first carried out by Wu and others in 1957) solves the Ozma problem. The fundamental assymmetries in particle physics remove the requirement that we and others we are trading signals with be able to refer to some common enantiomorphic object in order to co-ordinate our conventions for 'right' and 'left'.

Bennett (this volume, pp. 127—128) correctly sees these results as undermining his version of the Kantian hypothesis by providing a way to reveal to the Alphan his mistake without appeal to showing. Let us examine Gardner's suggestion about how to use a version of Wu's famous experiment to signal our convention for 'left'.

We say to the scientists of Planet X: "Cool the atoms of cobalt-60 to near absolute zero. Line up their nuclear axes with a powerful magnetic field. Count the number of electrons flung out by the two ends of the axes. The end that flings out the most electrons is the end we call 'south'. It is now possible to label the ends of the magnetic axis of the field used for lining up the nuclei, and this in turn can be used for labelling the ends of a magnetic needle. Put such a needle above a wire in which the current moves away from you. The north pole of this needle will point in the direction we call 'left' (Gardner 1979, p. 206; this volume, p. 94).

Clearly, we don't have to be able to make any ostensive or otherwise demonstrative reference to the actual runs of the experiment the Alphans (scientists from Planet X) use to align their interpretation of 'left' with ours.

On the other hand, and just as clearly, the Alphans have to appeal to *non de dicto* knowledge *de se* when they actually do use the experiment to align their interpretation with ours. To apply Gardner's suggestion one must orient the wire and the suspended magnet to one's body. This is an obvious example of just the sort of transformation of knowledge of relative positions among objects into knowledge of regions that Kant was talking about. It requires an essential appeal to intuition to locate the quarters toward which the parts of the various objects are directed relative to your point of view. This appeal to intuition also locates your point of view relative to the system of related objects. It is an example, indeed an obvious example, of *non de dicto* knowledge *de se*.[25]

Consider the contrasting case of the Betan's mistake about 'between'. Suppose you are sent a co-ordinate representation of the relations among a group of objects. You are able to read off the relative positions of these objects among themselves even if you have no knowledge whatsoever of the location and orientation of that reference frame with respect to your point of view. This holds also for any specification of the interpretation of the assignments of signs to axes that would align it with your embedding in space. Between is a paradigm example of a relation among relative positions. It is fixed independently of how local neighborhoods containing that system of relative positions are embedded in global space.[26] It is invariant with respect to alternative sign assignments to co-ordinate representations of that system of relative positions as well as invariant with respect to translations and rotations of such co-ordinate representations.

As we have seen, Nerlich regards global space (the space-time manifold) as a common object to which both parties can refer. He takes Bennett to task for formulating the constraint on admissible evidence so as to rule out such a joint reference to the common space-time manifold (Nerlich 1973, p. 350; this volume, p. 165).[27] I think this objection is a good one. The rather mundane examples of *non de dicto* knowledge *de se* generated when we do our run of the Wu experiment and they do theirs facilitate a more profound sort of *non de dicto* knowlege *de se*. We treat our separate runs of the experiment as showing a common alignment between our respective embeddings in the space-time manifold. This alignment is made possible by our respective demonstrative references to our embeddings in global space when we each place our experimental apparatus according to regions by locating them relative to our respective points of view.

POSTSCRIPT ON GEOMETRY

Kant's discussion of incongruent counterparts in the Inaugural Dissertation (Section 15c) makes it quite clear that the essential role for intuition in discriminating between shapes of incongruent counterparts is intended to be a further illustration of what he takes to be the essential appeal to intuition required for geometrical demonstrations. This elaborates upon Kant's claims in the 1768 paper (p. 41, AK vol. II, p. 381, lines 16—18; p. 43, AK vol. II, p. 383, lines 24, 25) that the difference between incongruent counterparts rests on a relation to space as geometers conceive it. Similarly, in the *Metaphysical Foundations of Natural Science* (1986) (Ellington, p. 23, AK vol. IV, p. 484) we have Kant's explicit claim that the concept of the difference in shape between incongruent counterparts can be *constructed*, even though it does not admit of being clarified by universal marks in any merely discursive method of knowing. These passages suggest that Kant intends mathematical constructions (at least what he calls the *ostensive* constructions in geometry A 717, B 745) to be sufficient to carry the sort of non-discursive information required to discriminate a left- from a right-handed helix. It would seem, therefore, that Kant would regard pure intuitions used in such constructions as making demonstrative reference to an embedding in space in much the same way that empirical intuitions do so when we orient things around us to the embedding fixed demonstratively by our point of view.

NOTES

[1] Here I am following the translation in Kerferd and Walford (1968) pp. 36—43. I shall also include volume, page, and line references to the *Akademie Ausgabe (Kants gesammelten Schriften)*, 1902—0000. Editor's note: The Handyside translation is reprinted in this volume, pp. 27—33.

[2] See Nerlich (1973, pp. 340—341; this volume, pp. 154—155) for additional discussion of the inadequacy of Remnant's interpretation of Kant's intentions.

[3] In the translation from Kerferd and Walford (p. 43) Kant's one sentence is broken up into three separate sentences, one for each conjunct.

[4] The reference to inner sense in this sentence suggests a possible alternative reading of the puzzling passages where Kant claims that the difference between incongruent counterparts is an inner difference and that it must be based on an inner principle. We have been reading these passages as asserting something like the relevant features, e.g. shape, that distinguish a hand from an incongruent counterpart of it must be intrinsic to the object — that the differences are real differences so that they are among the facts which an adequate account of space must account for. The alternative reading suggested by this reference to inner sense would be to assert that the differences are inner in that they are grounded in the sensibility of the human subject. This alternative reading was pointed out to me by Philip Caton.

[5] This interpretation of twisting and turning as rigid motions is even better supported by Kant's German. The German phrase translated as "Twist and turn as one will" is

"drehen und wanden, wie man will" (AK II, p. 382, line 31). The German word "drehen" is more closely associated with rigid rotations than is the English translation "twist". I owe this observation to Brigitte Sassen.

[6] There is a little problem here. Rigid motions require that the space support a free mobility condition. This requires that it have a geometry of constant curvature. It is clear that the non-orientable surfaces under consideration have variable curvatures. They do not afford any literally rigid motions taking a figure around into its counterpart. What we can say, however, is that the relevant motion can be made as close as we want to rigid by making the surface large enough with respect to the size of the figure to be moved. We can also note that the deformations undergone during transport are not the sort of deformations which by themselves would induce the relevant change of shape (such as e.g. twisting the curved part to point the other way).

[7] John Earman (1987; this volume, pp. 235—255) has recently suggested an interesting way to argue from the global universality of the laws requiring non-conservation of parity to the global orientability of our space.

[8] Van Cleve offers a modest revision of Poincaré's definition of dimension to support his claim that there would be an intact sphere with an unbroken path from inside to outside that doesn't penetrate it if space had a fourth dimension (pp. 65—67; this volume, pp. 228—229). The claim is quite correct. Indeed, it follows from the standard neighborhood definition of dimensionality, where Van Cleve appeals to the consequence of his definition:

> that in a four-dimensional space any two points would be separated by something three dimensional, but not in general by something two-dimensional.

The standard definition would appeal to the corresponding consequence:

> that in a four-dimensional space any point has arbitrarily small neighborhoods whose boundaries have dimension three, but not in general arbitrarily small neighborhoods whose boundaries have dimension two or less.

In each case we could get a pair of points such that no sphere or other closed two dimensional surface could separate one from the other, not even when the surface passed between two points.

[9] One of the salient results of this investigation is that the sort of story Kant provides to explain how it is that ostensive geometrical constructions provide warrant limits the *a priori* warrant to figures of a size that human observers could accurately take in and to tolerances corresponding to human perceptual capacities. As I see it the correct Kantian story about how these *a priori* local results are to be extended to very large or very small figures involves non-constructive conceptions. Such a conception (e.g. the full Euclidean parallels postulate) may at some time enjoy a *contextual-a priori* status as the only serious candidate for approximating an unrealizable ideal of pure reason (e.g. articulating the global geometry for space), but it can be made vulnerable by the development of a rival conception that provides a more successful way to approximate the ideal (see Harper 1986).

[10] There is a special difficulty for Van Cleve's application of such thought experiments to settle the global orientability of space. Van Cleve (p. 45; this volume, p. 212) suggests that:

> . . . a non-orientable space of n dimensions is possible only in an ambient space of $n + 1$ dimensions. (If you haven't got three dimensions to work in, you can't make the twist necessary to create a Möbius strip.) The possibility of three-dimensional non-orientable spaces is thus dependent on the possibility of four-dimensional spaces, . . .

This is a mistake. As Earman (1987) pointed out (and as is common knowledge among those who work on philosophy of space-time physics) three-dimensional non-orientable spaces can be specified without embedding them in any space of higher dimension.

[11] The point at issue here is due to John Earman (1987). I got the idea of formulating a local definition of incongruent counterpart from his local definition of enantiomorph. My definition of global incongruent counterpart here is just his definition of incongruent counterpart.

[12] It doesn't require any very fine tuning about what are to count as the relations of the parts among themselves to see that Kant's Premise 1 comes out true in our toy model space. Consider the figures

d b

on our strip of vinyl. They are local incongruent counterparts (if the vinyl is not twisted into a Möbius strip they are also global incongruent counterparts). They have the relevant difference. Now cut the strip between them and turn over the right hand piece so that we have the following configuration:

d q

Now, our figures no longer have the relevant difference. They are no longer local (or global) incongruent counterparts. (A simple rotation of the figure on the right hand side makes clear that we now have two d's rather than a d and a b as before.) I think it is clear that the change we made in no way altered the relations among themselves of the parts of either figure. We did nothing at all to the left hand figure. The right hand figure was moved, but this move could be as rigid a motion as we want. Imagine cutting a little piece of vinyl around the letter completely out and carefully turning it over so that there is no stress from twisting the strip. Our operation might be carried out without in any way disturbing the subsurface within the surface defining this neighborhood of the figure. If we have not disturbed anything within this neighborhood then we have not disturbed spatial relations among the parts of a figure which is wholly within that same neighborhood.

[13] It may be worth noting that the addition of contractions to rotations and translations as a method of establishing shape-congruence answers an objection Van Cleve (1987, p. 57; p. 53, note 36; this volume, pp. 218–219, 232) makes to Nerlich's assumption that congruence requires rigid motion superposability.

We should also note that contraction with preservation of shape requires that the space have locally flat curvature. Indeed, it is well known that a homogeneity postulate (that figures of the same shape can have arbitrary diferences in size) is equivalent to Euclid's parallels postulate (e.g. see Toretti 1978, p. 207).

[14] A standard criticism of Nerlich is that he has not done much to argue that a relationist could not manage to recover topological structure (e.g. Sklar 1974). Earman (1987) has suggested that multiple connectedness of a space might show up in

appropriate multiple relations among objects in it. This would allow a relational account of the orientability of a flat two-space, since non-orientability requires multiple connectedness. This would not settle the matter for a strip bent into a cylinder, however, since this would be a multiply connected orientable space. If multiple connection is to be represented by multiple relations then, perhaps, the mappings allowed by the space may be able to count as relations among the objects in it.

[15] See Harper (1984) for more on applications of supervaluation semantics to Kant's empirical realism. Bas van Fraassen invented supervaluation semantics. See his 1966 paper for a classic presentation of it.

[16] This is Gauss's famous *Theorema Egregium* (see Spivak, Vol. II, p. 112). The expositions by Sklar and Nerlich are extremely simplified. Spivak gives a rigorous and comprehensive development of Gauss's theory of surfaces which is also very clear. His treatment contains a historical account of the development of these ideas including a detailed exposition of the relevant parts of Gauss's original papers. He also uses diagrams to appeal to geometrical intuitions with great effect.

[17] We have been using the idea that the two-space observer looks at the figures on the surface of the strip from a point outside it, just as we would. This is, I think, the sort of intuition to which Nerlich, Earman and others who use two-space models to illuminate Kant's discussion of incongruent counterparts normally appeal. One might wonder whether the points argued would hold up if the observer were required to be right in the two-space with the figures. Key Dewdney's *Planiverse* is the most recent, and perhaps the most developed, of the accounts of two-dimensional worlds that have grown up since Abbott's classic *Flatland*. In these worlds objects are observed edge on, as lines. Enantiomorphic objects reveal right-left asymmetries in the lines exhibited to observers at appropriate perspectives. In Dewdney's *Planiverse* there are two cultures with mirror image versions of dot-dash codes for writing. With an appropriate twist available to change the embeddings of objects and characters this would provide just the sort of background to support neat versions of the Alice story. As far as I can tell, the points we have been arguing go through quite unproblematically. Rob DiSalle pushed me to check this out. I am grateful to him and to Key Dewdney for a fascinating discussion about how these issues work out in the context of Dewdney's *Planiverse*.

[18] One way to do this is to think of the vicinity around here as e.g. the vicinity of our solar system. Now even if this "vicinity" is, itself, travelling about, it can serve to locate which observers count as the stay-at-homes.

[19] Here I may be disagreeing with Martin Gardner (1979) when he suggests:

> Of course, if you imagine yourself looking at the hand, naturally you will see it as either left or right, but that is equivalent to putting yourself (with your sense of handedness) into 3-space. You must imagine the hand in space to be completely removed from all relationships with other geometrical structures. Clearly, it would be as meaningless to say that the hand is left or right as it would be to say that it is large or small, or oriented with its fingers pointing up or down (p. 141; this volume, p. 63).

One way to imagine yourself looking at the hand is to imagine your body as an additional object in the space. This *is* equivalent to putting yourself in the space. To imagine the hand from a virtual point of view need not be equivalent to putting yourself in the space in this sense, but it does import your sense of handedness. Gardner's suggestion about a way of imagining a lone hand that does not import your sense of handedness — that is not from any virtual point of view at all — seems to me to be confusing imagining with conceiving. I grant that a shape which satisfies Nerlich's intrinsic hand shape and yet is indeterminate between the shape of a left hand and that of a right hand is conceivable; but, I don't think even Martin Gardner can imagine something with such a shape.

[20] Ralf Meerbote suggested to me that the German word "Gegenden" which Kerferd and Walford have translated as "regions" would be more accurately translated as "directions" in the 1768 paper. This would fit in rather nicely with the interpretation I offer here.

[21] The translation here is based both on Beck's (1950, p. 33; this volume, p. 37) and Bennett's (1970, p. 177; this volume, p. 101). Bennett's translation of the passage these lines are from is missing the words "; and yet the one cannot be put in place of the other". Here is a more correct translation:

> For example, two spherical triangles on opposite hemispheres, which have an arc of the equator as their common base may be completely equal, in sides as well as angles, so that nothing is found in either, when it is described alone and completely, which does not also appear in the description of the other; *and yet the one cannot be put in the place of the other* (that is, upon the opposite hemisphere). Here, then, is an *inner* difference between the two triangles, which no understanding can given an account of [an geben] as inner and which only reveals itself through the outer relation in space (AK II, p. 285, lines 33—35, p. 286, lines 1—6).

The missing phrase (which I have underlined) helps make it clear that the inner difference which no understanding can give an account of as inner is that the one cannot be put in the place of the other. This helps make it clear that Kant intends the difference to be the sort of thing he takes it that geometrical construction can provide.

[22] I am here following Bennett's (1970, p. 177; this volume, p. 101) correction to Beck's (1950, p. 34; this volume, p. 38) translation of Kant's parenthetical remark. Kant's German is "(z.b. widersinnig gewundener Schnecken)". Beck translates this, "(for instance, two symmetric helices)". Bennett translates it, "(e.g. spirals winding opposite ways)". Bennett is correct to explicitly say that the spirals are oppositely wound; however, I think that context makes it clear that Kant intends three-dimensional spirals, i.e. cylindrical helices, here.

[23] Bennett was apparently unaware of Kant's discussion of incongruent counterparts in the *Metaphysical Foundations of Natural Science* (see his remark on p. 175; this volume, p. 100).

[24] Project OZMA is a systematic search for radio messages from extraterrestrial intelligences. The name comes from Frank Baum's Oz books (see Gardner, pp. 153—160; this volume, pp. 75—81).

[25] Bennett dismisses the obvious sort of appeal to intuition we have been considering as irrelevant to the discrimination of the difference between incongruent counterparts.

> It is commonly believed that the distinction between a pair of enantio-

> morphs, when properly spelled out, must refer to the "point of view" of an "observer"; but this is false if it goes beyond the general point that any empirical distinction must qua empirical, have conceptual background. The idea seems to be that we should describe A [an example enantiomorph Bennett introduces on p. 179] like this: "When the line from its small cut to its small insert face runs the same way as the line from the observers feet to his head ... etc. But if a human body is used in describing A, why should it be an observers body? A corpse would serve as well (p. 180; this volume, pp. 107—108).

As I see it Kant's idea is that the obvious sort of relation of a system of positions to *our* point of view *demonstratively* fixes the relation of that system of positions to our embedding in global space. This is not the idea Bennett seems to have in mind. Certainly the considerations he raises here in no way undercut this idea of an essential role that relation to our points of view plays in our discrimination of handedness.

[26] This independence of the betweenness relation with respect to positions from what happens outside local neighborhoods containing these positions answers an objection Nerlich raises against one of Bennett's examples. Bennett uses the statement

> "Since Baltimore is between Washington and New York [B-W-NY],and we were flying in a straight line, we passed over New York first, then Baltimore, then Washington" (Bennett 1970, p. 183; this volume, p. 114).

as evidence to reveal the Betan's mistake. Nerlich's complaint is that examples like this do not yield the result Bennett wants, since the Betan's interpretation of 'between' is compatible with the order in which the cities are passed over. Just let the plane go the other way around the geodesic (great circle) connecting the three cities (Nerlich 1973, p. 351). Bennett's implicit appeal to the shortest geodesic connecting the three cities is correct.

[27] Here is Nerlich's actual objection:

> Bennett states on page 178 and again on page 180 that in certain possible spaces (some of which we know) there may not *be* a difference between left and right. So how could one possibly discover "the difference" between left and right in terms of sentences that must leave it entirely open whether there *is* any difference to the discovered? That cannot be settled short of some statement about the over-all connectivity of the space in which the things live (p. 350).

Even if the idea of local incongruent counterparts makes Nerlich's specific objection less telling than it may have originally appeared, this idea that the space-time manifold itself may count as a common object raises difficulties for Bennett's formulation. If we take the formulation literally, then the evidence from the parity-violating experiments becomes inadmissible once we and the Alphans actually use our runs of the experiment to align our embeddings in the common space-time manifold. If, as Bennett seems to, we use the constraint only to rule out common reference to mundane sorts of objects such as hands, stars, and galaxies, then the test fails to show that knowledge of handedness is not profoundly *non de dicto*.

REFERENCES

Bennett, J. (1966) *Kant's Analytic*, Cambridge University Press, Cambridge.

Bennett, J. (1970) 'The difference between right and left,' *American Philosophical Quarterly* Vol. 7, No. 3, 175—191. This volume, pp. 97—130.

Buroker, J. V. (1981) *Space and Incongruence, the Origin of Kant's Idealism*, Reidel Publishing Co., Dordrecht.

Dewdney, A. K. (1984) *The Planiverse (computer contact with a two-dimensional world)*, Poseidon Press, New York.

Earman, J. (1971) 'Kant, incongruous counterparts and the nature of space and space-time,' *Ratio* **13**, 1—18. This volume, pp. 131—149.

Earman, J. (1987) 'On the other hand . . . : a reconsideration of Kant, incongruent counterparts, and absolute space.' This volume, pp. 235—255.

Friedman, M. (1983) *Foundations of Space-Time Theories*, Princeton University Press, Princeton, N.J.

Friedman, M. (1986) 'Kant on the Foundations of Newtonian Science,' in Butts, ed. (1986).

Gardner, M. (1964, 2nd edition 1969, reprinted 1979) *The Ambidextrous Universe*, Charles Scribner and Sons, New York.

Grünbaum, A. (1973) *Philosophical Problems of Space and Time* (2nd enlarged edition), D. Reidel, Dordrecht.

Harper, W. L. (1984) 'Kant on space, empirical realism and the foundations of geometry,' *Topoi* **3**, 143—161.

Harper, W. L. (1986) 'Kant on the a priori and material necessity,' in Butts, R. E. (1986) ed., *Kant's Philosophy of Physical Science*, Reidel Publ. Co., Dordrecht, pp. 239—272.

Kant, I. (1768) 'Concerning the ultimate foundations of the differentiation of regions in space,' in Kerferd, G. B. and Walford, D. E. (1968) eds. and trans., *Kant, Selected Pre-Critical Writings and Correspondence with Beck*, Manchester University Press, Manchester, pp. 36—43. Editor's note: The Handyside translation is reprinted in this volume, pp. 27—33.

Kant, I. (1770) (Inaugural Dissertation) *On the Form and Principles of the Sensible and Intelligible World*, in Kerferd, G. B. and Walford, D. E. (1968), pp. 46—92.

Kant, I. (1781) *Critique of Pure Reason*, N. K. Smith (1933) trans., Macmillan, London.

Kant, I. (1783) *Prolegomena to Any Future Metaphysics*, Beck, L. W. (1950) ed. and trans., Bobbs Merrill, New York.

Kant, I. (1786) *Metaphysical Foundations of Natural Science*, Ellington, J. (1970) trans., Bobbs Merrill, New York.

Körner, S. (1966) *Kant*, Penguin Books, Baltimore.

Lewis, D. (1979) 'Attitudes De Dicto and De Se,' *The Philosophical Review* **88**, pp. 513—43; reprinted in Lewis (1983), pp. 133—156.

Lewis, D. (1983) *Philosophical Papers*, Vol. I, Oxford University Press, Oxford.

Mortenson, C. and Nerlich, G. (1983) 'Space-time and Handedness,' *Ratio* **XXV**, Vol. 1, pp. 1—13.

Nerlich, G. (1973) 'Hands, knees, and absolute space,' *Journal of Philosophy* **70**, 337—351. Revised version reprinted in this volume, pp. 151—172.

Nerlich, G. (1976) *The Shape of Space*, Cambridge University Press, Cambridge.

Reichenbach, H. (1927) *Space and Time*, Reichenbach, M. trans. (1958), Dover Publ. Inc., New York.

Remnant, P. (1963) 'Incongruent counterparts and absolute space,' *Mind* **72**, 393–399. This volume, pp. 51–59.

Sklar, L. (1974) *Space, Time, and Space-Time*, University of California Press, Berkeley.

Spivak, M. (1979) *A Comprehensive Introduction to Differential Geometry*, Vol. 2, 2nd ed., Publish or Perish, Inc., Berkeley.

Toretti, B. *Philosophy of Geometry from Riemann to Poincaré.*

Van Cleve, J. (1987) 'Right, left, and the fourth dimension,' *Philosophical Review* **XCVI**, 33–68. This volume, pp. 203–234.

van Fraassen, B. (1966) 'Singular terms, truth value gaps, and free logic,' *Journal of Philosophy* **63**, 481–495.

van Fraassen, B. (1970) *An Introduction to the Philosophy of Time and Space*, Random House, New York.

Walker, R. C. S. (1978) *Kant*, Routledge and Kegan Paul, London.

Weyl, H. (1952) *Symmetry*, Princeton University Press, Princeton.

JILL VANCE BUROKER

THE ROLE OF INCONGRUENT COUNTERPARTS IN KANT'S TRANSCENDENTAL IDEALISM*

INTRODUCTION

Undoubtedly the most puzzling of Kant's Critical views is his thesis that virtually all aspects of our experience of objects are contributed by the perceiving subject rather than by the things experienced, and are not features of these things as they exist independently of sensible perceivers. This position, which Kant calls transcendental idealism, is striking because nothing could be less commonsensical than the belief that things as they appear to us have nothing in common with things as they are independently of being perceived. From a more technical point of view the doctrine is perplexing because Kant apparently does not support it very well. Beginning with Kant's contemporaries, critics have pointed out that among all the arguments in the *Critique of Pure Reason*, none apparently entails the conclusion that things in themselves *cannot* be like objects of sense experience in any way. So, for example, although Kant's theory of synthetic *a priori* knowledge provides some support for transcendental idealism, there is nothing in the analysis of the synthetic *a priori* ruling out the possibility that features contributed to experience by the perceiving subject may correspond to characteristics of things in themselves, although we might never know this to be so. And even though Kant views transcendental idealism as the solution to the Antinomies, this is at best indirect support for the position. Moreover, Kant asserts the merely subjective character of sensible representations in 1770, long before he developed the theory of the Antinomies. Because transcendental idealism is so radical, it seems that Kant should provide especially strong reasons in its support.

In the *Critique of Pure Reason* Kant first draws his idealistic conclusion in the Transcendental Aesthetic, following his arguments that the representations of space and time are synthetic and *a priori*. In the section titled 'Conclusions from the above Concepts' he says with respect to space:

(a) Space does not represent any property of things in themselves, nor does it represent them in their relation to one another. That is to say, space does not represent any determination that attaches to the objects themselves, and which remains even when abstraction has been made of all the subjective conditions of intuition. . . .

(b) Space is nothing but the form of all appearances of outer sense. It is the subjective condition of sensibility, under which alone outer intuition is possible for us. . . . ([3], A26 = B42)

This presentation suggests that Kant's idealism is founded in his analysis of sensibility, and in particular, in his views of space and time. In spite of the absence of compelling arguments in the *Critique*, then, the first place to look for the origin of Kant's idealism is in his thinking about the sensory aspects of experience.

A close look at Kant's writings before 1781 bears out this hunch, for we see that the first sketch of his Critical theory of sensibility, in the 1770 Inaugural Dissertation, resulted directly from objections to the Leibnizian analysis of sense experience. These objections were initially aimed at Leibniz's relational theory of space, and took Kant from there to Leibniz's account of sense perception and finally to the metaphysical status of space. In short, Kant's transcendental idealism makes sense when seen as a repudiation of the Leibnizian theory of ideas.

The key to transcendental idealism is a series of arguments appearing in Kant's writings from 1768 to 1786. In each case Kant claims that the existence of a certain kind of physical object, which he calls an 'incongruent counterpart,' demonstrates something about the nature of space. Counterparts are objects like left and right hands which are mirror image reflections of one another. They are 'incongruent' because, in spite of their similarities, they cannot be superimposed on one another, that is, made identical to one another, by any continuous motion in space. The first incongruent counterparts argument occurs in his short essay of 1768, *Concerning the Ultimate Foundation of the Differentiation of Regions in Space*, where he argues that left- and right-handed objects support the Newtonian theory of absolute space over Leibniz's relational theory, the theory Kant had adhered to for the preceding 20 years. The essay is noteworthy, however, because it signals not only a radical shift in Kant's thought about space, but also the prelude to the Critical philosophy. For one thing, *Regions in Space* appears just two years before the Inaugural Dissertation of 1770, in which Kant first sets out his Critical theory of the sensibility and argues that space is the form of outer sense. For another, subsequent versions of this argument appear in three of Kant's Critical writings, including the Dissertation, supporting conclusions quite different from that argued for in the *Regions* essay. In particular, the argument in the *Prolegomena to Any Future Metaphysics* of 1783 stands out for its conclusion that space is transcendentally ideal, that it is a merely subjective representation, and that things as they exist independently of sensible perceivers can in no way be spatial.

Commentators focusing on these arguments have generally claimed that this final conclusion, that space is merely subjective and ideal, flatly contradicts Kant's first conclusion from incongruent counterparts, that space has an absolute reality of its own. Norman Kemp Smith says that the argument in the *Prolegomena* "is made to yield to a very different conclusion from that drawn in the Dissertation and a *directly opposite conclusion* from that drawn in 1768". ([9], p. 164) In his introduction to the translation of *Regions of Space*, D. E. Walford agrees, claiming that "there is clearly an incompatibility of purpose to which Kant puts the paradox of incongruent counterparts in the 1768 treatise and the 1783 *Prolegomena*." ([5], p. xvii) It is hard to disagree with Jonathan Bennett's view that "Kant could not decide which if any of his doctrines about space can draw strength from special facts about the right/left distinction." ([1], p. 176)

However tempting this appraisal, it is not supported by a closer look at the texts. Despite the apparent incompatibility among Kant's various conclusions there is a coherent line of thought running through the different versions of the argument, so than rather than conflicting with the first argument, the later conclusions actually incorporate it. The key to the progression in Kant's incongruent counterparts arguments is the relation between the theory of space and the analysis of sensibility in Leibnizian thought. Once Kant's reasoning is charted against this Leibnizian background, we can see not only that the different versions of the argument are consistent, but also that they provide the missing support for Kant's idealism.

To say that the incongruent counterparts arguments are the source of Kant's idealism is not to imply that his doctrine had no other motivation. But it is to say that among all the considerations favoring that doctrine, incongruent counterparts provided its *strongest justification*. That is, the phenomenon of incongruent counterparts produced the *direct support* for the analysis of space and sensibility which culminated in his transcendental idealism. To see how this happens, we must look at the successive steps in Kant's thinking about space, beginning with the earliest version of the argument. Part I summarizes the argument in *Regions in Space*, showing how incongruent counterparts first led Kant to reject Leibniz's relational theory of space. In Part II we see how this first conclusion raised serious questions for Leibniz's epistemology, and so brought Kant to formulate a competing theory of sense perception. And finally, Part III shows how this latter analysis results in the transcendental ideality of space.

I. *PHASE I*: THE ARGUMENT FOR ABSOLUTE SPACE

The debate between the Newtonians and Leibnizians over the nature of space concerned, first and foremost, the proper ontological relation of space to the objects which occupy it. For the Newtonians it was a 'prejudice' to think of space and spatial properties such as distance or direction as derived from the properties and relations of material objects. A more careful analysis results in the distinction between absolute and relational space:

> Absolute space, in its own nature, without relation to anything external, remains always similar and immovable. Relative space is some movable dimension or measure of the absolute spaces; which our senses determine by its position to bodies. . . . Absolute and relative space are the same in figure and magnitude; but they do not remain always numerically the same. ([14], vol. 1, p. 6)

For Newton, absolute space is an actual existing thing which, although not itself material, serves as a container for all material objects and a necessary condition of their existence. In this sense space is *real* and *objective*, since its existence and nature do not depend on the prior existence of the bodies occupying it. It is also *prior* to the objects it contains in two senses, since, first, they must exist in space although space could exist without them, and second, the spatial relations among bodies are derived from the relations among parts of space.

Leibniz objected vigorously to this theory, arguing on metaphysical grounds that there can be no *real* entity such as absolute space. In particular, he claimed, Newton's concept of space violates the metaphysics of substance and property in addition to the fundamental principles of the Identity of Indiscernibles and Sufficient Reason. Absolute space cannot be a substance since, for Leibniz, substances are characterized on the physical level by their dynamical relations to one another. And Newton himself granted that space and physical objects exert no forces on one another. But space cannot be a property either: "What substance will that bounded empty space be an affection or property of . . .?" ([10], p. 37) Moreover, Newton's conception of absolute space as an infinite set of indistinguishable mathematical points runs afoul of the Principle of the Identity of Indiscernibles ("there cannot be two individual things in nature which differ only numerically" ([12], p. 268)) and the Principle of Sufficient Reason ("there is nothing without a reason, or no effect without a cause" ([12], p. 268)). According to Newton it makes sense to distinguish the state in which all material objects occupy one part of absolute space, from that in which the same objects with the same relations among them occupy a different part of space. But Leibniz points out in reply:

> . . . without the things placed in it, one point of space does not absolutely differ in any respect whatsoever from another point of space. Now from hence it follows (supposing space to be something in itself, besides the order of bodies among themselves), that 'tis impossible there should be a reason, why God, preserving the same situations of bodies among themselves, should have placed them in space after one certain particular manner, and not otherwise. . . . ([10], p. 26)

For Leibniz space is nothing more than the set of all possible positions objects can occupy relative to one another. If there were no objects, there would be no space. Space is just the way of ordering coexisting things: "abstract space is that order of situations when they are conceived as being possible. . . ." ([10], p. 89) With respect to the ontological status of space, then, Leibniz's relational theory maintains that physical objects are *ontologically independent of* and *prior to* space. Not only the application, but even the meanings of the terms 'space,' 'place' and 'motion' require reference to existing material frameworks.

In his pre-Critical writings from 1747 up to 1768, Kant was firmly committed to this Leibnizian theory as well as to a form of Leibnizian metaphysics. Although Kant's first published essay, *Thoughts on the True Estimation of Living Forces*, focuses for the most part on the analysis of force and motion, it says enough about space to show where his loyalties lay, as in the following remark: "It is easy to prove that there would be no space and no extension if substances had no force to act outside themselves. For without this force there is no connection, without this no order and without this, finally, no space." ([4], p. 23) And similarly, in explaining why space has three dimensions, Kant argues: "the properties of extension and consequently their threefold dimension itself will be grounded in the properties of the force which substances possess in relation to the things with which they are connected." ([4], p. 24) Since, according to Kant, gravitation is the force defining relations among physical objects, and since it obeys the inverse square law, the space which results from this dynamical matrix must be three-dimensional.

In several essays appearing between 1755 and 1758 Kant presents essentially this same relational analysis of space. He remarks in the *New Elucidation of the First Principles of Metaphysical Knowledge* that "position, situation, space are relations of substances by which . . . they have reference to other substances actually distinct from themselves. . . ." ([7], p. 414) And in the *Universal Natural History and Theory of the Heavens*, where he develops his famous hypothesis concerning the origin of the universe, he again claims that the properties of space depend on the nature of gravity. A few years later, in his 1758 essay, *A New System of Motion and Rest*, he defends the relational theory by arguing that the concepts of motion and rest signify relations among physical objects rather than relations of objects to space. Up to 1768, then, Kant adheres firmly to the Leibnizian ontology of space: space is not an entity existing independently of material objects, but merely the set of possible positons derived from their actual and possible relations. In metaphysical terms, physical objects are prior to space; in linguistic terms, the meaning of spatial concepts presupposes reference to existing material frameworks. Kant does not depart from these views until the first incongruent counterparts argument of 1768.

Kant begins that argument in *Regions in Space* by distinguishing between the 'positons' of objects and the 'regions' according to which these positions are related. He says:

> In the most abstract sense, region does not consist of the relation of one thing in space to the next. That would really be the concept of position. Region really consists rather in the relation of the system of these positions in absolute space. ([5], pp. 36—37)

> The region, however, to which this order of the parts is directed, is related to space outside, but not with reference to localities, for this would be nothing else than the position of just those parts in an external relation; region is related rather to space in general as a unity, of which each extension must be regarded as a part. ([5], p. 37)

In setting off positions from regions in space, Kant is drawing attention to the difference between a set of relations between objects or parts of objects in space, and the way in which such a system of relations is situated in its encompassing space. The references in these passages to "absolute space" and "space in general as a unity" indicate that Kant is thinking of a region as a part of global space.

The difference between a region and a position is best illustrated by certain kinds of objects, which he calls "incongruent counterparts." All species of hops twist around poles from left to right, while beans wind in the opposite direction. With some exceptions, the shells of snails all coil in the same direction. But the most common example of the incongruent counterpart is the human hand. When we examine objects such as left and right hands, the first thing we notice is that in virtually all respects they are similar: "in respect both of size and proportion and even of the situation of the parts relative to each other, they may be in perfect agreement." ([5], p. 39) And if one considers merely these features — size, proportion, relative position of the parts — "a complete description of the one must apply, in all respects, to the other." ([5], p. 41) But in spite of their similarities, these objects "cannot be included exactly within the same limits." ([5], p. 41) Or, as Kant puts it elsewhere, "the surface that includes the one could not possibly include the other." ([5], p. 42) Left and right hands are counterparts, then, because in terms of certain physical characteristics they are identical. But they are incongruent because there is no continuous rigid motion which can bring them into coincidence in space.

Let us consider first Kant's claim about the ways in which such counterparts are alike.[1] It is clear that the size and proportion of left and right hands can be identical, but it is not so clear that this is true of the relations among the parts. When we hold our hands so that the thumbs and fingers of both hands are in the same order, for example, then the direction of the back-to-palm relation differs. This suggests that the relations among the parts of left and right hands can be considered the same as long as each relation (say the distance and angle between two parts) is considered in any given plane, that is, in two dimensions. The reason the hands are not substitutable in spite of these identical two-dimensional relations is that, taken together, all of these distances and angles constitute a *three-dimensional* system of relations. In other words, the relevant differences between incongruent counterparts arise only when one considers the relations among the parts of hands in all three dimensions.

This interpretation is supported by Kant's remarks about figures lying on a flat surface:

When two figures, drawn on a surface, are like and similar, they cover each other. But it is often different with physical extension or even with lines and surfaces not lying on a flat surface. They can be perfectly like and similar and yet be in themselves so different that the limits of the one cannot at the same time be the limits of the other. ([5], p. 41)

Figures 'on a flat surface,' then, are two-dimensional figures. If two-dimensional asymmetrical triangles, which have angles of the same size and sides of the same length, are placed facing different directions, they can be made to coincide by rotating one outside the plane in three-dimensional space. This illustrates the principle that an n-dimensional asymmetrical object can be turned into its counterpart if it is rotated in a space of $n + 1$ dimensions. Now it follows from this principle that objects like hands could be made congruent if they were rotated in a space of at least four dimensions. But since our space is only three-dimensional, they can never be made to coincide, no matter how they are moved. Thus Kant recognized a fact about three-dimensional asymmetrical objects which sets them apart from two-dimensional counterparts. The incongruence of counterparts such as hands, then, is a function jointly of the properties of the objects themselves and those of the space in which they are located.

In addition to its dimensionality, there is a second property of space connected with incongruence which Kant undoubtedly did not conceive as such: this is the *orientability* of space. Contrary to the principle established above, it is possible for an object of n dimensions to be transformed into its counterpart in an n-dimensional space if that space is *non-orientable*. Orientable spaces are those in which two figures which are locally incongruent in Kant's sense are also globally incongruent; if they cannot be made to coincide in the local region, then taking them outside that region into larger space of the same dimensions will not make them congruent. This is true, for example, of all Euclidean spaces. But there are spaces which do not behave in this way. For these non-orientable spaces, two figures which are incongruent locally might be made congruent by moving one through space outside the region and back. The Möbius band is one such non-orientable space. Now although Kant did not understand orientability as such, he evidently realized that the Euclidean nature of space precludes turning incongruent three-dimensional objects into their counterparts. Orientability is relevant to the analysis of incongruent counterparts because it shares an important feature with the three-dimensional nature of space: namely, both are features of *global space*.

The clearest statement of Kant's argument in favor of absolute space in *Regions in Space* occurs in this thought experiment: he instructs the reader to try to imagine a universe consisting of only one object, a human hand. Now if the Leibnizians were correct, so that space consisted merely of the relations of parts of the hand to one another, and since, Kant thinks, these relations are identical for right and left hands, "the hand would be completely indeterminate with respect to such a quality, that is, it would fit on either side of the human body." ([5], p. 43) But that is, of course, impossible; the hand must be either left or right. Therefore the Leibnizians are wrong about the nature of space.

If we imagine a universe consisting only of a human hand and its "handspace", can we determine whether the hand could be made to coincide with its counterpart? Here I think Kant is right. The answer is no, because we would have to know the nature of the space in which the hand and its handspace are contained; more precisely, we would have to know the dimensions and orientability of global space. In pointing out that their incongruence depends both on the properties of hands and the topology of space as a whole, Kant has raised a serious challenge to the Leibnizian view of space. If incongruent counterparts are not *intrinsically* incongruent, then incongruence is a relation between the objects and the space containing them.[2] And this implies that space is ontologically independent of physical objects.

Kant's argument for absolute space in *Regions in Space* can be summarized in the following steps:

1. The contrast between two-dimensional and three-dimensional counterparts demonstrates that there is a property of spatial objects which depends partly on the nature of the global space in which they are located.
2. If any properties of objects are dependent on the nature of space, then space must exist independently of objects.
3. Therefore, the space of our experience exists independently of and prior to the objects located in it.
4. The ontological priority of space to spatial objects is compatible with the absolute theory of space but not with the relational theory.

5. Therefore, experience refutes the Leibnizian view and supports the absolute theory of space.

Thus the incongruent counterparts argument, in its first appearance, provides the evidence forcing Kant to reject the relational view of space. It is not the case that the spatial properties of physical objects can be accounted for entirely by the relations between such objects: in particular, the incongruence of counterparts depends equally on the topology of the global space in which objects are located. And from this it follows that the nature of global space is *independent of* the objects it contains. The first step towards transcendental idealism, then, is Kant's conviction that the relational view that objects are ontologically prior to space cannot adequately account for all aspects of our experience of physical objects.

II. *PHASE 2*: THE NATURE OF SENSIBILITY

Although Kant's purpose in *Regions in Space* is to refute the derivative ontological status of space, the essay contains two passages suggesting that he was beginning to appreciate the implications of his argument for Leibniz's theory of sense perception. Near the beginning of the essay Kant describes his purpose as "to see whether, in the *intuitive* judgments of extension, such as include geometry, a clear proof can be found" that space is absolute. ([5], p. 37, emphasis added) And he concludes, near the end, that incongruent counterparts require us to "regard the concept of space in the way geometers regard it . . . as other than a mere entity of reason." ([5], p. 43) Since an entity of reason is a kind of mental construction, this latter quote implies that if the Leibnizians are wrong about the ontology of space, their epistemology of space must similarly be mistaken. To see why this is so, we need to examine the way in which the Leibnizians connect the relational theory of space with the metaphysics of monads and the theory of "well-founded phenomena."

For Leibniz, physical objects and the space and time they exist in are only *phenomenal* manifestations of metaphysical substances. These fundamental substances, called monads, are the absolutely simple elements out of which everything is composed. Since monads have no parts they are not spatially or temporally extended, and hence are not material; Leibniz calls them "intelligible" or "noumenal" entities, and conceives them on an analogy with minds. Properly speaking, monads are not objects of sense perception, but it is possible to know about them by reason or the understanding. Since monads are not material objects, Leibniz must explain how our experience of such objects comes about. In recognition of this demand he proposes the doctrine of "well-founded phenomena." Well-founded phenomena, which include both physical objects and space and time, result from the incomplete or "confused" perception of monads by the imagination. When in perception the imagination fails to distinguish all the properties of monads, in effect it abstracts from the properties distinguishing them from other monads, and in this way unites them into groups or aggregates. Thus for Leibniz our experience of objects extended spatially and temporally arises because the imagination erroneously combines distinct substances into *phenomenal unities*. Unlike monads, which are "substantial unities," aggregates of monads have only a "unity in thought, which is insufficient for the reality within the phenomena." ([12], p. 343)

This is not to say, however, that such phenomena are illusory. In terms of their relations to monads, well-founded phenomena are those which express, however confusedly, properties which monads actually have. As Leibniz remarks to Arnauld, "they have as much of reality or substantiality as there is of true unity in that which enters into their composition." ([11], p. 191) Well-founded phenomena, then, have both *objective* and *subjective* aspects. The objective aspect obtains because these phenomena correspond in certain exact ways to the states of monads; that is, there is a basis in monads which allows them to be perceived as aggregates. But the apparent unity of well-founded phenomana is subjective: it comes about only because monads are perceived incompletely by the imagination.

Leibniz's account of the process of spatial and temporal representation has two stages. As a consequence of the first act of confused perception, the perceiver experiences objects which are extended and which have certain relations to one another. In this sense extension and duration are attributes of the physical object produced by the experience, but not metaphysical attributes of anything. Space and time are then created in a second stage of the process when the perceiver abstracts from physical objects all characteristics but their relative positions, thus representing the actual locations occupied by objects. From this set of positions the perceiver then constructs a matrix which includes not only these actual positions but also those which objects may possibly occupy. These constructed matrixes are space and time.

This theory includes two features which become especially prominent in the incongruent counterparts arguments. First is Leibniz's doctrine of the ideality of relations. The second is what I shall call Leibniz's Correspondence Principle and its basis in his theory of sensibility. Let us look briefly at each of these features.

As for the nature of relations, all that exists fundamentally in the noumenal world of monads are individual substances and their accidents. Strictly speaking, there are no relations between monads because there is no system functioning as a framework for relating monads in the way that Newtonian absolute space functions with respect to physical objects. In his fifth letter to Clarke, Leibniz says that only substances and their accidents are concrete, and relations are neither of these. For if one views a relation as an accident shared by two substances,

> . . . we should have an accident in two subjects, with one leg in one, and the other in the other; which is contrary to the notion of accidents. Therefore we must say, that this relation, in this third way of considering it, is indeed out of the subjects; but being neither a substance nor an accident, it must be a mere ideal thing, the consideration of which is nevertheless useful. ([10], p. 71)

This theory of relations consists of two main theses. The first is that all relations between substances are derived from non-relational properties. That is, if we separate the terms predicable of substances into intrinsic or internal denominations, those referring to a single substance, and external or extrinsic denominations, those requiring reference to more than one substance, then Leibniz holds that "there are no purely extrinsic denominations which have no basis at all in the denominated thing itself." ([12], p. 268) The second thesis is that relations among monads are *reducible* to their non-relational properties. This is usually interpreted as the claim that relational facts can be analyzed into conjunctions of non-relational facts, or that relational propositions are replaceable by equivalent subject-predicate propositions. G. H. R. Parkinson claims that Leibniz held this view in both a weak and a strong sense:

> . . . when one says that *A* has a certain relation to *B*, the proposition asserted is reducible to subject-predicate propositions whose subjects are *A* and *B* respectively. . . . However, Leibniz also takes the statement that there are no purely extrinsic denominations to mean that in order to assert a relational proposition about (say) *A*

and *B*, it is in principle enough to know either the predicates of *A* alone or the predicates of *B* alone. ([15], p. 45)

In sum, Leibniz's view that relations are ideal implies not only that nonrelational properties are ontologically prior to relations, but also that relational propositions are, in principle, eliminable.

The second part of Leibniz's theory especially relevant to the incongruent counterparts arguments concerns his account of the relation between sense experience and intellectual representations. As we have seen, well-founded phenomena "represent" or express, in a confused manner, the real properties of noumenal substances. Although they differ from these noumenal properties by virtue of their sensory character, spatial properties of physical objects nonetheless correspond in some exact way to these more fundamental, "intelligible" properties. Thus the representation of space is *objective* in the sense that it is derived from properties actually belonging to monads. But it is also *subjective* insofar as its character as represented depends on the nature of the perceptual process. According to this analysis, then, sensory representations are nothing more than confused or incomplete concepts. For Leibniz the basic elements of all knowledge are concepts or general ideas, not sense impressions: sense impressions of phenomenal things are only composites of concepts not completely or clearly grasped. Were it not for the capacity of the imagination to distort intellectual representations, sense perceptions and their spatial context would never arise.

In the two most fully-developed versions of the incongruent counterparts argument appearing after 1768, Kant's focus shifts from the ontological status of space to the nature of spatial experience. Following his rejection of the relational theory, Kant goes on to argue in the Inaugural Dissertation and the *Prolegomena to Any Future Metaphysics* that the representation of space is a pure intuition. As opposed to Leibniz's account of sense perception, Kant will show that our knowledge of space and spatial objects is *irreducibly sensible*, that it cannot be explained in terms of defective concepts. Furthermore, as we shall see, this argument presupposes the conclusion from *Regions in Space*.

In the Dissertation Kant introduces incongruent counterparts after arguing that space is a pure intuition, apparently taking counterparts as examples or illustrations of his thesis. After remarking that the fact that space is a pure intuition can be seen in the axioms of geometry (e.g., space contains no more than three dimensions, between any two points there is only one straight line), Kant says this:

> Which things in a given space lie towards one quarter and which things incline towards the opposite quarter are things that *cannot be described discursively or reduced to intellectual marks by any mental acuteness.* Thus between solids which are perfectly similar and equal but not congruent, in which genus are the left hand and the right hand (in so far as they are conceived solely according to their extension), or spherical triangles from two opposite hemispheres, there is a diversity *which makes it impossible for the boundaries of their extension to coincide although they could be substituted for one another as far as concerns all the things which may be expressed in marks intelligible to the mind in speech.* And so it is clear that in these cases the diversity, I mean the discongruity, *can only be noticed by a certain act of pure intuition.* ([5], p. 69, emphases added)

From this passage we can extract the following three claims about incongruent counterparts:

D1. The differences between them "cannot be described discursively or reduced to intellectual marks by any mental acuteness."

D2. Despite their incongruence, they are substitutable "as far as concerns all the things which may be expressed in marks intelligible to the mind in speech."

D3. The incongruity between counterparts "can only be noticed by a certain act of pure intuition."

From the discussion in the *Prolegomena*, which is too long to reproduce in its entirety, these claims stand out as essential:

P1. If things are equal in all respects ascertainable, "quantitatively and qualitatively it must follow that the one can in all cases and under all circumstances replace the other. . . ."

P2. But incongruent counterparts "exhibit, notwithstanding a complete internal agreement, such a difference in their external relation that the one figure cannot possibly be put in the place of the other."

P3. There is nothing about an incongruent counterpart, "if it be described for itself alone and completed, that would not equally be applicable to both. . . ."

P4. This is evidence of "an internal difference" between counterparts, "which difference our understanding cannot describe as internal and which only manifests itself by external relations in space."

P5. Between left and right hands there are "no internal differences which our understanding could determine by thinking alone."

P6. "Yet the differences are internal as the senses teach" for the one hand cannot be superimposed on the other.

P7. Therefore those objects are not things "as some mere understanding would know them, but sensuous intuitions. . . ." ([8], p. 33)

In the first place, Kant evidently wants to show that certain facts about incongruent counterparts depend on the intuition of space, and cannot be known or verified on the basis of *merely conceptual* knowledge. When he denies, as at P7, that these facts can be known by the 'mere understanding,' he does not mean to claim that they cannot be conceptualized or expressed verbally — all judgments require concepts — but that the basis of these facts resides in the irreducibly sensible nature of space. In order to establish their truth one must be *acquainted* with the nature of space, which is possible only through sensible intuition, and not through the mere concept of spatiality. This is also the point of D1, D2, and P5.

Now we must consider *which* facts Kant has in mind. D1 implies that these facts include a description of left- and right-handed objects, a description linking their properties with their incongruence. Putting this together with the previous point, we can restate Kant's conclusion this way: *The fact that objects such as left and right hands (three-dimensional, equal in size and proportion, having the same distances and angles between parts in any two dimensions, the parts in the third plane being mirror-image reflections, etc.), are incongruent can be known only by means of the (pure) sensible intuition of space.* This accounts for D3, since if this is true it would follow that to *recognize or notice* the incongruence of such objects would presuppose the intuition of space. Of course Kant realizes that we can describe incongruent counterparts by means of concepts; he wants to show that we could not formulate or justify the judgment that such objects are incongruent by *conceptual knowledge alone.* Consequently such knowledge must be *irreducibly sensible.*

As stated in the *Prolegomena* the argument takes the form of a paradox involving Leibniz's principle of the Identity of Indiscernibles and theory of well-founded phenomena. The Identity of Indiscernibles asserts that things that are qualitatively identical are also numerically identical; all numerical distinctions among objects are accompanied by differences in their properties. Now one criterion for qualitative identity is substitutivity; two things have the same properties if and only if they are completely substitutable. Leibniz applied this principle to both phenomena and noumena, so that no two things that exist, material or intelligible, differ merely numerically. As we have seen, according to the theory of well-founded phenomena, objects known by the senses represent, albeit confusedly, properties and relations of intelligible substances. As a metaphysical principle this postulates a correspondence between phenomena and noumena. Epistemologically it asserts a *reduction* of sensory representations to confused or incomplete concepts. In addition, since space is 'ideal' or constructed from properties and relations of phenomena, we can describe spatial objects completely without referring to space. All references to space are in principle eliminable.

Kant derives his first conclusion in the Critical period, that knowledge of space is *irreducibly sensible*, by showing that these Leibnizian views, taken in conjunction with the analysis of incongruent counterparts from *Regions in Space*, result in a paradox:

1. Two things are completely substitutable if and only if they have identical properties. (Identity of Indiscernibles)
2. The properties and relations of phenomena correspond in an exact way to the properties and relations of noumena. (Leibniz's Correspondence Principle)
3. Space is ideal; phenomena can be completely described without referring to space. (Ideality of Relations)
4. In a three-dimensional Euclidean space, objects *L* and *R* are not subsitutable and hence (by 1) are not identical.
5. Therefore, noumena corresponding to *L* and *R* must differ in some respect corresponding to their incongruence. (2, 3, 4)
6. In a four-dimensional (or non-orientable) space, *L* and *R* are substitutable and therefore have identical properties.
7. Therefore, noumena corresponding to *L* and *R* do not differ in any respect giving rise to incongruence. (2, 3, 6)

8. Therefore, noumena corresponding to *L* and *R* both do and do not differ in some respect relevant to the congruence or incongruence of *L* and *R*. (5, 7)

Lines 4 and 6 above come from the argument in *Regions in Space*, since they express the view that whether left and right hands could be made to coincide depends on the space in which the hands are located. But if, as the Leibnizians believe, space has no status independent of spatial objects, reference to the nature of space can be eliminated from complete descriptions of phenomena. Thus we can assert *simpliciter*, given the Correspondence Principle and the analysis of incongruence, that noumena corresponding to the hands have contradictory properties. Since this is impossible, the Leibnizian theory must be wrong.

Before seeing how Kant then derives his conclusions about the nature of sensibility, let us use the paradox to interpret his statements in the Dissertation and the *Prolegomena*. For one thing, we can now understand claims D1, D2, P4, and P5 as depending on the Leibnizian distinction between internal and external properties (or denominations). Internal properties are, first of all, the *non-relational* properties, those whose description requires reference to a single substance. But since, for Leibniz, relational properties are *reducible* to non-relational properties, internal properties are also the *essential, irreducible* properties of things. With respect to Leibniz's theory of ideas, these properties of monads are *intelligible* or *purely conceptual* rather than sensory. So internal properties are those which are (1) non-relational, (2) essential, (3) irreducible, and (4) intellible.

In his discussion of incongruent counterparts Kant apparently *identifies* the intelligible (or conceptual) and non-relational aspects of internal properties. We can see how this might come about. In the first place, he agrees with Leibniz that space and time are systems of possible relations among phenomenal objects. But, as he argues in the Dissertation and the *Critique of Pure Reason*, they are also necessary conditions of all experience of sensible objects. Consequently, insofar as sensory experience is spatial and temporal, it is inherently relational. By implication, the purely non-relational properties of things must be non-sensory or intelligible. In this way Kant collapses the distinction between relational and non-relational properties into the distinction between sensible and intelligible properties. As he reads Leibniz, properties divide neatly into two groups, with non-relational and intelligible properties on the noumenal or fundamental side, and relational and sensible properties on the phenomenal or derived side. This explains his claims in the Dissertation and the *Prolegomena* that incongruent counterparts do not differ conceptually. He means to say that since in some spatial circumstances counterparts are substitutable, the *intelligible entities* underlying them must be identical in the relevant respects. In Leibnizian terms, objects which are manifested phenomenally as incongruent counterparts are identical noumenally, *insofar as they are known purely by the intellect.*

The tie connecting Kant's paradox with the view that knowledge of space is irreducibly sensible is the Leibnizian theory of relations. As we saw earlier, this involves two claims: first, that all relational properties presuppose or depend on non-relational properties of substances; and second, that relations among substances are *reducible* to sets of non-relational properties. Kant apparently shares both views. In the section of the *Critique* titled the 'Amphiboly of the Concepts of Reflection,' in which he criticizes the Leibnizians, he contrasts their use of the matter—form distinction with his own. This is what he says:

> In any judgment we can call the given concepts logical matter (i.e., matter for the judgment), and their relation (by means of the copula) the form of the judgment. In every being the constituent elements of it (*essentialia*) are the matter, the mode in which they are combined in one thing the essential form. ([3], A266 = B322)

Just as in logic the matter of a judgment refers to its contents and the form to the way they are related, an analogous use applies in metaphysics. With respect to things that exist, 'matter' signifies the essential, non-relational properties of a thing, and 'form' the way these properties are related, both within a substance and to other substances. What is the status of relations?

> The understanding, in order that it may be in a position to determine anything in definite fashion, demands that something be first given, at least in concept. Consequently in the concept of the pure understanding matter is prior to form; and for this reason Leibniz first assumed things (monads), and within them a power of representation, in order afterwards to found on this their outer relation and the community of their states (i.e., of the representations). ([3], A267 = B323)

So according to reason form always presupposes matter; relations presuppose non-relational properties. This is clear from the concept of relation alone: if there were no prior relata to be related, there would be no relations. Moreover, since this truth can be known from the *mere concept* of relation, this knowledge about relations can be attributed to the understanding or the intellect. It is a *conceptual* truth about relations that they presuppose non-relational properties of existing things. Kant repeats this point later in the Amphiboly when he says,

> Through mere concepts I cannot, indeed, think what is outer without thinking something that is inner; and this for the sufficient reason that concepts of relation presuppose things which are absolutely [i.e., independently] given, and without these are impossible. ([3], A284 = B340)

Leibniz had that much right. That is why, as Kant remarks, he postulated monads and their properties as the foundation of all that exists, and space and time as systems derived ultimately from those substances. According to the theory of relations, then, if space represented intelligible substances — Kant says if it were a 'determination' of things in themselves — it could not exist *independently* of substances or their phenomenal manifestations. But Kant believes, thanks to incongruent counterparts, that our space *does* exist independently of spatial objects. In the world of sensible phenomena there are relations (among the parts of space) that cannot be derived from the non-relational properties of objects in space alone. The nature of our space conflicts with the conceptual analysis of relations.

This conclusion, that the nature of space contradicts the theory of relations, is the basis for Kant's claims that knowledge of spatial objects is, *qua* spatial, irreducibly sensible and, later, that space is transcendentally ideal. Let us consider the first conclusion. If relations as conceived by the understanding are dependent solely on the monadic properties of real things, as the theory of relations maintains, and if spatial relations are not so dependent, as the analysis of incongruence shows, then spatial experience does not correspond to the properties and relations of intelligible substances. So Leibniz's Correspondence Principle is false. The properties and relations of sensible phenomena do *not* correspond to or represent, even incompletely, the properties and relations of things known by the intellect alone. But if that is so, then sensations are not just confused concepts. Sensible experience cannot be analyzed as incomplete knowledge by the understanding because some sensible (spatial) relations are *independent* of the things so related. Although I think this line of reasoning does not establish the *a priori* nature of our knowledge of space, it does entitle Kant to conclude that knowledge of spatial objects is irreducibly sensible.

III. *PHASE 3*: SPACE AND THINGS IN THEMSELVES

The final conclusion of Kant's incongruent counterparts arguments, that space is a merely subjective representation which in no way represents properties or relations of things in themselves, follows immediately from the previous conclusion. We have seen that the conclusion from *Regions in Space*, taken in conjunction with Leibniz's analysis of relations, entails that Leibniz's Correspondence Principle is false. If the independent status of space relative to the objects in it conflicts with the dependent status of relations as conceived by the intellect, then sensible relations cannot correspond, even in a confused fashion, to relations among purely intelligible substances. Thus the sensibility is not derived from the intellect; it must be an independent source of representation. As a system of relations standing independently of the things related in it, space 'applies' only to things as they appear to the senses, and not at all to the reality which underlies them. Consequently, this underlying reality, which Kant calls things in themselves, is in no way spatial. By showing how the formal character of sensible phenomena conflicts with the formal features of things as defined by reason, the incongruent counterparts argument provides the *direct support* for the non-spatiality of things in themselves that is otherwise lacking.

As I remarked earlier, there is no mention of incongruent counterparts in the *Critique of Pure Reason*. This seems odd, given the significance I have assigned to the argument and the fact that the *Critique* contains the fullest treatment of space and sensibility in Kant's writings. Despite the lack of explicit reference to incongruent counterparts, however, there are passages in the *Critique* which express Kant's reasoning in the incongruent counterparts arguments.

The argument appears in a more general form in the Transcendental Aesthetic. Rather than reasoning from the existence of objects like left and right hands to the independent status of sensibility and the ideality of space, Kant argues directly in terms of the analysis of relations. The first relevant passage in the Aesthetic occurs in the Metaphysical Exposition where Kant points out how part-whole relations differ for concepts and intuitions. Here is what Kant says:

3. Space is not a discursive or, as we say, general concept of relations of things in general, but a pure intuition. For, in the first place, we can represent to ourselves only one space; and if we speak of diverse spaces, we mean thereby only parts of one and the same unique space. Secondly, these parts cannot precede the one all-embracing space, as being, as it were, constituents out of which it can be composed; on the contrary, they can be thought only as *in* it. Space is essentially one; the manifold in it, and therefore the general concept of spaces, depends solely on [the introduction of] limitations. ([3], A24—25 = B39)

Here Kant is claiming that the difference between sensible and intellectual representations is analysable in terms of two features characterising their part-whole structures.[3] For concepts to be general in form means that they can be organized into hierarchies, that they may bear a genus-species relation to one another. Consider the following three concepts: 'physical object', 'animal', and 'left-handed.' 'Physical object' is the most general or highest concept of the three; the concept 'animal' stands under it as a species to its genus. The same relation holds between the concepts 'animal' and 'left-handed,' since the latter concept is in turn subordinated to the former. Now considered from the other direction, the concepts 'animal' and 'left-handed' both *presuppose* the concept 'physical object,' because the content of these concepts includes the content of the concept 'physical object.' How are lower concepts obtained from higher concepts? On Kant's view the extension of the higher concept (that is, the set of instances to which the concept is applied) is divided on the basis of some differentia, or some characteristic. One might notice, for example, that some physical objects are living and capable of locomotion. These would then be the characteristics or differentiae distinguishing animals from other physical objects. Thus the concept 'animal' has a smaller extension than the concept 'physical object', but a *greater content or intension.* In this way we obtain lower-order concepts — more complex concepts — by *adding* content to higher-order concepts. In this case, for example, the complex concept 'animal' can be analysed as composed of the *logically prior* concepts 'physical object', 'living,' and 'capable of locomotion.' For concepts, then, the wholes — the complex concepts — are logically dependent on their simpler parts.

For sensible representations, those that *reproduce* an individual object, this part-whole ordering is quite different. For one thing, the relation of the parts to the whole is not hierarchical. Instead, the (proper) part is contained *within* the whole. My perception of my thumb, for example, is a part of my perception of my hand, and it is contained *in* that larger perception rather than falling *under* it. In other words, my perception of my hand is not a genus of which my perception of my thumb is a species. Moreover — and this is the key part — in sensible representations of spatial—temporal objects the whole is *logically prior* to the parts. This is because there are no smallest parts of space (or time) which, when added together, constitute larger spaces (or times). For sense perceptions the logical priority works in the other direction; parts of space (and time) are created by drawing boundaries in the global framework. Thus Kant's third argument in the Metaphysical Exposition consists in the claim that the original representation of space is irreducibly sensible because the representation of the whole logically precedes the representations of its parts.

The different part-whole structures of intuitions and concepts show that spatial and temporal experience cannot be reduced to or explained in terms of confused concepts. Adding to this Kant's argument from the Amphiboly, that relations as understood by reason always presuppose simple, non-relational parts, it follows that spatial relations cannot represent the relations among things in themselves. In fact, Kant makes this very argument in a section of the General Observations on Transcendental Aesthetic added in the B edition:

> II. In confirmation of this theory of the ideality of both outer and inner sense, and therefore of all objects of the senses, as mere appearances, it is especially relevant to observe that everything in our knowledge which belongs to intuition . . . contains nothing but mere relations; namely, of locations in an intuition (extension), of change of location (motion), and of laws according to which this change is determined (moving forces). What it is that is present in this or that location, or what it is that is operative in the things themselves apart from change of location, is not given through intuition. Now a thing in itself cannot be known through mere relations; and we may therefore conclude that since outer sense gives us nothing but mere relations, this sense can contain in its representation only the relation of an object to the subject, and not the inner properties of the object in itself. ([3], B66—67)

Although Kant does not mention incongruent counterparts explicitly, he does argue for the transcendental ideality of space based on the logical priority of the whole of space to its parts. And this is equivalent to the third and final conclusion derived from his analysis of incongruent counterparts.

CONCLUSION

I hope to have shown that the standard reading of Kant's incongruent counterparts arguments is mistaken. Kant was not confused about what he could infer from the existence of left- and right-handed objects; there are no incompatibilities among the different arguments. We have seen, rather, that the later arguments, that experience of spatial objects is irreducibly sensible and merely subjective, are based on the first conclusion, that space exists independently of the objects in it. Moreover, Kant's Critical theory of space is Newtonian in so far as he views space as logically prior to the objects in it. But with respect to the relation of space to perceivers, and hence to things in themselves, Kant rejects both the Newtonian and Leibnizian positions.

Incongruent counterparts are the key to transcendental idealism. In Kant's break from the Leibnizian tradition, incongruent counterparts first convinced him that space is an entity independent of phenomenal objects. From there he was required to reject Leibniz's epistemology of space, and, ultimately, the views that sensibility is a defective kind of intellection and that sensible phenomena are confused representations of things in themselves. But this is so only because throughout the progression of Kant's thought he never questioned the theory of relations at the center of Leibnizian metaphysics.

However important Hume's influence on Kant, it is Leibniz who shapes the Critical theory of sensibility, both as ancestor and adversary. The radical subjectivity of transcendental idealism can be understood only as a revision of the qualified subjectivity in Leibniz's idealism. This, I think, is the importance of the incongruent counterparts arguments. In drawing our attention to the role of relations in Kant's theory of sensibility, and the connection between this analysis and transcendental idealism, incongruent counterparts illuminate aspects of Leibnizian thought fundamental to Kant's Critical philosophy.

NOTES

* This article is extracted from portions of my book, *Space and Incongruence: The Origin of Kant's Idealism* (Dordrecht: D. Reidel Publishing Company, 1981), especially Chapters 3 through 5. Fuller discussions of the issues raised here as well as the literature on incongruent counterparts are presented there.

[1] This discussion owes much to articles by John Earman, Graham Nerlich, and

Lawrence Sklar. In Chapter 3 in *Space and Incongruence* I discuss the differences among their interpretations of Kant's argument in more detail. See [2], [13], and [16].

[2] A recent article by James Van Cleve has convinced me that this issue deserves more attention than my discussion here or in *Space and Incongruence* suggests. Unfortunately I did not have time to take into account Van Cleve's interesting analysis of Kant's "single-hand experiment" in [17].

[3] Kant's theory of concepts is outlined in his *Logic*, paragraphs 5—16, and 110. See [6], pp. 99—106. My understanding of this theory was greatly aided by Kirk Wilson's analysis in [18], pp. 252—54.

REFERENCES

1. Bennett, Jonathan. 'The Difference Between Right and Left,' *American Philosophical Quarterly* **7** (1970): 175—91. This volume, pp. 97—130.
2. Earman, John. 'Kant, Incongruous Counterparts and the Nature of Space and Space-Time,' *Ratio* **13** (1971): 1—18. This volume, pp. 131—149.
3. Kant, Immanuel. *Critique of Pure Reason*. Translated by Norman Kemp Smith. London: Macmillan and Co., 1929.
4. Kant, Immanuel. *Gedanken von der wahren Schätzung der lebendigen Kräfte* (1747). In *Kants Werke*, Vol. 1: 1–182. Preussischen Akademie der Wissenschaften edition. Berlin: Georg Reimer, 1902.
5. Kant, Immanuel. *Kant, Selected Pre-Critical Writings*. Edited and translated by G. B. Kerferd and D. E. Walford. Manchester: Manchester University Press, 1968.
6. Kant, Immanuel. *Logic*. Translated by Robert S. Hartman and Wolfgang Schwarz. Indianapolis: Bobbs-Merrill Co., 1974.
7. Kant, Immanuel. *Principiorum primorum cognitionis metaphysicae nova dilucidatio* (1755). In *Kants Werke*, Vol. 1: 385—416.
8. Kant, Immanuel. *Prolegomena to Any Future Metaphysic*. Edited by Lewis White Beck. Indianapolis: Bobbs-Merrill Co., 1950.
9. Kemp Smith, Norman. *A Commentary to Kant's Critique of Pure Reason*. New York: Humanities Press, 1962.
10. Leibniz, Gottfried Wilhelm. *The Leibniz—Clarke Correspondence*. Edited by H. G. Alexander. Manchester: Manchester University Press, 1965.
11. Leibniz, Gottfried Wilhelm. *Discourse on Metaphysics, Correspondence with Arnauld and Monadology*. Translated by George R. Montgomery. La Salle, Illinois: Open Court, 1968.
12. Leibniz, Gottfried Wilhelm. *Philosophical Papers and Letters*. Translated and edited by Leroy E. Loemker. Dordrecht: D. Reidel Publishing Co., 1969.
13. Nerlich, Graham. 'Hands, Knees and Absolute Space,' *Journal of Philosophy* **70** (1973): 337—51. Revised version reprinted in this volume, pp. 151—172.
14. Newton, Isaac. *Mathematical Principles of Natural Philosophy and His System of the World*. Translated by Florian Cajori. 2 volumes, Berkeley: University of California Press, 1966.
15. Parkinson, G. H. R. *Logic and Reality in Leibniz's Metaphysics*. Oxford: The Clarendon Press, 1965.

16. Sklar, Lawrence. 'Incongruous Counterparts, Intrinsic Features and the Substantiviality of Space,' *Journal of Philosophy* **71** (1974): 277–90. This volume, pp. 173–186.
17. Van Cleve, James. 'Right, Left, and the Fourth Dimension,' *The Philosophical Review* **XCVI**, No. 1 (Jan., 1987): 33–68. This volume, pp. 203–234.
18. Wilson, Kirk. 'Kant on Intuition,' *Philosophical Quarterly* **25** (1975): 247–65.

JAMES VAN CLEVE

INCONGRUENT COUNTERPARTS AND THINGS IN THEMSELVES

> Those who cannot yet rid themselves of the notion that space and time are actual qualities inherent in things in themselves may exercise their acumen on the following paradox. When they have in vain attempted its solution and are free from prejudices at least for a few moments, they will suspect that the degradation of space and time to mere forms of our sensuous intuition may perhaps be well founded.
>
> Immanuel Kant: *Prolegomena to Any Future Metaphysics*

Kant called upon incongruent counterparts during the course of his career to establish three very different conclusions: in 1768, that space is absolute, i.e., exists independently of matter; in 1770, that our knowledge of space and spatial configurations is by way of intuition, not concept; and in 1783, that space is ideal, i.e., belongs only to our form of intuition and not to things in themselves.[1] It is this third conclusion that is the most startling and has the least apparent connection with the premises. My aims here are (i) to assess briefly three different reconstructions of Kant's argument and (ii) to consider the argument in relation to three different conceptions of things in themselves.

I. THE ARGUMENT FROM INTELLIGIBILITY

Norman Kemp Smith interprets the argument given in Kant's *Prolegomena* (1783) along the following lines[2]

1. The difference between two incongruent counterparts (e.g., a left and a right hand) cannot be known conceptually. (That is, a purely conceptual description of a hand will not determine whether it is left or right.)

2. Things in themselves *can* be adequately known conceptually.
3. Therefore, objects like left and right hands are not things in themselves.

From this conclusion one can presumably generalize and say *no* spatial objects are things in themselves.[3]

The argument above is not explicit in the *Prolegomena*. The materials for it are rather to be found in the Inaugural Dissertation of 1770, where Kant does indeed assert each of the premises and the conclusion. Curiously, however, he does not assemble them all into one argument; he asserts the conclusion on independent grounds, and the only moral he draws from incongruent counterparts is that our knowledge of space is intuitive rather than conceptual.

Could the argument above be the argument Kant meant to give in the *Prolegomena*? The first premise is indeed there, and there is a hint of the second in Kant's use of the phrase "things as they are in themselves *and as some pure understanding would know them*" (emphasis mine). It is extremely puzzling, however, to find Kant using such an argument in 1783. As Kemp Smith notes, the second premise is pre-Critical. It amounts to saying that things in themselves are noumena in the positive sense (things adequately knowable by nonsensible means), not just in the negative sense (things not adequately knowable by sensible means), and this is a doctrine Kant rejects in the *Critique of Pure Reason* (at least so far as human beings are concerned). Kemp Smith thus concludes that the argument is a remnant of a repudiated view, taking this to explain why there is no trace of the argument in the second edition of the *Critique* in 1787.

II. THE ARGUMENT FROM INTERCHANGEABILITY

What is the 'paradox' of incongruent counterparts? Kant states it as follows:

> If two things are quite equal in all respects as much as can be ascertained by all means possible, quantitatively and qualitatively, it must follow that the one can in all cases and in all circumstances replace the other, and this substitution would not occasion the least perceptible difference. This in fact is true of plane figures in geometry; but some spherical figures [for example, spherical triangles in opposite hemispheres with an arc of the equator as common base and corresponding sides equal] exhibit, notwithstanding complete internal agreement, such a difference in their external relation that the one figure cannot possibly be put in the place of the other.[4]

Let me use 'internally alike' to abbreviate "quite equal in all respects, etc." (in other words, alike in all nonrelational respects) and 'interchangeable' to abbreviate "the one can in all cases, etc." The paradox can then be stated more briefly as follows:

1. Things that are internally alike must be interchangeable.
2. Some spatial figures are internally alike, but *not* interchangeable.

This is indeed a paradox, since as they stand 1 and 2 contradict each other. What is the way out?

According to Kant, we must first of all recognize that 1 is true only with a qualification; it must be rewritten as

1′. Things *in themselves* that are internally alike must be interchangeable.

This restores consistency – *provided* we do not shrink from drawing the conclusion that now follows:

3. Some spatial figures are not things in themselves.

This is just the conclusion Kant draws: the figures in question are not things in themselves, but only "appearances, whose possibility rests upon the relation of certain things unknown in themselves to something else, namely, to our sensibility." As before, we can presumably generalize to all spatial configurations.

My criticism of this argument is that the first premise seems perfectly plausible *without* the qualification. If that is so, the only way to avoid contradiction is to deny the *second* premise. This would mean either denying that incongruent counterparts are internally alike or holding that they are interchangeable after all. I shall mention a supporting ground for each option.[5]

(i) Suppose we take seriously the idea of *absolute direction*.[6] Then it would be false that incongruent counterparts are internally alike: Kant's spherical triangles would differ in that the direction from middle to largest to smallest angle would be clockwise in one and counterclockwise in the other.

(ii) Suppose we take seriously the idea of a *fourth spatial dimension*. Then it would be false that incongruent counterparts are noninterchangeable: although we cannot interchange spherical triangles in *our* space, spaces would be possible in which they *could* be interchanged.[7] One could resist this objection to the second premise by holding that 'interchangeable' must mean 'interchangeable in the *actual* space', but then the *first* premise would become questionable.

III. THE ARGUMENT FROM REDUCIBILITY

According to Jill Vance Buroker, the gist of Kant's incongruent counterparts argument is that "the nature of space is incompatible with the analysis of relations."[8] I agree completely as to gist, but would like to give a different account of the details. Here, then, is a third reconstruction of Kant's argument:

1. Incongruent counterparts (e.g., Kant's two spherical triangles) are different in virtue of their differing relations to space as a whole. (Thus triangle 1 bears R to Space and triangle 2 does not.)
2. All relations among things in themselves are reducible to nonrelational characters (qualities) of the relata.
3. Therefore, if Space and figures within it are things in themselves, triangle 1 must differ internally from triangle 2.
4. But in fact triangle 1 does *not* differ internally from triangle 2.
5. Therefore, Space itself and figures within it are not things in themselves.[9]

Now for a series of comments to elucidate this argument.

A. The premise that comes to the fore in this version is the *reducibility of relations* among things in themselves. It is a merit of Buroker's book to bring out the importance of this Leibnizian element in Kant's philosophy.

Kant does not explicitly state this premise in the *Prolegomena* argument, but he does endorse it at least implicitly elsewhere. It underlies two of his assumptions in the 'Amphiboly' section of the *Critique of Pure Reason*: that things in themselves cannot have their natures exhausted by relations and that they cannot differ numerically without also differing qualitatively.[10]

B. Strictly speaking, it is not the *reducibility* of relations that is needed in the argument above, but only their *supervenience*. That is, a relational fact aRb need not be *equivalent* to any conjunction Fa & Gb, but there must be some such conjunction that *entails* it. I shall nonetheless stick with the language of reducibility in what follows.

C. In accordance with the reducibility principle, if we have

triangle 1 *R* Space & ~ (triangle 2 *R* Space)

we must also have

F(triangle 1) & *G*(Space) & ~ [*F*(triangle 2) & *G*(Space)]

which implies

F(triangle 1) & ~ *F*(triangle 2).

Thus, the triangles must differ internally. This is the justification for step 3 above.

D. Note that the first premise in the present argument is the conclusion of Kant's 1768 argument: the difference between incongruent counterparts is a difference in their relations to Space. This bears out Buroker's contention that Kant's later arguments incorporate, rather than contradict, his earlier conclusion.[11]

E. Note, however, that it would not have been necessary for Kant to make Space a term of relations in the argument above. He could simply have noted that the relation of incongruent counterparthood, a relation holding directly between the triangles, is an irreducible relation. If relations among things in themselves must be reducible, it would then follow that the triangles are not things in themselves. So I disagree with Buroker's view that if Kant's conclusion of 1768 were denied, the 1783 version of his argument would be undercut.[12]

F. Moreover, once armed with the reducibility principle, Kant need not have resorted to anything so recondite as incongruent counterparts. He could simply have said: *distance* is not a reducible relation; therefore, nothing in the field of the distant-from relation (which is to say, nothing in space) is a thing in itself.

G. The big question raised by this interpretation of the argument is this: Why should Kant have thought that the Leibnizian reducibility principle holds for things in themselves, but not for appearances? I shall consider three possible answers to this question, each invoking a different conception of the contrast between appearances and things in themselves.

IV. THINGS IN THEMSELVES AS INTELLIGIBILIA

The first answer is given by the following syllogism:

1. Intelligible entities (entities adequately knowable by the intellect alone) can stand in none but reducible relations.
2. Things in themselves are intelligible entities.
3. Therefore, things in themselves can stand in none but reducible relations.

Appearances, as sensible entities, would not fall under the major premise.

This answer to our question (which seems in places to be Buroker's [13]) is unsatisfying for two reasons. First, the minor premise is the Dissertation doctrine all over again, so we are thrown back to the argument from intelligibility. Is there an answer to our question the *Critical* Kant can give? Second, why should even the pre-Critical Kant have thought the major premise true? That is, why should he have thought irreducible relations a bar to intelligibility? To use James's famous figure, why may the life of the intellect not include flights as well as perchings?

I will note in this connection that according to some philosophers, the nature of a number is exhausted by its relations to other numbers.[14] Yet numbers are intelligible entities *par excellence*.[15]

V. THINGS IN THEMSELVES AS THINGS APART FROM RELATION

In some contexts Kant's phrase 'in itself' seems to contrast with 'in relation to other things (or to us)'. If we understand the phrase this way, the unknowability of things in themselves is no longer the utter unknowability of things of a special kind, but the unknowability in certain respects (namely, nonrelational respects) of things in general.[16] The way is then open for holding that what present themselves to our senses are the very things that are unknowable "in themselves;" what is denied us is not any access whatever to these things, but just knowledge of their nonrelational or qualitative aspects. In support of such an interpretation one might cite the following passage:

Now a thing in itself cannot be known through mere relations; and we may therefore conclude that since outer sense gives us nothing but mere relations, this sense can contain in its representation only the relation of an object to the subject, and not the inner properties of the object in itself.[17]

Can this interpretation of the 'in itself' be used to answer our question? The idea would be that if 'in itself' means 'apart from relation', it becomes a tautology to say that things in themselves stand in no relations (or none but reducible relations). but there is clearly a fallacy here. 'No knowledge of things in themselves' may mean 'no knowledge of things apart from relation', but we cannot infer that 'things in themselves' means 'things apart from relation'. Or if we do, we have no right to assume that there *are* any things in themselves; that would be to make an illicit shift from 'in itself' as adverb to 'in itself' as adjective.

VI. THINGS IN THEMSELVES AS REAL EXISTENTS

In various of his writings Wilfrid Sellars has interpreted Kant's distinction between things in themselves and appearances as the distinction between things having real being or 'formal reality' and things merely having intentional being or 'objective reality' (in the medieval and Cartesian sense of this term).[18] The idea is that to be a thing in itself is to exist *simpliciter*, whereas to be an appearance is to exist only as a content of thought or awareness. I believe this interpretation to be essentially correct,[19] and in what follows I wish to see what light it throws on the incongruent counterparts argument.

It will be useful to begin with a little paradox. Consider first the following principle:

1. A genuinely existent entity must be *fully determinate* — that is, one predicate out of every possible pair of contradictorily opposed predicates must belong it.

To me this principle seems self-evident.[20] Kant endorses it at A573 = B601, and I believe it also underlies his assumption in Section 7 of the Antinomies chapter that the spatiotemporal world, if existent as a thing in itself, would have to be either finite or infinite.

So far all is well, but now let's confront our principle with the following recalcitrant fact:

2. The dragon I dreamt of last night was *not* fully determinate. It had teeth, but no definite number of them.

Here is the 'paradox': 1 and 2 both seem true, yet seem to conflict. The solution is obvious: 1 and 2 do *not* conflict, but can both be true provided the following is also true:

3. The dragon I dreamt of was not a genuinely existent entity.

The dragon existed only in the dreaming of it, and therefore falls outside the scope of the determinacy principle.

Now what I would like to suggest is that Kant's incongruent counterparts argument should be viewed as paralleling the argument 1—3:

1′. Genuinely existent entities must stand in none but reducible relations.
2′. Spatial figures stand in some irreducible relations.
3′. Therefore, spatial figures are not genuinely existent entities.

This is exactly what Kant's argument would become on the suggested interpretation of things in themselves. If you wish to verify this, please refer back to comment E of Section III, and if you think 3′ a most unKantian conclusion, please read on.

One thing I wish to bring out with the parallel is the following possibility: that Kant does not regard the reducibility principle as holding for things in themselves in virtue of some special feature they possess, but instead simply regards it as holding for existents as such. Like the determinacy principle, it may be intended as a general logical or ontological principle. That being so, the more pressing half of our question would not be why things in themselves must be subject to the reducibility principle, but how appearances can be exempt from it.[21]

At one level the answer is that appearances, as things having only intentional being, do not really exist, and hence are not there to be exceptions to any principle. But that cannot be the whole story. There is an important sense in which spatial items *do* exist (they are "empirically real," as Kant insists), and we must inquire how items with their mode of being can be exempt from laws governing the *an sich*.

My answer to this has two parts. (I): *Things that exist only in relation to consciousness are logical constructions out of conscious states.* I think this assumption is both plausible in its own right and plausibly attributable to Kant. If we do not make this assumption, we must evidently suppose that "intentional being" or "objective reality" is a separate ontological status, a second way of being in the world, and it has always been hard to understand what such a status could amount to. Moreover, that Kant himself thinks of intentional items as constructions is strongly suggested by many of his remarks about "objects" in the Transcendental Deduction and elsewhere.[22] (II): *If A's are logical constructions out of B's, A's need not obey all the same laws as B's, however self-evident these laws may be.* Here is one example: matrices are logical constructions out of ordinary numbers; ordinary multiplication is commutative; matrix multiplication is not.

Putting I and II together, we see that intentionalia need not obey all the same laws as the realia out of which they are constructed. (I assume that conscious states are realia.) For an illustration, let's return to the dragon of my dream. The dragon is a logical construction out of dream states; this explains not only how it is possible for the dragon to have no existence outside the dream, but also how it is possible for it to be indeterminate. The construction is governed by the following rule: if 'p' states a fact about the dragon internal to the dream, then p iff Dp (i.e., one dreams that p).[23] Now since reality is determinate, we must have either Dp or $\sim Dp$; but we need not have either Dp or $D\sim p$. It follows that the dragon may be indeterminate: there need not be an answer to the question, "Did it have more than 100 teeth or not?"

Lest the dream example mislead, I must insert two caveats. I do not mean to suggest that appearances in general have for Kant the same status as dream objects. They do not: an empirically real beast has relations to the rest of our experience that the dragon does not.[24] But appearances in general are nonetheless *like* dream objects in that they exist only in relation to consciousness. Nor do I mean to suggest that the basis in reality for any fact about appearances must always consist in a subject's apprehending that very fact. To take one of Kant's famous examples, the basis for saying that opposite sides of a house exist simultaneously is not that anyone perceives them as existing simultaneously; it is rather that the perceptions of the sides individually can be obtained in any order.[25]

I have not addressed the question how the reducibility principle in particular might fail for appearances,[26] but I think I have at least made it intelligible how appearances might fail to obey laws that are deemed valid for things in themselves.

NOTES

[1] 1768: 'On the First Ground of the Distinction of Regions in Space,' in *Kant's Inaugural Dissertation and Early Writings on Space*, trans. by John Handyside (Chicago: Open Court, 1929); 1770: *Dissertation on the Form and Principles of the Sensible and Intelligible World*, in Handyside; 1783: *Prolegomena to Any Future Metaphysics*, trans. by L. W. Beck (Indianapolis: Bobbs-Merrill, 1950). Relevant parts of all three of these works are reprinted in this volume. The *Prolegomena* argument is briefly repeated on pp. 23—24 of *Metaphysical Foundations of Natural Science*, trans. by James Ellington (Indianapolis: Bobbs-Merrill, 1970).

[2] *A Commentary to Kant's 'Critique of Pure Reason'* (New York: Humanities Press, 1962; reprint of 1923 edition), pp. 161—66. This volume, pp. 43—48.

[3] Or would anyone dare suggest that perhaps *symmetrical* objects, which lack incongruent counterparts, might for all the argument shows be things in themselves? To take this suggestion seriously would be to countenance the possibility that while my fingers are things in themselves, my hand as a whole is appearance!

[4] *Prolegomena*, p. 33. This volume, p. 37.

[5] I discuss the two points that follow in more detail in 'Right, Left, and the Fourth Dimension,' *The Philosophical Review* **96** (1987), 33—68. This volume, pp. 203—234.

[6] That is, suppose we maintain that one of a thing's part's lying in a certain direction from another part does not tacitly involve a relation to something outside the whole.

[7] Spherical triangles, though themselves objects of two-dimensions, can exist only in a space of three dimensions and would require for their interchange a space of four dimensions.

[8] Jill Vance Buroker, *Space and Incongruence* (Dordrecht, Holland: D. Reidel, 1981), p. 85.

[9] All that strictly follows is that *either* Space *or* one of the triangles is not a thing in itself — a conclusion that leaves room for someone to maintain that the triangles are things in themselves although Space is not. But this position will be ruled out in comment E below.

[10] It may also underlie premise 1 in the argument from interchangeability.

[11] Buroker, pp. 4 and 69.

[12] Buroker, p. 87.

[13] See especially p. 83.

[14] See Paul Benacerraf, 'What Numbers Could Not Be,' *The Philosophical Review* **74** (1965), 47—73.

[15] I do not say that Kant would agree. He holds our knowledge of arithmetic to be related to our temporal form of sensibility in a way that apparently requires that numbers be sensible.

[16] See D. P. Dryer, *Kant's Solution for Verification in Metaphysics* (London: George Allen & Unwin Ltd, 1966), ch. 11, sec. vi, especially pp. 513—14.

[17] Immanuel Kant, *Critique of Pure Reason*, trans. by N. K. Smith (New York: St. Martin's Press, 1965), B67. (All further page references are to the Kemp Smith edition.) A competing interpretation of this passage is also possible, and in light of the 'Amphiboly' section recommends itself: things in themselves cannot "consist solely of relations;" appearances *do* consist solely of relations; therefore, no appearance is a

thing in itself. Since Kant defines 'appearance' as the object of intuition, it would follow that we have no intuition of things in themselves.

[18] See, for example, *Science and Metaphysics: Variations on Kantian Themes* (London: Routledge and Kegan Paul, 1968), ch. II. A similar interpretation is illuminatingly discussed by Philip Cummins in 'Kant on Outer and Inner Intuition,' *Nous* **2** (1968), 271–92.

[19] Section 6 of the Antinomies chapter (to mention just one passage) seems to me to leave very little room for doubt on this matter.

[20] At any rate, it seems self-evident to me that every genuine existent must be fully determinate; perhaps it may be questioned whether the law of excluded middle is a sufficient test for such determinacy.

[21] Of course, I still find it perplexing why Kant was so convinced of the reducibility principle. For me it has nothing like the evidence of the determinacy principle – certainly not enough evidence to make me give up the reality of things in space!

[22] "If we enquire what new character *relation to an object* confers upon our representations, what dignity they thereby acquire, we find that it results *only* in subjecting the representations to a rule, and so in necessitating us to connect them in some one specific manner." (A197 = B242; second emphasis mine.)

[23] Cf. Kant's observation at p. 54 of the *Metaphysical Foundations of Natural Science*: "For with regard to what is actual only by its being given in representation, there is not more given than is met with in the representation."

[24] More accurately: the experiences constituting the real beast have relations to one another and to other experiences that are not possessed by the experiences constituting the dragon.

[25] A192–93 = B237–38.

[26] Here is one suggestion though, which I owe to Ernest Sosa. Suppose that a fact about intentionalia holds iff it is apprehended as holding, and suppose further that I apprehend a relational fact aRb about intentionalia without apprehending any grounding facts Fa and Gb: it will follow that R is an irreducible or ungrounded relation. As Sosa notes, this explanation is not altogether satisfying, since it seems that the absence of grounding properties for spatial relations is not due simply to my *de facto* failure to perceive any. I *could* perceive in the dragon properties enough to make it determinate, but what properties could I perceive in two incongruent counterparts that would ground their incongruence?

BIBLIOGRAPHY

The following bibliography is divided into four sections: works by Kant relating to space and incongruent counterparts, discussions of the nature of space (including dimensionality and orientability), discussions of incongruent counterparts, and discussions of parity. In the section on incongruent counterparts, we have tried to provide a comprehensive bibliography of items in English. In the other sections, we have been highly selective and have for the most part not listed highly technical works.

I. WORKS BY IMMANUEL KANT

Kant, Immanuel. *Kant's Inaugural Dissertation and Early Writings on Space.* Translated by John Handyside. Chicago: Open Court, 1929.

Kant, Immanuel. *Kant: Selected Pre-Critical Writings.* Edited by G. B. Kerferd and D. E. Walford. Manchester: Manchester University Press, 1968.

Kant, Immanuel. *The Critique of Pure Reason.* Translated by Norman Kemp Smith. London: Macmillan, 1970.

Kant, Immanuel. *Prolegomena to Any Future Metaphysics.* Translated by L. W. Beck. Indianapolis: Bobbs-Merrill, 1950.

Kant, Immanuel. *Metaphysical Foundations of Natural Science.* Translated by James Ellington. Indianapolis: The Bobbs-Merrill Co. Inc., 1970.

II. WORKS ON THE NATURE OF SPACE

Abbott, Edward A. *Flatland.* New York: Barnes and Nobel, 1963.

Alexander, H. G. *The Leibniz—Clarke Correspondence.* Manchester: Manchester University Press, 1965.

Broad, C. D. 'Leibniz's Last Controversy with the Newtonians.' *Theoria* **12** (1946): 143—68.

Dewdney, A. K. *The Planiverse (Computer Contact With a Two Dimensional World).* New York: Poseidon Press, 1984.

Earman, John. *World and Space—Time Enough: Absolute vs. Relational Theories of Space and Time.* Cambridge Mass.: MIT Press, 1989.

Euler, Leonhard. 'Reflexions sur l'Espace et le Temps.' *Opera Omnia*, Series 3, Volume 2. Edited by E. Hoppe, K. Matter, and J. Burckhardt. Geneva: Societatis Scientiarum Naturalium Helveticae, 1942. Pp. 376—83. This article was originally published in 1748.

Grünbaum, Adolf. *Philosophical Problems of Space and Time.* 2nd. ed.; Dordrecht: D. Reidel, 1973.

Hilbert, David and Stephan Cohn-Vossen. *Visual Geometry.* Berlin: J. Springer, 1932.

Hinckfuss, I. E. *The Existence of Space and Time.* Oxford: Oxford University Press, 1975.
Manning, Henry P. *The Fourth Dimension Simply Explained.* Gloucester, Mass.: Peter Smith, 1977.
Nerlich, Graham. *The Shape of Space.* Cambridge: Cambridge University Press, 1976.
Reichenbach, Hans. *The Philosophy of Space and Time.* New York: Dover, 1958.
Russell, Bertrand. *Essay on the Foundations of Geometry.* Cambridge: Cambridge University Press, 1897.
Sklar, Lawrence. *Space, Time, and Space-Time.* Berkeley: University of California Press, 1974.
Smart, J. J. C., ed. *Problems of Space and Time.* New York: Macmillan, 1964.
Swinburne, Richard. *Space and Time.* London: Macmillan, 1968.
Van Fraassen, Bas C. *An Introduction to the Philosophy of Space and Time.* New York: Random House, 1970.

III. DISCUSSIONS OF INCONGRUENT COUNTERPARTS

Alexander, Peter. 'Incongruent Counterparts and Absolute Space.' *Proceedings of the Aristotelian Society* **85** (1984—85): 1—21.
Allison, Henry. *Kant's Transcendental Idealism.* New Haven, Conn.: Yale University Press, 1986. Pp. 99—102.
Beck, Lewis W. *Early German Philosophy.* Cambridge, Mass.: Harvard University Press, 1969. Pp. 446—51.
Bennett, Jonathan. 'The Difference Between Right and Left.' *American Philosophical Quarterly* **7** (1970): 175—91. Reprinted in this volume.
Block, N. J. 'Why Do Mirrors Reverse Right/Left and Not Up/Down?' *Journal of Philosophy* **71** (1974): 259—277.
Borel, Emile. *Space and Time.* New York: Dover, 1960. Pp. 85—89.
Broad, C. D. *Kant: An Introduction.* Cambridge: Cambridge University Press, 1978. Pp. 37—44.
Buchdahl, Gerd. *Metaphysics and Philosophy of Science.* Cambridge, Mass.: MIT Press, 1969. Pp. 598—606.
Buroker, Jill Vance. 'The Role of Incongruent Counterparts in Kant's Transcendental Idealism.' Pp. 315—40 in this volume.
Buroker, Jill Vance. *Space and Incongruence: The Origin of Kant's Idealism.* Dordrecht: Reidel, 1981.
Butts, Robert E. *Kant and the Double Government Methodology.* Dordrecht: D. Reidel, 1984. Chapter 5.
Caird, Edward. *The Critical Philosophy of Immanuel Kant.* Glasgow: Maclehose and Sons, 1889. Vol. I, pp. 165—67.
Curd, Martin. 'Showing and Telling: Can the Difference Between Left and Right Be Explained in Words?' *Ratio* **26** (1984): 63—69. Reprinted in this volume.
Earman, John. 'Kant, Incongruous Counterparts, and the Nature of Space and Space—Time.' *Ratio* **13** (1971): 1—18. Reprinted in this volume.

Earman, John. *World and Space-Time Enough*. Cambridge, Mass.: MIT Press, 1989. Chapter 8. Reprinted in this volume except for minor changes.
Fritsch, Vilma. *Left and Right in Science and Life*. London: Barrie and Rockliffe, 1968.
Gardner, Martin. *The Ambidextrous Universe: Left, Right, and the Fall of Parity*. New York: Basic Books, 1964. 2nd ed.; New York: Penguin Books, 1982. 3rd ed.; San Francisco: W. H. Freeman and Co., 1989. Chapter 17 reprinted in this volume.
Garnett, C. P. *The Kantian Philosophy of Space*. New York: Columbia University Press, 1939. Pp. 112—118, 127—28, 152, 174, and 207.
Grünbaum, Adolf. *Philosophical Problems of Space and Time*. 2nd ed.; Dordrecht: D. Reidel, 1973.
Harper, William. 'Kant on Incongruent Counterparts.' Pp. 263—313 in this volume.
Jammer, Max. *Concepts of Space*. New York: Harper, 1960. Pp. 129—32.
Korner, Stephan. *Kant*. Baltimore: The Penguin Press, 1960. Pp. 33—34.
Lange, H. 'Uber den Unterschied der Gegenden im Raume.' *Kant-Studien* **50** (1958—59): 479—99.
Lucas, J. R. *Space, Time and Causality*. Oxford: Oxford University Press, 1984. Pp. 143—155.
Mayo, Bernard. 'The Incongruity of Counterparts.' *Philosophy of Science* **25** (1958): 109—115.
Mortensen, Chris and Graham Nerlich. 'Spacetime and Handedness.' *Ratio* **25** (1983): 1—13.
Nerlich, Graham. 'Hands, Knees, and Absolute Space.' *Journal of Philosophy* **70** (1973): 337—351.
Nerlich, Graham. *The Shape of Space*. Cambridge: Cambridge University Press, 1976. Chapter 2. Reprinted in this volume.
Pears, D. F. 'The Incongruity of Counterparts.' *Mind* **61** (1952): 78—81.
Remnant, Peter. 'Incongruent Counterparts and Absolute Space.' *Mind* **72** (1963): 383—399. Reprinted in this volume.
Robinson, Hoke. 'Incongruent Counterparts and the Refutation of Idealism.' *Kant-Studien* **72** (1981): 391—97.
Scott-Taggart, M. J. 'Recent Work on the Philosophy of Kant.' *American Philosophical Quarterly* **3** (1966): 178—80.
Sklar, Lawrence. 'Incongruous Counterparts, Intrinsic Features, and the Substantiviality of Space.' *Journal of Philosophy* **71** (1974): 277—90. Reprinted in this volume.
Smart, J. J. C. *Between Science and Philosophy*. New York: Random House, 1968. Pp. 217—18.
Smith, Norman Kemp. *A Commentary on Kant's Critique of Pure Reason*. New York: Humanities Press, 1962; reprint of 1918 edition. Pp. 161—166. Reprinted in this volume.
Van Cleve, James. 'Right, Left, and the Fourth Dimension.' *The Philosophical Review* **96** (1987): 33—68. Reprinted in this volume.
Van Cleve, James. 'Incongruent Counterparts and Things in Themselves.' *Proceedings: Sixth International Kant Congress*. Edited by G. Funke and Thomas M. Seebohm. Washington, D.C.: University Press of America, 1988. Reprinted in this volume.
Walker, R. C. S. *Kant*. London: Routledge and Kegan Paul, 1978. Pp. 44—51. Reprinted in this volume.

Weyl, Hermann. *Symmetry*. Princeton: Princeton University Press, 1952. Especially pp. 16—38.

Winterbourne, A. T. 'Incongruent Counterparts and the Intuitive Nature of Space.' *Auslegung* **1** (1982): 85—98.

Wittgenstein, Ludwig. *Tractatus Logico-Philosophicus*. Translated by D. F. Pears and B. F. McGuinness. London: Routledge and Kegan Paul, 1961. Pp. 141—142. Reprinted in this volume.

IV. DISCUSSIONS OF PARITY

Adair, Robert K. 'A Flaw in a Universal Mirror.' *Scientific American*, February, 1988: 50—56.

Bouchiat, Marie-Anne and Lionel Pottier. 'An Atomic Preference Between Left and Right.' *Scientific American*, June, 1984: 100—111.

Feynman, Richard P. *The Feynman Lectures on Physics*. Edited by Richard Feynman, Robert Leighton, and Matthew Sands. Reading, Mass.: Addison-Wesley, 1963—65. Vol. I, Ch. 52.

Frisch, O. R. 'Parity Not Conserved: A New Twist To Physics?' *Universities Quarterly* **2** (1957): 235—44.

Gardner, Martin. *The Ambidextrous Universe: Left, Right, and the Fall of Parity*. New York: Basic Books, 1964. 2nd ed.; New York: Penguin Books, 1982. 3rd ed.; San Francisco: W. H. Freeeman and Co., 1989. Portions of chapters 18, 20, and 22 reprinted in this volume.

Haber, Howard W. and Gordon L. Kane. 'Is Nature Supersymmetric?' *Scientific American*, June, 1986: 52—75.

Lee, T. D. and C. N. Yang. 'Question of Parity Conservation in Weak Interactions.' *Physical Review* **104** (1956): 254—258.

Morrison, Philip. 'The Overthrow of Parity.' *Scientific American*, April, 1957: 45—53.

Overseth, Oliver E. 'Experiments in Time Reversal.' *Scientific American*, October, 1969: 88—101.

Treiman, S. B. 'The Weak Interactions.' *Scientific American*, March, 1959: 72—84.

Wigner, Eugene. 'Violations of Symmetry in Physics.' *Scientific American*, December, 1965: 28—36.

Wigner, Eugene. *Symmetries and Reflections*. Bloomington: Indiana University Press, 1967. Pp. 57—62, 74—75.

Wilczek, Frank. 'The Cosmic Asymmetry Between Matter and Antimatter.' *Scientific American*, December, 1980: 82—90.

CONTEMPORARY CONTRIBUTORS

LEWIS WHITE BECK, Department of Philosophy, University of Rochester.
JONATHAN BENNETT, Department of Philosophy, Syracuse University.
JILL VANCE BUROKER, Department of Philosophy, California State University.
MARTIN CURD, Department of Philosophy, Purdue University.
JOHN EARMAN, Department of Philosophy, University of Pittsburgh.
MARTIN GARDNER, Hendersonville, NC.
WILLIAM HARPER, Department of Philosophy, University of Western Ontario.
GRAHAM NERLICH, Department of Philosophy, University of Adelaide.
PETER REMNANT, Department of Philosophy, University of British Columbia.
LAWRENCE SKLAR, Department of Philosophy, University of Michigan.
RALPH WALKER, Magdalen College, Oxford University.

INDEX

THE UNIVERSITY OF WESTERN ONTARIO
SERIES IN PHILOSOPHY OF SCIENCE

A Series of Books in Philosophy of Science, Methodology, Epistemology, Logic, History of Science, and Related Fields

1. J. Leach, R. Butts, and G. Pearche (eds.), *Science, Decision and Value*. 1973, vii + 219 pp.
2. C. A. Hooker (ed.), *Contemporary Research in the Foundations and Philosophy of Quantum Theory*. 1973, xx + 385 pp.
3. J. Bub, *The Interpretation of Quantum Mechanics*. 1974, ix + 155 pp.
4. D. Hockney, W. Harper, and B. Freed (eds.), *Contemporary Research in Philosophical Logic and Linguistic Semantics*. 1975, vii + 332 pp.
5. C. A. Hooker (ed.), *The Logico-Algebraic Approach to Quantum Mechanics*. 1975, xv + 607 pp.
6. W. L. Harper and C. A. Hooker (eds.), *Foundations of Probability Theory, Statistical Inference, and Statistical Theories of Science*. 3 Volumes. Vol. I: *Foundations and Philosophy of Epistemic Applications of Probability Theory*. 1976, xi + 308 pp. Vol. II: *Foundations and Philosophy of Statistical Inference*. 1976, xi + 455 pp. Vol. III: *Foundations and Philosophy of Statistical Theories in the Physical Sciences*. 1976, xii + 241 pp.
7. C. A. Hooker (ed.), *Physical Theory as Logico-Operational Structure*. 1979, xvii + 334 pp.
8. J. M. Nicholas (ed.), *Images, Perception, and Knowledge*. 1977, ix + 309 pp.
9. R. E. Butts and J. Hintikka (eds.), *Logic, Foundations of Mathematics, and Computability Theory*. 1977, x + 406 pp.
10. R. E. Butts and J. Hintikka (eds.), *Foundational Problems in the Special Sciences*. 1977, x + 427 pp.
11. R. E. Butts and J. Hintikka (eds.), *Basic Problems in Methodology and Linguistics*. 1977, x + 321 pp.
12. R. E. Butts and J. Hintikka (eds.), *Historical and Philosophical Dimensions of Logic, Methodology and Philosophy of Science*. 1977, x + 336 pp.
13. C. A. Hooker (ed.), *Foundations and Applications of Decision Theory*. 2 volumes. Vol. I: *Theoretical Foundations*. 1978, xxiii + 442 pp. Vol. II: *Epistemic and Social Applications*. 1978, xxiii + 206 pp.
14. R. E. Butts and J. C. Pitt (eds.), *New Perspectives on Galileo*. 1978, xvi + 262 pp.
15. W. L. Harper, R. Stalmaker, and G. Pearce (eds.), *Ifs. Conditionals, Belief, Decision, Chance, and Time*. 1980, ix + 345 pp.

16. J. C. Pitt (ed.), *Philosophy in Economics*. 1981, vii + 210 pp.
17. Michael Ruse, *Is Science Sexist?* 1981, xix + 299 pp.
18. Nicholas Rescher, *Leibniz's Metaphysics of Nature*. 1981, xiv + 126 pp.
19. Larry Laudan, *Science and Hypothesis*. 1981, x + 258 pp.
20. William R. Shea, *Nature Mathematized*. 1983, xiii + 325 pp.
21. Michael Ruse, *Nature Animated*. 1983, xiv + 266 pp.
22. William R. Shea (ed.), *Otto Hahn and the Rise of Nuclear Physics*. 1983, x + 252 pp.
23. H. F. Cohen, *Quantifying Music*. 1984, xvii + 308 pp.
24. Robert E. Butts, *Kant and the Double Government Methodology*. 1984, xvi + 339 pp.
25. James Robert Brown (ed.), *Scientific Rationality: The Sociological Turn*. 1984, xiii + 330 pp.
26. Fred Wilson, *Explanation, Causation and Deduction*. 1985, xviii + 385 pp.
27. Joseph C. Pitt (ed.), *Change and Progress in Modern Science*. 1985, viii + 398 pp.
28. Henry B. Hollinger and Michael John Zenzen, *The Nature of Irreversibility*. 1985, xi + 340 pp.
29. Kathleen Okruhlik and James Robert Brown (eds.), *The Natural Philosophy of Leibniz*. 1985, viii + 342 pp.
30. Graham Oddie, *Likeness to Truth*. 1986, xv + 218 pp.
31. Fred Wilson, *Laws and Other Worlds*. 1986, xv + 328 pp.
32. John Earman, *A Primer on Determinism*. 1986, xiv + 273 pp.
33. Robert E. Butts (ed.), *Kant's Philosophy of Physical Science*. 1986, xii + 363 pp.
34. Ian B. MacNeill and Gary J. Umphrey (eds.), *Advances in the Statistical Sciences*, Vol. I, *Applied Probability, Stochastic Processes, and Sampling Theory*. 1987, xxv + 329 pp.
35. Ian B. MacNeill and Gary J. Umphrey (eds.), *Advances in the Statistical Sciences*, Vol. II, *Foundations of Statistical Inference*. 1987, xvii + 287 pp.
36. Ian B. MacNeill and Gary J. Umphrey (eds.), *Advances in the Statistical Sciences*, Vol. III, *Time Series and Econometric Modelling*. 1987, xix + 394 pp.
37. Ian B. MacNeill and Gary J. Umphrey (eds.), *Advances in the Statistical Sciences*, Vol. IV, *Stochastic Hydrology*. 1987, xv + 225 pp.
38. Ian B. MacNeill and Gary J. Umphrey (eds.), *Advances in the Statistical Sciences*, Vol. V, *Biostatistics*. 1987, xvi + 283 pp.
39. Ian B. MacNeill and Gary J. Umphrey (eds.), *Advances in the Statistical Sciences*, Vol. VI, *Actuarial Science*. 1987, xvi + 250 pp.
40. Nicholas Rescher, *Scientific Realism*. 1987, xiii + 169 pp.
41. Brian Skyrms and William L. Harper (eds.), *Causation, Chance, and Credence*. 1988, xii + 284 pp.
42. William L. Harper and Brian Skyrms (eds.), *Causation in Decision, Belief Change and Statistics*. 1988, xix + 252 pp.
43. R. S. Woolhouse (ed.), *Metaphysics and Philosophy of Science in the Seventeenth and Eighteenth Centuries*. 1988, x + 362 pp.
44. R. E. Butts and J. R. Brown (eds.), *Constructivision and Science. Essays in Recent German Philosophy*. 1989, xxv + 287 pp.
45. A. D. Irvine (ed.), *Physicalism in Mathematics*. 1990, xxvi + 365 pp.

46. J. van Cleve and R. E. Frederick (eds.), *The Philosophy of Right and Left*. 1991, x + 358 pp. + index
47. F. Wilson, *Empiricism and Darwin's Science*. 1991 (forthcoming)
48. G. G. Brittan, Jr. (ed.), *Causality, Method, and Modality. Essays in Honor of Jules Vuillemin. With a Complete Bibliography of Jules Vuillemin*. 1991, viii + 238 pp.

GPSR Compliance

The European Union's (EU) General Product Safety Regulation (GPSR) is a set of rules that requires consumer products to be safe and our obligations to ensure this.

If you have any concerns about our products, you can contact us on ProductSafety@springernature.com

In case Publisher is established outside the EU, the EU authorized representative is:

Springer Nature Customer Service Center GmbH
Europaplatz 3
69115 Heidelberg, Germany

Batch number: 10815974

Printed by Printforce, the Netherlands